The Handbook of Environmental Chemistry

Editors-in-chief: O. Hutzinger · D. Barceló · A. Kostianoy

Volume 5 Water Pollution
Part J

The Handbook of Environmental Chemistry

Recently Published and Forthcoming Volumes

Polymers: Chances and Risks
Volume Editors: P. Eyerer, M. Weller and C. Hübner

Environmental Specimen Banking
Volume Editors: P. R. Becker, S. A. Wise, E. W. Gunter, and S. Tanabe

Biosensors for the Environmental Monitoring of Aquatic Systems
Bioanalytical and Chemical Methods for Endocrine Disruptors
Volume Editors: D. Barceló and P.-D. Hansen
Vol. 5/J, 2009

Environmental Consequences of War and Aftermath
Volume Editors: T.A. Kassim and D. Barceló
Vol. 3/U, 2009

Degradation of Synthetic Chemicals in the Environment
Volume Editor: A. Boxall
Vol. 2/P, 2009

The Black Sea Environment
Volume Editors: A. Kostianoy and A. Kosarev
Vol. 5/Q, 2008

Emerging Contaminants from Industrial and Municipal Waste
Removal Technologies
Volume Editors: D. Barceló and M. Petrovic
Vol. 5/S/2, 2008

Emerging Contaminants from Industrial and Municipal Waste
Occurrence, Analysis and Effects
Volume Editors: D. Barceló and M. Petrovic
Vol. 5/S/1, 2008

Fuel Oxygenates
Volume Editor: D. Barceló
Vol. 5/R, 2007

The Rhine
Volume Editor: T. P. Knepper
Vol. 5/L, 2006

Persistent Organic Pollutants in the Great Lakes
Volume Editor: R. A. Hites
Vol. 5/N, 2006

Antifouling Paint Biocides
Volume Editor: I. Konstantinou
Vol. 5/O, 2006

Estuaries
Volume Editor: P. J. Wangersky
Vol. 5/H, 2006

The Caspian Sea Environment
Volume Editors: A. Kostianoy and A. Kosarev
Vol. 5/P, 2005

Marine Organic Matter: Biomarkers, Isotopes and DNA
Volume Editor: J.K. Volkman
Vol. 2/N, 2005

Environmental Photochemistry Part II
Volume Editors: P. Boule, D. Bahnemann and P. Robertson
Vol. 2/M, 2005

Air Quality in Airplane Cabins and Similar Enclosed Spaces
Volume Editor: M.B. Hocking
Vol. 4/H, 2005

Environmental Effects of Marine Finfish Aquaculture
Volume Editor: B.T. Hargrave
Vol. 5/M, 2005

The Mediterranean Sea
Volume Editor: A. Saliot
Vol. 5/K, 2005

Biosensors for Environmental Monitoring of Aquatic Systems

Bioanalytical and Chemical Methods for Endocrine Disruptors

Volume Editors: Damià Barceló · Peter-Diedrich Hansen

With contributions by

J. Adrián · D. Barceló · S. Belkin · A. Biran · S. Boronat
R. Brix · S. Buchinger · M. Casado · V. Dethlefsen
H. Dizer · M. Farré · F. Fernández · T. Grummt
P.-D. Hansen · M. Kuster · M. López de Alda
M. J. López de Alda · M.-P. Marco · A. Muriano
R. Obregon · A. Olivares · R. Pedahzur · B. Piña
C. Postigo · J. Ramón-Azcon · G. Reifferscheid
P. Rettberg · S. Rodriguez-Mozaz · H. Rosenthal
J. Sherry · N. Tort · H. Tüg · E. Unruh · M. Villagrasa
P. Waldmann · H. von Westernhagen · E. Wittekindt
J. Zipperle

Springer

Environmental chemistry is a rather young and interdisciplinary field of science. Its aim is a complete description of the environment and of transformations occurring on a local or global scale. Environmental chemistry also gives an account of the impact of man's activities on the natural environment by describing observed changes.
The Handbook of Environmental Chemistry provides the compilation of today's knowledge. Contributions are written by leading experts with practical experience in their fields. The Handbook will grow with the increase in our scientific understanding and should provide a valuable source not only for scientists, but also for environmental managers and decision-makers.
The Handbook of Environmental Chemistry is published in a series of five volumes:

Volume 1: The Natural Environment and the Biogeochemical Cycles
Volume 2: Reactions and Processes
Volume 3: Anthropogenic Compounds
Volume 4: Air Pollution
Volume 5: Water Pollution

The series Volume 1 The Natural Environment and the Biogeochemical Cycles describes the natural environment and gives an account of the global cycles for elements and classes of natural compounds.
The series Volume 2 Reactions and Processes is an account of physical transport, and chemical and biological transformations of chemicals in the environment.
The series Volume 3 Anthropogenic Compounds describes synthetic compounds, and compound classes as well as elements and naturally occurring chemical entities which are mobilized by man's activities.
The series Volume 4 Air Pollution and Volume 5 Water Pollution deal with the description of civilization's effects on the atmosphere and hydrosphere.
Within the individual series articles do not appear in a predetermined sequence. Instead, we invite contributors as our knowledge matures enough to warrant a handbook article.
Suggestions for new topics from the scientific community to members of the Advisory Board or to the Publisher are very welcome.

The Handbook of Environmental Chemistry, Subseries 5 ISSN 1433-6863
ISBN 978-3-642-05548-5 e-ISBN 978-3-540-36253-1
DOI 10.1007/978-3-540-36253-1
Springer Dordrecht Heidelberg London New York

Cover design: SPi Publisher Services

Printed on acid-free paper

Springer is part of Springer Science+Business Media (www.springer.com)

The Handbook of Environmental Chemistry Also Available Electronically

For all customers who have a standing order to The Handbook of Environmental Chemistry, we offer the electronic version via SpringerLink free of charge. Please contact your librarian who can receive a password or free access to the full articles by registering at:

springerlink.com

If you do not have a subscription, you can still view the tables of contents of the volumes and the abstract of each article by going to the SpringerLink Homepage, clicking on "Browse by Online Libraries", then "Chemical Sciences", and finally choose The Handbook of Environmental Chemistry.

You will find information about the

– Editorial Board
– Aims and Scope
– Instructions for Authors
– Sample Contribution

at springer.com using the search function.

Color figures are published in full color within the electronic version on SpringerLink.

Preface

This book on "Biosensors for the Environmental Monitoring of Aquatic Systems" is based on the scientific developments and results achieved within a group of European Union (EU) funded projects that will be briefly discussed below. Indeed, everything started with the development of biosensors for environmental measurements in 1991 as part of the EC Environment Programme (1991–1994) under Area II: Technologies and engineering for the environment, section 1: Assessment of environmental quality and monitoring. In the context of this Programme, a biosensor was defined as a compact analytical device containing a biological or biologically derived sensing element (e.g., enzymes antibodies, microorganisms, or DNA) either integrated with or in intimate contact with a physicochemical transducer (e.g., electrochemical, optical, thermometric, or piezoelectric). The first European Workshop on Biosensors for Environmental Monitoring took place at the Technische Universität Berlin (BIT - Berlin Institute of Technology) in February 1993, organized by Prof. P.D. Hansen and Dr. J. Büsing. This was the first focused workshop of a series of workshops in London (1994) organized by P. Bennetto and J. Büsing, Florence (1995), Barcelona (1996), Freising (1997), Kinsale (1998), Paris (1999), Cascais (2000), and Alcala de Henares (2001). All these workshops covered a large variety of technical and scientific topics related to biosensors for environmental monitoring with additional presentations of the progress reports of the specific EU projects. The workshop in Florence had already presented ten biosensor projects supported under the EC Environment Programme, dealing with pesticide detection, microbiological contamination, and xenobiotics in water and air. This workshop involved some 100 scientists from 15 European countries and from Japan. The objective of the workshop in Florence (1995) was to present the state of European research activities supported under the EC Environment and Climate Programme (1995–1998) in the field of biosensor development and to identify priorities for future Research & Development.

During that period the European Commission's Environment and Climate Research Programme supported a three-year co-ordination project to guide technological developments in Biosensors for Environmental Technology (BIOSET). The aim of BIOSET was to enhance the development of biosensors for practical applications in monitoring pollutants in the environment. BIOSET maintained close contact with European industry and explored possibilities for collaboration with the commercial sector. This ensured eventual transfer of the technology demonstrated at the workshops and Technical Meetings to marketable devices. BIOSET provided assistance for researchers from European laboratories to meet to exchange ideas, use equipment, and establish a basis for new joint projects. The secretariat of the Concerted Action BIOSET supported the Technical Meetings.

There were three Technical Meetings held, two in Berlin in 1997 and 1998, and the third in Barcelona, in April 2000. The goal of these technical meetings was to join different research and industrial teams to evaluate the performance of their biosensor technology in field conditions with common and standardized surface and waste waters.

As a result of these field experiments, the additional information that biosensors can offer to environmental monitoring was also evaluated. Thus, these three Technical Meetings were useful accompanying measures and practical additions to the currently organized yearly workshops. The concerted action BIOSET was followed by the SENSPOL network. The 1st SENSPOL Workshop was held on the 9–11 May 2001 on Sensing Technologies for Contaminated Sites and Groundwater at the University of Alcala. There was one special Workshop on "Genotoxicity Biosensing (TECHNOTOX)" supported by the European Commission DG XII D-1 and BIOSET in the year 2000. The TECHNOTOX meeting at the Flemish Institute for Technological Research (VITO) in Mol was organized by Phillippe Corbisier (VITO), Peter-D. Hansen (TU Berlin) and Damia Barcelo (CSIC Barcelona). Thirty scientists participated in this meeting and 14 genotoxicity tests were performed on-site simultaneously on a maximum of 11 samples (chemical compounds and environmental samples). The AMES Test (ISO 16240) and the well-accepted umuC Test (ISO 13829) were performed as reference genotoxicity tests. The panel of genotoxicity tests included six prokaryotic tests, one eukaryotic test, four mammalian tests, one DNA test, and one bacterial test. Four assays were commercially available. It was demonstrated that no single test is capable of detecting all relevant genotoxins. Therefore, a Test Battery for genotoxicity is recommended (see the chapter by T. Grummt et al., this volume). The concept was to transform the standardized and harmonized microplate assay into an automated on-line assay and/or finally into a biosensor. The EuCyanobacteria Electrode was one of the first biosensors developed under the EC Environment Programme (1991–1994) initiated by Jürgen Büsing. The EuCyanobacteria Electrode was developed and optimized as an on-stream biosensor for pesticide detection (EV5V-CT92-0104). The EuCyanobacteria Electrode, in principle a so-called Rawson sensor, was transformed into a single rod electrode. The EuCyanobacteria Electrode with detection limits for atrazine <10 microgram per liter and diuron <2.5 microgram per liter was implemented in the monitoring programme on the river Rhine after the SANDOZ disaster in 1986 and optimized in several research projects in the following years. Special catch system devices were coupled with secondary electrochemical methods of signal transduction based on enzyme or whole-cell reactions. With the aid of mass spectrometry it was possible to evaluate the bio-recognition and detection components.

The Technical Meetings showed that biosensors are useful tools for controlling the degree of pollution of surface waters and of WWTP (wastewater treatment plants).

All biosensors evaluated showed much higher "alarm" values for the influent waste waters as compared to the final treated WWTP effluent. This is already remarkable, since the main advantage of the different biosensors evaluated, like BOD, ToxAlert, or phenols is that the measurements are being carried out very rapidly—a maximum of 20 minutes—so the information is available immediately. This is certainly an advantage over conventional methods that require several days, either for the BOD-5 or for conventional chromatographic/mass spectrometric methods.

The contributions in this book intend to summarize some of the practical achievements in the field of biosensors for environmental monitoring in Europe focusing, mainly, on examples of chemical groups of analytes (or effects) in different water matrices and more specific aspects of biosensor technology. The target objective of this book is to provide an overview of biosensors as a practical alternative and complementary or additional measurement methodology to traditional chromatographic techniques. Emphasis will also be given to the validation of the applied technology and its application to real-world environmental samples. The various chapters cover examples in different areas that are listed below:

General parameters of water quality. Biochemical Oxygen Demand (BOD) which is usually determined by measuring the oxygen consumption of a mixed culture over 5 days (BOD5). In the Third Technical Meeting held in Barcelona, the commercial firm Biosensores presented an on-line biosensor technique for the determination of BOD. This on-line biosensor device was installed in the la Llagosta wastewater treatment plant and after the Technical meeting, was installed in two more WWTP for a period of one month. This is an example of the implementation of biosensors into real-world situations, since the authorities are already using two of their devices for routine control.

Trace organics in surface water. The RIver ANAlyzer (RIANA), an optical immunosensor coupled to a FIA system based on a competitive solid-phase fluorescence immunoassay with immobilized analyte derivative and free fluorescence-labeled anti-atrazine or anti-simazine antibodies has been developed and applied for the measurement of chlorotriazine pesticides in river water samples. A pilot monitoring survey in the Ebre area (Tarragona, Spain) was carried out over a three-month period. Simazine and atrazine were detected at levels ranging from 0.5 to 1 microgram per liter and in soil samples from 19 to 40 g kg^{-1}. Phenylurea herbicides and paraquat were also determined by this immunosensor approach. The validation of the RIANA biosensor for the determination of phenylurea herbicides was carried out by LC-MS whereas in the case of paraquat capillary electrophoresis(CE)-UV detection was used. Enzymatic biosensors based on catalytic inhibition have been widely used and applied for the analysis of organophosphorus and carbamate pesticides in water samples. A cholinesterase biosensor using acethylthiocholine chloride as substrate was able to distinguish between different batches of water containing 3–10 ppb levels of pesticides.

Toxic and endocrine-disrupting effects. There are different methods of measuring toxicity. Two of the most commonly used methods are Microtox® and the recently developed ToxAlert@ system. Although they are not strictly biosensors, they are a kind of bioprobe used in a standardized way to measure toxicity of the effluents. They are widely accepted and internationally recognized. During the Third Field Experiment held in Barcelona, Spain more data was generated and a variety of cytotoxic compounds were identified in the wastewater treatment plant during this field experiment. The efficiency of the WWTP was evaluated by collecting and analyzing samples at various stages of the water treatment process. The chemical compounds identified could explain the toxicity observed at the different steps of the WWTP. Endocrine-disrupting effects are today widely studied. One of the most current measurements is the so-called vitellogenin (VTG) level in male fish. If VTG levels are high as compared to control values, exposure to endocrine-disrupting chemicals can be concluded. This approach has been used for the development of a biosensor called VITELLO within the EU projects PRENDISENSOR and SANDRINE. Within the EU project SANDRINE different biosensors, bioassays, and receptor assays are being developed for detecting endocrine-disrupting compounds in WWTP. In this respect, MCF-7 and vitellogenin by the ELRA and Vitello sensor of the SANDRINE project are a few of the different assays in use. A practical example was the determination of the presence and the effects of endocrine-disrupting compounds in two tributaries of the Llobregat river (NE Spain). Water samples and carp, *Cyprinus carpio*, were collected from selected sites along a transect, for chemical and biological determination, respectively. Western blot analysis of male carp plasma using a polyclonal antibody raised in the cyprinid Koi carp, detected this protein in all samples, the VTG increase being more evident at the vicinity of the treatment plants. A certain correlation was also found between NP in water and VTG induction in fish ($r = 0.75$). Within the EU project ALGAETOX a bio-analytical system was developed with the immunotoxic relevant endpoint. Phagocytosis was measured by oxidative burst (ROS) and luminescence. There is now a recent application of this system in the COLUMBUS/ BIOLAB programme. The bio-analytical system contributes to risk assessment concerning immunotoxicity under space-flight conditions. A schematic work flow demonstrates the close connection between environmental chemistry and the development and validation of a biosensor.

All these promising results demonstrate some of the European efforts to implement biosensor technology and other assays, including toxicological relevant assays and ELISA, in environmental analysis. The developments achieved during the last years by the different EU Environment Programs have permitted us to show some key examples of the application of biosensors for environmental monitoring, demonstrating not only cost-effective alternatives for screening but also effect-related readouts where new and unpredictable approaches are expected. Biosensor technology has reached maturity for BOD, pesticides, and phenols incorporating whole cells, affinity or catalytic events with electrochemical or optical transduction. Complex and difficult analytes (e.g., surfactants) are progressively being approached by biosensors. The large variety of analytes, the complexity of environmental matrices, the low limits of detection, and the legislative compliance should not shadow the progress achieved in this area. On the contrary, biosensor research is offering new strategies for the evaluation of biological effects where cocktails of compounds may show synergistic, additive, or antagonistic effects that will remain unsolved by chromatographic or spectroscopic techniques. The challenge for environmental biosensors remains open and further research in this area will only prove that new parameters are needed for safer and more complete environmental risk assessment. It is expected that collaborative work, focused research, and interdisciplinary approaches may lead to biosensor indexes that will certainly contribute to a better knowledge and real-time characterization of environmental samples. The book is organized to provide information on relevant principles, novel biosensor techniques and their applications. It is hoped that it will stimulate much discussion and bring new ideas to one of the most exciting and environmentally beneficial areas of environmental chemistry and eco-toxicology. It is hoped that the book and its methods contribute to the broader goal of conserving and protecting our aquatic ecosystems and their biota for future generations.

The book will be of interest to a broad audience of analytical chemists, biologists, environmental chemists, water management operators, and technologists working in the field of wastewater treatment, or newcomers who want to learn more about the topic using new measurement devices such as biosensors. Finally, we would like to thank all the contributing authors of this book for their time and effort in preparing this comprehensive compilation of research papers. Special thanks to Dr. Juergen Büsing for "pushing" this topic since 1991 at the EU level thus allowing excellent scientific and technological developments. The book that you have in your hands is part of such achievements.

February 2009

Damiá Barcelò
Peter-Diedrich Hansen

Contents

Hdb Env Chem (2009) 5J: 1–32
DOI:10.1007/698_2008_11

Biosensors for Environmental Monitoring at Global Scale and the EU Level

Marinella Farré, Sara Rodriguez-Mozaz, Miren López de Alda, Damià Barceló, and Peter-Diedrich Hansen

Abstract Biosensor technology is based on a specific biological recognition element in combination with a transducer for signal processing. Since their inception, biosensors have been expected to play a significant analytical role in environmental monitoring. The use of biological data to complement chemical analysis, as well as the development of biosensors and other biological approaches, has grown steadily in recent years. However, the commercialization of biosensor technology has significantly lagged behind their search output as reflected by a plethora of publications and patenting activities. The rationale behind the slow and limited technology transfer could be attributed to cost considerations and some key technical barriers.

During the last 15 years a relevant support has been provided by public institutions for biosensor research in the USA, Japan and, especially, in Europe. In addition, biomonitoring is an essential tool for rapid and cost-effective environmental monitoring. For this reason, biosensor technology has been considered as a key tool for the complete implementation of the new European Union (EU) directives, such as the Water Framework Directive (WFD), and other related directives such as the recent Marine Strategy Framework Directive.

Analytical chemistry has changed considerably, driven by automation, miniaturization and system integration with high throughput for multiple tasks. Such requirements pose a great challenge in biosensor technology where the biosensor is

M. Farré, M.L. de Alda, and D. Barceló (✉)
Department of Environmental Chemistry, IDAEA-CSIC,
c/Jordi Girona, 18-26, 08034, Barcelona, Spain
e-mail: dbcqam@iiqab.csic.es

S. Rodriguez-Mozaz and D. Barceló
Catalan Institute of Water Research, ICRA, C/Pic de Peguera, 15, 17003,
Girona, Spain

P.-D. Hansen
Department of Ecotoxicology, Technische Universitaet Berlin,
Franklinstrasse 29, 10587, Berlin, Germany

often designed to detect a single or a few target analytes. Successful biosensors must be versatile to support interchangeable biorecognition elements, and, in addition, miniaturization must be feasible to allow automation for parallel sensing with ease of operation at a competitive cost. A significant up front investment in research and development is a prerequisite in the commercialization of biosensors. The progress in such endeavours is incremented with limited success, thus, the market entry for a new venture is very difficult unless a niche product can be developed with a considerable market volume.

Recent technological developments in the miniaturization of electronics and wireless communication technology have led to the emergence of Environmental Sensor Networks (ESN). These will greatly enhance monitoring of the natural environment and in some cases open up new techniques for taking measurements or allow previously impossible deployments of sensors.

Herein we present the principles, advantages and limitations of biosensor technology for environmental diagnosis, with special emphasis on those based on nanomaterials and technologies for remote biosensing developed under the support provided by public institutions for research in USA, Japan and Europe.

Keywords Biosensor, Environmental monitoring, Network, Sensor, Water Framework Directive

Contents

1 Introduction

Because of the difficulty of predicting the collective effects of the increasing number of pollutants in receiving ecosystems, there is a need for screening methods in environmental monitoring.

The implementation of safety programs calls for environmental analysis comprising two parts: screening methods based on high-throughput analysis and capable of continuous monitoring on-line at site with low costs, and analysis of the positive samples with confirmatory analytical techniques.

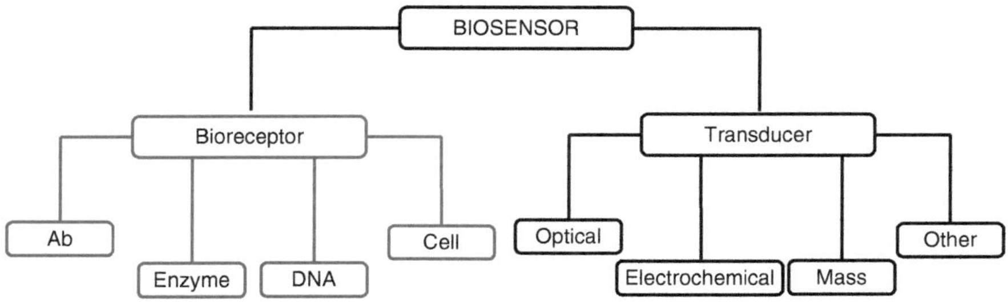

Fig. 1 General scheme of biosensors

Bioresponse-linked instrumental analysis combines two processes, biological recognition initiating a biological effect, and chemical analysis. The biomolecular components are used as targets for active substances. Although it is possible to apply, for example, binding assays that provide effects-linked equivalents, information on the responsible contaminants is only accessible by chemical analysis. Therefore a subsequent step is provided by the chemical analysis of the substances that are bound by the biorecognition components and that are therefore bio-effective.

Biosensors are usually classified either according to the transduction element (e.g. electrochemical, optical, piezoelectrical, or thermal), or according to the biorecognition principle (enzymatic, immunoaffinity recognition, whole-cell sensors, or DNA) (see Fig. 1).

The main advantages of biosensors are short times of analysis, low cost of assays, and possibility of portability, real-time measurements and remote control. These new technologies have been applied equally in the quantitative analysis of target analytes and in ecotoxicological measurements.

At present, biomonitoring is an essential tool for the complete implementation of the EU directives, such as the Water Framework Directive (WFD), and the Marine Strategy Framework directive. Nevertheless, many of these new technologies are still being developed and rely on combined efforts from diverse scientific fields. In Europe, the European Commission (EC) has been the flagship for the development of these new technologies in environmental analysis. Under the 3rd, 4th, 5th and 6th Framework Programs (FP) a great number of funded research projects were dedicated to the development of new analytical tools, and more than 300 projects on sensors, biosensors and biological techniques have been carried out [1].

Several reviews on biosensors, optical sensors for different analytical fields, or biological techniques for environmental analysis have been published during the last years: see Rodriguez-Mozaz [1], González-Martínez [2], Kurosawa [3], Farré [4].

Herein we aim to present the principles, advantages and limitations of biosensor technology for environmental monitoring, and summarize the advances obtained in this field under the support provided by public institutions.

2 Biosensor Research Funding Schemes

Europe, USA and Japan have been trend leaders in biosensor research and manufacture, especially for health, food and environmental analysis.

The European research funding in the area of biosensors for environmental monitoring as a specific topic started in 1992 and the first workshop was held in Berlin in 1993. A selection of projects developed in the past years under different framework programs is listed in Table 1.

During the first EU Framework Programmes, FP4 and FP5, funding was devoted to supporting research projects related to the determination of pesticides in water, in support to the European Union Drinking Water Directive (98/83/EC).

Table 1 Research projects involving biosensors supported by European Union funding

Title, acronym	Contract no.	Duration
Optical biosensing techniques for monitoring organic pollutants in the aquatic environment, BIOPTICAS	EV5V-CT92-0067	1993–1996
Development of new multi-sensing biosensors for the detection of bioavailable heavy metals in solid matrices, BIOMETSENSOR	ENV4-CT95-0141	1996–1999
Integrated immuno extraction sampling and portable biosensor prototype for in field monitoring, INExSPORT	ENV4-CT97-0476	1997–2000
Prediction of the behaviour of potential endocrine disruptors in soils using vitellogenin ELISA as biosensors, PRENDISENSOR	ENV4-CT97-0473	1997–2000
The application of integrated biosensors with antibody and macrocyclic receptor libraries in the measurement of algal cells and toxins in water, ALGAETOX	ENV4-CT98-0784	1999–2002
Development of improved detection systems for monitoring of toxic heavy metals in contaminated groundwaters and soils, DIMDESMOTOM	EVK1-CT-1999-00002	2000–2003
Protection of groundwater resources at industrially contaminated sites, PURE	EVK1-CT-1999-00030	2000–2003
Automated water analyzer computer supported system, AWACSS	EVK1-CT-2000-00045	2001–2004
Artificial recharge demonstration project, ARTDEMO	EVK1-CT-2002-00114	2002–2005
Water catchment areas: tools for management and control of hazardous compounds, WATCH	EVK1-CT-2000-00059	2001–2004
A biological nanoactuator as a molecular switch for biosensing, BIONANO-SWITCH	Specific Targeted Research Project 43288	2006–2009
Tools and technologies for the analysis and synthesis of nanostructures TASNANO	NMP-2003-3.4.1.4-1	2005–2007
DIagnostic NAnotech and MICrotech Sensors, DINAMICS	Integrated Project 26804	2007–2010

A good example of such projects was BIOPTICAS, which was followed up by RIANA, AWACSS and SANDRINE being the most successful projects at the EU level in that area. In the frame of these projects biomolecular receptors were obtained, and different biosensor configurations were developed and applied to real water samples. Under the RIANA and AWACSS projects optical immunosensors were developed and validated against several groups of organic pollutants, for example, for the simultaneous determination of pesticides and estrogens in natural water samples [5]. It is important to highlight BIOSET and SENSPOL concerted actions, which have been organizing various annual workshops and technical meetings on different areas of biosensors for environmental monitoring in order to stimulate research in that topic in Europe and to demonstrate the applicability of biosensors under real-world conditions.

At the same time, in the USA, "biodefence" and research with potential contribution to national security was an issue of particular concern, and biosensors for biohazards detection were developed [6, 7]. In addition, important initiatives were carried out under the auspices of the National Science Foundation (NSF), which funded biosensing network proposals, and "The National Centre For Environmental Research" (NCER), which funds research grants and graduate fellowships in numerous environmental science and engineering disciplines. The Office of Naval Research also promoted the science and technology programs through government laboratories, such as the "Naval Research Laboratory", which is one of the leading groups in biosensors research in the USA. They have developed several single and multi-analyte biosensors for toxins, bacteria and virus detection in clinical and environmental samples. Another remarkable research group, the biosensor group at UC Davis undertakes research into the development of miniaturized, fast and sensitive biosensors for use in environmental research and monitoring.

Regarding biosensor research in Japan, it is important to mention that biocomputing and nanotechnology are areas of increasing interest. One of the main research groups devoted to biosensor development is Karube's group in the "Research Centre for Advanced Science and Technology", at the University of Tokyo. Japan Fund for Global Environment (JFGE) provides grants for environmental conservation activities. For example, support is provided for activities, such as conserving air and water quality, and organizing symposiums, seminars and workshops. For example, Biosensors 2002, the Seventh World Congress on Biosensors, was held in Kyoto, Japan.

More recently, in Europe under FP6 less funding was directly devoted to biosensors for environmental monitoring; however, other important fields related to the development of environmental biosensing were boosted, such as nanotechnology. Some examples of research funded projects in this period are BIONANO-SWITCH, BIOMED-nano, OPTANANOGEN, TASNANO, DINAMICS and ISAMCO, most of them related to nanotechnology and material science advancements. For example, BIONANO-SWITCH is still on-going, and the main aim is the development of a nanoactuator/biosensor system.

The BIOMED-nano STREP-project aimed at improving enzymatic electron transfer reactions for application towards integrated bio-powered biosensing systems for diagnosis and healthcare. Such improvements were based on novel enzymes; modification of enzymes; design of novel nano-structured scaffolds for enzyme immobilization, in order to obtain biosensing devices with improved stability and electron transfer efficiency. Another example was OPTONANOGEN: Integrated opto-nanomechanical biosensor for functional genomic analysis. The aim of this project was to develop a portable genotoxicity biosensor microsystem able to detect hybridization of nucleic acids with sensitivity to distinguish single base substitutions, deletions and insertions.

In addition to initiatives for biosensing development, an important restriction is the successful marketing of biosensors for environmental applications. In order to overcome implementation barriers and bridge the gap between innovative and accepted techniques for environmental monitoring in Europe an Environmental Technology Verification (ETV) system has been developed following the model of the US and Canada. Therefore, Europe is now heading towards an Environmental Technology Verification System. Under the 6th FP several projects have been devoted to the development of this verification system including PROMOTE-ETV, Eurodemo, Testnet, AirETV and Tritech-ETV. A proposal for an EU framework on ETV is planned to be published by the European Commission in autumn 2008.

Research based on nanotechnology has been extremely important worldwide, and has been a new motor for biosensors development. In addition to the use of nanomaterials to improve analytical properties of new biosensors, the development of nanobiosensing devices is one of the main promising perspectives for future developments.

In this sense, during the last years in the US the National Science Foundation (NSF), through the Directorate for Engineering and the Directorate for Computer and Information Science and Engineering, carried out a broad interdisciplinary program of research and education in the area of advanced sensor and biosensors development. An example of successful achievements has been the creation of the world's first all-integrated sensor circuit based on nanowire arrays, combining light sensors and electronics made of different crystalline materials, by Scientists at the U.S. Department of Energy's Lawrence Berkeley National Laboratory and the University of California at Berkeley [8, 9]. Several projects are under development such as MEMS-Based Magnetic Probe Microscopy Project, in which a scanning-probe magnetic microscope based on a high-resolution magnetic tunnel junction (MTJ) sensor is under construction. The completed microscope will be used to characterize the magnetic properties of ferromagnetic metal samples.

In the same way the Japan Science and Technological Agency has been promoting a number of initiatives focused on nanobiosensors.

In Europe the 7th FP presents a thematic area on Nanosciences, nanotechnologies, materials and new production technologies. Also under the 7th FP, the first call has included funding for projects to develop sensors and biosensors platforms for remote environmental monitoring. The last tendency in Europe is research in new autonomous platforms able to operate on-line in an autonomous manner.

3 Biosensors

Biosensors can be defined as an integrated receptor-transducer device, which is capable of providing selective quantitative or semi-quantitative analytical information using a biological recognition element. A biosensor converts a biological event into a detectable signal by the action of a transducer and a processor.

The automation and the miniaturization of biological analytical techniques, as well as the development of on-line and remote measurement equipment, can be achieved through biosensor technology. In recent years a growing number of biosensor applications have been developed for environmental analysis [10] for targeted analysis and for monitoring whole biological effects such as toxicity or estrogenicity [11].

3.1 Classes and Fundamentals

Biosensors can be classified according to the bioreceptor elements involved in the recognition and according to the physicochemical transduction elements.

The main classes of bio-receptors applied in environmental analysis of organic pollution are: enzymes, antibodies, DNA and whole cells.

On the other hand, there are four basic groups of transduction elements: electrochemical, optical, mass sensitive and thermal sensors, with most of the biosensors described in the literature being electrochemical biosensors. The following sections discuss the main classes of biosensors according to the transduction element.

3.1.1 Electrochemical Transduction

Electrochemical sensors measure the electrochemical changes that occur when chemicals interact with a sensing surface of the detecting electrode. The electrical changes can be based on a change in the measured voltage between the electrodes (potentiometric), a change in the measured current at a given applied voltage (amperometric), or a change in the ability of the sensing material to transport charge (conductometric). Such effects may be stimulated electrically or may result in a spontaneous interaction at the zero-current condition. Voltammetric, amperometric, potentiometric, potentiometric solid electrolyte gas sensor and chemically sensitized field effect transistor (CHEMFET) are subgroups that may be distinguished.

The principle operation of amperometric biosensors is defined by a constant potential applied between a working and a reference electrode. The imposed potential promotes a redox reaction, which produces a current. The magnitude of this current is proportional to the concentration of electro active species present in solution.

Oxidase enzymes have been the most frequently investigated and applied. A number of amperometric biosensors are based on the monitoring of oxygen consumption, or hydrogen peroxide generation. Both are electrochemically active; oxygen can be electrochemically reduced, and hydrogen peroxide can be oxidized. The current generated is proportional to the concentration of the enzyme substrate present in a sample.

The use of mediators should permit one to replace oxygen as an electron acceptor and to operate at a much lower potential, reducing the effects of other electrochemically active species found in complex matrices.

Potentiometric biosensors are based on the monitoring of the potential at a working electrode, with respect to a reference electrode.

During the last few years this type of transduction has achieved a great development. Recent research has led to powerful nanomaterial-based electrical biosensing devices and examines future prospects and challenges. New nanoparticle-based signal amplification and coding strategies for bioaffinity assays are in use, along with carbon-nanotube molecular wires for achieving efficient electrical communication with redox enzyme and nanowire-based label-free DNA sensors.

3.1.2 Optical Transducers

Optical transducers are based on various technologies of optical phenomena, which are the result of an interaction of an analyte with the receptor part. Optical transduction may employ linear optical phenomena, including adsorption, fluorescence, phosphorescence, polarization, rotation, interference, etc., or non-linear phenomena, such as second harmonic generation.

Reflectometric techniques are based on a white incident light passing the interface between different refractive indices, which will be reflected in part. These reflected beams superimpose and build a characteristic interference spectrum. The binding of a biological receptor such as an antibody to the surface changes the thickness of the toggling layer, which causes a change in the reflectance spectrum (Fig. 2).

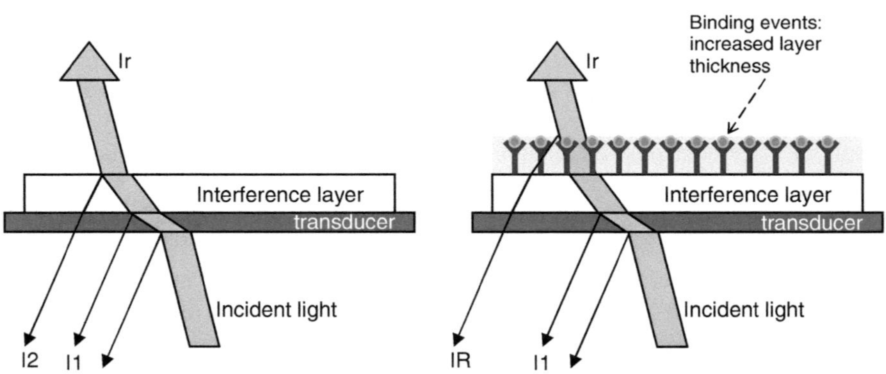

Fig. 2 Reflectometric based biosensors scheme

Thus, the interaction process between the bioreceptor and the analyte can be detected time-resolved [12].

Interferometry has also been exploited for biosensor development. When an immunoreaction takes place on the waveguide surface, it produces a change in the refractive-index profile within the evanescent field volume; correspondingly, the effective refractive index of the waveguide system is changed. In Mach-Zehnder interferometry (MZI), an optical waveguide is split into two arms and after a certain distance they are recombined again. The sensor arm will be exposed to a variation of the refractive index due to a biorecognition reaction such as an immunoreaction in the sensor channel. Over this distance, light travelling in the sensing arm will experience a phase shift in comparison with guided light in the reference arm [13].

Optical waveguide lightmode spectroscopy (OWLS) is a new sensing technique using an evanescent field for the in situ and label-free study of surface processes at molecular levels. It is based on the precise measurement of the resonance angle of linearly polarized laser light, diffracted by a grating and incoupled into a thin waveguide layer. The incoupling is a resonance phenomenon that occurs at a defined angle of incidence that depends on the refractive index of the medium covering the surface of the waveguide. In the waveguide layer, light is guided by total internal reflection to the edges where it is detected by photodiodes. By varying the angle of incidence of the light, the spectrum is obtained from which the effective refractive indexes are calculated for both electrical and magnetic field waves [14].

Total internal reflection fluorescence (TIRF) has been used with planar and fibre optic waveguides as signal transducers in a number of reported biosensors. In these transducers, light is propagated down a waveguide which generates an electromagnetic wave (evanescent wave) at the surface of the optically denser medium of the waveguide and the adjacent less optically dense medium. The amplitude of the standing wave decreases exponentially with distance into the lower refractive index material. The fluorescence of a fluorophore excited within the evanescent field can be collected.

At low angles, total internal reflection results when light propagating within a dense medium (e.g. quartz) reaches an interface with a less dense medium (e.g. aqueous solution). Although the light is fully reflected, an evanescent field is generated that extends beyond the interface and into the aqueous solution. Typically, the penetration depth (or the thickness of evanescent field) is in the range of half the wavelength of the light. The evanescent field provides the surface selectivity of TIRF. Only fluorophores adsorbed, adhered, or bound to the surface will be excited and therefore fluoresce. Conversely, fluorophores in bulk solution will not be excited. Therefore, if the surface is made biologically active so that one may "trap" fluorescently labelled compounds of interest, one can detect analytes within complex sample solutions. Because the excitation light is totally reflected away from the detection, one can easily discriminate the fluorescence signal from the excitation light and achieve high sensitivities and low detection limits. TIRF systems provide measurement of real time kinetics of a bioanalyte's binding to a surface immobilized sensor molecule. TIRF is a fast, non-destructive, sensitive and versatile technique that is well suited for monitoring biomolecular interactions. TIRF allows monitoring conformational changes, orientation changes and lateral mobility of biomolecules. TIRF has been used in a high number of biosensors. Examples of application are RIANA and AWACSS devices [10, 15, 16, 17].

The measurement based on evanescent waves has been exploited for environmental biosensor development. Evanescent waves propagating from waveguides can be used to probe only the sensing film to decrease the optical background signal from the sample.

Another important type of transduction is *surface plasmon resonance* (SPR). A surface plasmon wave can be described as a light-induced collective oscillation in electron density at the interface between a metal and a dielectric. At SPR, most incident photons are either absorbed or scattered at the metal/dielectric interface and, consequently, reflected light is greatly attenuated. The resonance wavelength and angle of incidence depend upon the permittivity of the metal and dielectric. SPR biosensors are based on the special electromagnetic waves—surface plasmon polaritons—to probe interactions between an analyte in solution and a biomolecular recognition element immobilized on the SPR sensor surface. Under the frame of different EU projects different SPR biosensors have been developed for environmental applications. Under the MENDOS project one of the main objectives was the development of an SPR optical sensor system and application for its assessing and screening endocrine disrupting compounds (EDCs).

Fluorescence/luminescence occurs when a valence electron is excited from its ground state to an excited singlet state. The excitation is produced by the absorption of light of sufficient energy [18]. The common principle of luminescence immunosensors is that an indicator or chemical reagent placed inside or on an immunoreactor is used as a mediator to produce an observable optical signal. Typically, conventional techniques, such as spectrometers, are employed to measure changes in the optical signal. This approach has been reviewed by Schobel et al. [19]

The general advantages of the optical techniques are the speed and the reproducibility of the measurement, and the main drawback is the high cost of the apparatus. Several applications [20, 21, 22, 23] have been described during the last years for environmental analysis; some of them are revised in the next section.

Optical biosensors offer different advantages such as low limits of detection and robustness and, are, in general, suitable for screening a large number of samples simultaneously; however, as a main limitation they cannot be easily miniaturized.

3.1.3 Mass-Sensitive Sensors

Measurements of small changes in mass is a transduction form that has been used for biosensor development [24, 25]. Piezoelectric devices and surface acoustic wave devices can be grouped under this category.

The vibration of piezoelectric crystals produces an oscillating electric field in which the resonant frequency of the crystal depends on its chemical nature, size, shape and mass. These crystals can be made to vibrate at a specific frequency of oscillation which is dependant on the electric frequency. The frequency of oscillation is therefore dependant on the electrical frequency applied to the crystal as well as the crystal's mass. When the mass increases due to the binding of analytes, the oscillation frequency of the crystal changes and the change can be measured. The general equation of crystal microbalances can be summarized as follows when the change in mass (*m*) is very small compared to the total mass of the crystal:

$$\Delta f = Cf^2 \Delta m/A,$$

where *f* is the vibration frequency of the crystal in the circuit, *A* is the area of the electrode and *C* is a constant determined in part by the crystal material and thickness. Piezoelectric crystals, sometimes referred to as quartz crystal microbalances (QCM), are typically made of quartz and operate at frequencies between 1 and 10MHz. These devices can operate in liquids with a frequency determination limit of 0.1Hz; the detection limit of mass bound to the electrode surface is about 10^{-10} to 10^{-11}g.

Acoustic wave devices, made of piezoelectric materials, are the most common sensors, which bend when a voltage is applied to the crystal. Acoustic wave sensors are operated by applying an oscillating voltage at the resonant frequency of the crystal, and measuring the change in resonant frequency when the target analyte interacts with the sensing surface.

Limitations for this transduction method involve format and calibration requirements, which are time consuming.

3.1.4 Thermometric Sensors

Thermometric sensors are based on the measurement of the heat effects of a specific chemical reaction or adsorption which involve the analyte. In this group, the heat effects may be measured in various ways, for example in the catalytic sensors the heat of a combustion reaction or an enzymatic reaction is measured by use of a thermistor. Calorimetric biosensors detect variations of heat during a biological reaction.

3.2 *Biosensors for Environmental Monitoring*

The following sections discuss the biosensors developed and applied in environmental monitoring, classified according to the biological recognition element.

3.2.1 Enzyme Biosensors

Enzyme electrodes have been widely used for monitoring a wide range of clinically or environmentally important substrates, especially using oxido-reductases (as tyrosinase, peroxidase and laccase) [26, 27], and hydrolases (choline estearases)[28].

Enzyme biosensors can be distinguished into those that measure the inhibition of a specific enzyme due to the presence of target analytes, and those that measure the catalytic transformation of target analytes by a specific enzyme.

Most of the transduction elements associated to enzyme-based biosensors are electrochemical: amperometric or potentiometric. Some examples of environmental applications of electrochemical biosensors are reported in Table 2.

Enzymes typically used in amperometric biosensors are oxidases. The main advantage of this class of transducer is the low cost; it is also highly reproducible and a wide variety of disposable electrodes are often available. The instrumentation is also very easy to obtain and can be inexpensive and compact; this allows for the possibility of on-site measurements. Limitations for amperometric measurements include potential interferences to the response if electroactive compounds are present in the sample, because they can generate false current values.

The development of electrochemical DNA and protein sensors on screen-printed electrodes (SPEs) based on the catalytic activity of hydrazine has been described by Shiddiky et al. [51] A high sensitive flow enzymatic biosensor for the detection of 2,4-D, atrazine and ziram has been reported by Kim et al. [52]. The same group [53] has also recently reported the preparation of an optimized coverage of gold nanoparticles at a tyrosinase electrode for the measurement of the pesticide in various water samples. Screen-printed electrodes containing immobilized acetylcholine esterase (AChE) were used for the electrochemical determination of organophosphorous and carbamate pesticides [54]. In this work, the extent of AChE deactivation by the pesticides was determined in the presence of acetylcholine (AChCl) substrate. In this approach AChE was attached to the SPE by in situ bulk polymerization of acrylamide to ensure efficient adherence within the membrane with minimal losses in enzyme activity. Responses were observed for the pesticides monocrotophos, malathion, metasystox and lannate over the concentration range 0–10 $\mu g\ L^{-1}$.

Most of the recent advances in electrochemical biosensors have been based on nanotechnology. The huge interest in nanomaterials is driven by their many desirable properties. In particular, the ability to tailor the size and structure and hence the properties of nanomaterials offers excellent prospects for designing novel sensing systems and enhancing the performance of the bioanalytical assay.

An extremely important challenge in amperometric enzyme electrodes is the establishment of satisfactory electrical communication between the active site of the enzyme and the electrode surface. The redox centre of most oxidoreductases is electrically insulated by a protein shell. Because of this shell, the enzyme cannot be oxidized or reduced at an electrode at any potential. In general, there is a need for co-substrates or mediators to achieve efficient transduction of the biorecognition event. Aligned carbon nanotubes (CNT), prepared by self assembly, can act as molecular wires to allow electrical communication between the underlying electrode and redox proteins (covalently attached to the ends of the single wall nanotubes (SWNT)[55, 56].

The catalytic properties of metal nanoparticles have also facilitated the electrical contact of redox centres of proteins with electrode surfaces. For example, gold nanoparticles were shown to be extremely useful as electron relays for the alignment of glucose oxidase on conducting supports and wiring its redox centre [57].

A wide range of enzyme electrodes based on dehydrogenase or oxidase enzymes rely on amperometric monitoring of the liberated NADH or hydrogen peroxide products. The anodic detection of these species at ordinary electrodes is often hampered by the large over voltage encountered for their oxidation. The greatly enhanced redox

Table 2 Examples of electrochemical biosensors reported for environmental analysis

Analyte/effect	Biosensor: Type AND Basis	Matrix	Limit of detection	Reference
Carbamates	AchE, BChE. Potentiometric detection	Synthetic samples	1.5×10^{-5}–2.5×10^{-3}M	[29]
Dimethyl-and diethyl-dithiocar-bamates	Tyrosinase amperometric detection	Aqueous solution	0.2mM	[30]
2,4,6-Trichloroanisole	Amperometric immunosensor	Water	1ng mL^{-1}	[31]
2,4-Dichlorophenoxyacetic acid	Amperometric immunosensor	Water	0.1μg L^{-1}	[32]
Bisphenol A	Tyrosinase immobilized on different carbon carriers with amperometric detection	Drinking and surface water	0.02μM	[33]
Bisphenol A	Potentiometric immunosensor	Water	0.6ngmL^{-1}	[34]
Bisphenol A, 17-β-estradiol	Tyrosinase-modified boron-dped diamond electrodes with amperometric detection	Water	10^{-6}M for bisphenol A	[35]
Catechol, *p*-cresol, phenol, *p*-chlorophenol, and *p*-methylcatechol	Tyrosinase immobilized in titania sol-gel with amperometric detection	Spring and surface water	9×10^{-8}M	[36]
Dichlovos	AchE inhibition amperometric detection	River water	70nM	[37]
Fenitrothion and ethyl *p*-nitrophenol thiobenzenephosphonate	Amperometric whole-cell biosensor	Water	1.4ppb of fenitrothion and 1.6ppb of EPN	[38]
Genotoxicity	DNA amperometric detection	Wastewater		[39]
Genotoxicity	Voltammetric DNA			[40]
Genotoxicity	DNA electrochemical biosensors using four types of DNA (calf thymus ssDNA, calf thymus dsDNA, salmon testis ssDNA and salmon testis dsDNA) were constructed using graphite screen printed electrodes	Water		[41]
Heavy metals	Amperometric DNA	Water	4.0×10^{-11}, 1.0×10^{-10}, 1.0×10^{-9}, and 5.0×10^{-9}M for Cu(II), Pb(II), Cd(II), and Fe(III), respectively	[42]
Organophosphates	Single-walled carbon nanotube (SWNT) amperometric whole-cell biosensor	Water	Phosphate levels in the marine system were 0.04mgL^{-1}	[43]

(continued)

Table 2 (continued)

Analyte/effect	Biosensor: Type AND Basis	Matrix	Limit of detection	Reference
Parathion	Parathion hidrolase enzyme action with amperometric detection	River water	10ngmL^{-1}	[44]
PAT gene	Nano-SiO2/p-aminothiophenol self-assembled films. DNA Impedance measurement	Water	1.0×10$^{-11ML-1}$	[45]
Phenol, *o*-cresol, *p*-cresol, *m*-cresol, catechol, dopamine, and ephinephrine	Multi–walled carbon nanotube-Nafion–Tyrosinase nanobiocomposites with amperometric detection	Aqueous solution	303μAmM^{-1}	[46]
Polycyclic aromatic hydrocarbons	Amperometric inmunosensor based on screen-printed electrodes	Water	Phenanthrene LOD of 2ngmL^{-1}	[47]
Progesterone	Colloidal gold–graphite–Teflon–Tyrosinase composite with amperometric detection	Milk	0.43ngmL^{-1}	[48]
Surfactants	Amperometric bacterial biosensor	Water	0.25mgL^{-1}	[49]
Toxicity	Amperometric whole-cell biosensor	Water and wastewater		[50]

Acetylcholinesterase (AChE) and butyrylcholinesterase (BChE)
Phosphinothricin acetyltransferase (PAT) gene

activity of hydrogen peroxide [58] and NADH [59] at CNT-modified electrodes addresses these over voltage limitations. The ability of CNT to promote electron transfer reactions is attributed to the presence of edge plane defects at their end caps.

The deposition of platinum nanoparticles onto CNT has led to further improvements in the detection of the enzymatically liberated peroxide species [60]. In addition to CNT films, it is possible to use CNT-based inks [61]. The excellent electrocatalytic properties of metal nanoparticles (compared to bulk metal electrodes) can also benefit amperometric enzyme electrodes. For example, You et al. [62] dispersed iridium nanoparticles in graphite-like carbon for improved amperometric biosensing of glutamate.

Another promising and controllable route for preparing conducting-polymer nanowire enzyme sensors involves electro-deposition within the channel between electrodes [63].

Recently, a lactate electrochemical biosensor with a titanate nanotube as a direct electron transfer promoter has been reported by Yang et al. [64]. In this work hydrogen titanate ($H_2Ti_3O_7$) nanotubes were synthesized by a one-step hydrothermal processing. Lactate oxidase (LOx) enzyme was immobilized on the three-dimensional porous nanotube network to make an electrochemical biosensor for lactate detection. Cyclic voltammetry and amperometry tests revealed that the LOx enzymes, which were supported on tunnelling nanotubes (TNTs), maintain their substrate-specific catalytic activity. The nanotubes offer the pathway for direct electron transfer between the electrode surface and the active redox centres of LOx, which enables the biosensor to operate at a low working potential and to avoid the influence of the presence of O_2 on the amperometric current response.

In another recent study Kochana et al. [36] reported the development of a titania sol-gel-derived tyrosinase-based amperometric biosensor for determination of phenolic compounds in water samples.

3.2.2 Immunosensors

The recognition elements are immunochemical interactions between antibody and antigen. This type of device combines the principles of solid-phase immunoassay with physicochemical transduction elements (electrochemical, optical, piezoelectric, evanescent wave and surface plasmon resonance).

Because most small molecular weight organic pollutants in the environment have few distinguishing optical or electrochemical characteristics, the detection of stoichiometric binding of these compounds to antibodies is typically accomplished using competitive binding-assay formats. The main limitation of these techniques is the electrochemical detection of the immunoreaction, because it is necessary to use enzymes that will generate electrochemically active compounds.

Electrochemical immunosensors have been widely used for environmental analysis in amperometric, potentiometric and conductimetric configurations. Amperometric immunosensors measure the current generated by oxidation or reduction of redox substances at the electrode surface, which is held at an appropriate electrical potential. A high number of successful applications have been reported during the last decade. Among them Wilmer et al. determined 2,4-dichlorophenoxyacetic (2,4-D) acid in water by an amperometric immunosensor with a limit of detection of 0.1 μgL^{-1} [32], and the same group developed also a flow injection system [65]. Screen-printed electrodes also have been extensively used. An example is the system developed by Fahnrich et al. [47] for the detection of polycyclic aromatic hydrocarbons (PAHs). Moore et al. [31] described the development of a biosensor for the quantitative detection of 2,4,6-trichloroanisole.

Another configuration used in electrochemical immunosensors for environmental analysis has been based on the use of recombinant single-chain antibody (scAb) fragments, this approach was used for atrazine determination [66]. The label-free detection of bisphenol A using a potentiometric immunosensor was reported by Piao et al. [34] with a detection limit of 0.6 ngmL^{-1}

The use of nanomaterials in order to improve electrochemical properties of immunosensors has also been studied. A nanostructured progesterone immunosensor using a tyrosinase-colloidal gold-graphite-Teflon biosensor as the amperometric transducer has been developed by the Pingarron group [48].

Acoustic transducers have also been applied in immunosensors for water analysis [67, 68]. The resonant frequency of an oscillating piezoelectric crystal can be affected by a change in mass at the crystal surface. Piezoelectric immunosensors are able to measure a small change in mass. Last publications have been based on immunosensors using a quartz crystal microbalance for the detection of trace amounts of chemical compounds, such as dioxins [3].

In general, a vast number of optical transduction techniques can be used for biosensor development. These may employ linear optical phenomena (e.g. adsorption, fluorescence, phosphorescence and polarization) or non-linear phenomena (e.g. second harmonic generation). The choice of a particular optical method depends on the analyte and the sensitivity needed.

Total internal reflection fluorescence has been used with planar and fibre-optic wave-guides as signal transducers in a number of biosensors. In these devices, light is propagated down a wave-guide, which generates an electromagnetic wave (evanescent wave) at the surface of the optically denser medium of the wave-guide and the adjacent less optically dense medium. The amplitude of the standing wave decreases exponentially with distance into the lower refractive index material. The fluorescence of a fluorophore excited within the evanescent field can be collected either outside the wave-guide, or by coupling the emission frequencies back into the wave-guide. The biological sensing element is immobilized on the side rather than at the end of the wave-guide. On the basis of the evanescent wave transducing principle, atrazine was detected at concentrations around 0.1 μgL^{-1} [69] and cyclodiene insecticides in the μgL^{-1} range [70].

Binding of the bioreceptor to the transducer is an important aspect of immunosensor fabrication. A method for the immobilization of hytidine-target antibodies onto a gold surface for surface plasmon resonance measurements has been reported by different authors [71].

A significant high number of SPR immunoassays have been recently used for rapid analysis of pollutants in different water matrices. In these assays the incident light is reflected from the internal face of a prism in which the external face has been coated with a thin metal film. At a critical angle, the intensity of the reflected light is lost to a resonant oscillation in the electrons created at the surface of the metal film. This method has been used to measure the binding of antibodies to antigens immobilized at the sensor surface because the critical angle depends on the refractive index of the materials of the metal surface. An assay approach was exploited by Nabok et al. [72] for in situ registration of low molecular weight toxins with specific antibodies immobilized onto the gold surface via a (poly)allylamine hydrochloride layer using an electrostatic self-assembly (ESA) technique. A simple and rapid detection method using surface plasmon resonance for dioxins, polychlorinated biphenyls and atrazine was described by Shimomura et al. [73, 74], with limits of detection ranging between 0.1 and 5ngmL^{-1}, requiring 15min for a single sample measurement. Mauriz et al. [75] presented an immunoassay based on a binding inhibition test using a portable surface plasmon resonance immunosensor for the continuous monitoring of the organophosphate pesticide chlorpyrifos in real water samples with limits of detection from 45 to 64ngL^{-1}. Sensor reusability was assured through the formation of alkanethiol self-assembled monolayers (SAMs). Using the same principle Mauriz et al. [76] developed single and multi-analyte surface plasmon resonance assays for simultaneous detection of cholinesterase inhibiting pesticides.

An immunosensor chip utilizing surface plasmon resonance was fabricated for detecting carcinoembryonic antigen based on protein A-conjugated [77].

In another example, a SPR immunosensor was developed for the detection of 2,4-dinitrophenol at ultra-low concentrations (immunoassay range between 1ppt and 1ppb). The sensor strategy was based on a competitive immunoreaction between 2,4-dinitrophenol and a 2,4-dinitrophenol-protein conjugate, namely DNP-bovine serum albumin conjugate (DNP-BSA). Anti-2,4-dinitrophenol monoclonal antibody was immobilized on a gold thin-film coated SPR-sensor chip by means of a chemical coupling process [78].

Farré et al. [22] described the development of a part-per-trillion immunoassay also based on SPR for atrazine determination in natural waters.

Among this variety of different immunosensor systems two optical immunosensor devices for environmental applications must be mentioned: RIANA and AWACSS. These immunosensors are based on solid-phase fluoro-immunoassays combined with an optical transducer chip chemically modified with an analyte derivative. The RIANA immunosensor was successfully applied in different environmental applications [5], and, along the same lines, a new analytical system, AWACSS, was developed and evaluated to measure several organic pollutants at low nanogram per litre level in a few minutes of analysis without pre-treatment steps [10, 15, 16].

A summary of recent applications of optical biosensors for environmental analysis is presented in Table 3.

3.2.3 DNA Biosensors

DNA's structure is very sensitive to the influence of environmental pollutants, such as heavy metals [89, 90], polychlorinated biphenyls (PCBs) [91, 92] or polyaromatic

Table 3 Examples of optical biosensors for environmental monitoring

Analyte	Type and basis	Sensitivity	Reference
Pesticides and herbicides			
Atrazine	Inhibition assay optical detection	1–10 μg L^{-1}	[79]
Isoproturon	Indirect assay and TIR-fluorescence detection	0.01–0.14μgL^{-1}	[17]
Triazines	Inhibition assay SPR detection	0.16μgL^{-1}	[80]
Triazines	Inhibition assay surface refractive index change detection	15μgmL^{-1}	[81]
Pesticides	Competitive immunoassays fluorescence detection		[10]
Carbaril	Competitive immunoassay SPR detection	54–64ngL^{-1} 1.38μgL^{-1}	[75]
DDT and related compounds	Competitive immunoassay SPR detection	15ngL^{-1}	[82]
Phenols			
Phenols	Indirect assay and TIR-fluorescence detection	0.9, 1.0, 1.0, 3.0μM and 0.038, 0.037, 0.056, 0.201mM for 4-chlorophenol, phenol, *m*-cresol and *p*-cresol, respectively	[83]
Phenols, organic acids	Whole cell optical biosensor with immobilized bioreporter Pseudomonas fluorescens HK44	1.2 and 0.5mgL^{-1} for naphthalene and salicylate	[84]
EDCs			
Hormones	Optical immunosensor based on fluorescence	Sub ngL^{-1}	[85]
Bisphenol A	Optical immunosensor based on fluorescence		[11]
Bisphenol A	SPR immunosensor		[86]
Surfactants			
Alkylphenols and their ethoxylates	Immunosensor using electrochemical transduction	μgL^{-1} range	[87]
Nonylphenol	Immunosensor using electrochemical transduction	10μgL^{-1}	[88]

compounds (PAHs) [93]. These substances are characterized by a great affinity to DNA, causing mutagenesis and carcinogenesis. So, it is very attractive to use DNA-containing systems, for example DNA-based biosensors [94, 95] to perform genotoxic assays, or for rapid testing of pollutants for mutagenic and carcinogenic activity. Lazarides et al. [96] presented gold colloidal-nanoparticle aggregates that were linked by short pieces of DNA [96] and that exhibited a colour change due to electromagnetic coupling between the gold non-spherical nanoparticles after DNA hybridization.

Electrochemical biosensors based on immobilized DNA and its parts integrate detection sensitivity with a high specificity of biomolecules. This reduces the consumption of DNA and gives rise to the development of modern methods of analysis of DNA effectors, including toxic ones, in environmental and biological effects [97, 39, 98, 42, 40].

The development of electrical DNA hybridization biosensors has attracted considerable research efforts [99]. Such DNA sensing applications require high sensitivity through amplified transduction of the oligonucleotide interaction.

Nanoparticle-based amplification schemes have led to improved sensitivity of bioelectronic assays by several orders of magnitude. The groups of Wang [100] and Limoges [101], both in 2001, reported the use of colloidal gold tags for electronic detection of DNA hybridization. This protocol relies on capturing the nanoparticles to the hybridized target, followed by highly sensitive anodic stripping electrochemical measurement of the metal tracer. Electronic DNA hybridization assays have been extended to other metal tracers, including silver [102] and iron [103]. Commonly the coupling of the biorecognition element to surfaces of magnetic beads offers an effective minimization of non-specific binding.

The hybridization of probe-coated magnetic beads with the metal-tagged targets results in three-dimensional network structures of magnetic beads, cross-linked through the DNA and gold nanoparticles. The placing of such a magnetic-bead/DNA/metal-label assembly onto the electrode leads to direct contact of the metal label and the surface and enables chronopotentiometric measurements without dissolving the metal tag [104].

Several amplification processes can be used to dramatically enhance the sensitivity of particle-based bioelectronic assays.

Treatment of the gold-linked DNA-hybrid assembly with silver ions in the presence of hydroquinone results in catalytic deposition of silver on the gold tracer (acting as a catalyst), leading to a dramatic signal amplification [105].

Wang et al. [106] described a triple amplification bioassay, coupling the carrier-sphere amplifying units (loaded with numerous gold nanoparticle tags) with the "built-in" pre-concentration of the electrochemical stripping detection and a catalytic enlargement of the multiple gold-particle tags. The success of these and other nanoparticle-based amplification strategies depends on maintaining a low background response.

Other strategies are based on the use of inorganic nanocrystals offering an electro diverse population of electrical tags as needed for designing electronic coding. Inorganic-nanocrystal tracers have been used for a multi-target electronic detection of DNA [103] or proteins [107]. Three encoding nanoparticles (zinc sulphide, cadmium sulphide and lead sulphide) have thus been used to differentiate the signals of three protein targets in connection with a sandwich immunoassay and stripping voltammetry of the corresponding metals. Each binding yields a distinct voltammetric peak corresponding to a different antigen. This concept can be scaled up and multiplexed.

Nanoparticle-induced changes in the conductivity across a microelectrode gap can also be exploited for highly sensitive and selective electronic detection of DNA hybridization [108]. One-dimensional nanowires can also be used for bridging two closely spaced electrodes for label-free DNA detection. For example, a p-type silicon nanowire functionalized with PNA probes has been shown to be extremely useful for real-time label-free conductometric monitoring of the hybridization event [109]. This relies on the binding of the negatively charged DNA target that leads to an increase in conductance, reflecting the increased surface charge.

Similar improvements have been reported in connection with nanowires and carbon nanotubes functionalized with other receptor molecules.

Ramanathan et al. [63] demonstrated the real-time monitoring of nanomolar concentrations of biotin at an avidin-embedded polypyrrole nanowire.

Carbon nanotubes can also lead to ultrasensitive bioelectronic detection of DNA hybridization [110, 111]. For example, carbon nanotubes can be used as carriers for several thousands of enzyme tags and for accumulating the a-naphthol product of the enzymatic reaction. Such a carbon nanotube-derived double-step amplification pathway (of both the recognition and transduction events) allows the detection of DNA down to the 1.3zmol level.

Several works developed different electrochemical DNA systems for environmental analysis. Disposable electrochemical DNA-based biosensors have been used for the determination of low molecular weight compounds with affinity for nucleic acids and for the detection of the hybridization reaction. The rapid screening of genotoxic compounds using the molecular interaction between surface-linked DNA and the target pollutants or drugs has been applied in different configurations.

Genotoxic compounds were measured by their effect on the oxidation signal of the guanine peak of calf thymus DNA immobilized on the electrode surface and investigated by chronopotentiometric analysis by Mascini et al. [111]. This type of DNA biosensor is able to detect known intercalating compounds, such as daunomycin, polychlorinated biphenyls, aflatoxin B1 and aromatic amines. Applicability to river and waste water samples is also demonstrated [112]. Disposable electrochemical sensors for the detection of a specific sequence of DNA were realized by immobilizing synthetic single-stranded oligonucleotides onto a graphite screen-printed electrode. The probes became hybridized with different concentrations of complementary sequences present in the sample. The hybrids formed on the electrode surface were evaluated by chronopotentiometric analysis using daunomycin as the indicator of the hybridization reaction.

Recently, Feng et al. [113] constructed gold nanoparticle/polyaniline nanotube membranes on a glassy carbon electrode (Au/nanoPAN/GCE) for sensing of DNA hybridization. The properties of the Au/nanoPAN/GCE and the characteristics of the immobilization and hybridization of DNA were studied by cyclic voltammetry, differential pulse voltammetry and electrochemical impedance spectroscopy. The synergistic effect of the two kinds of nanomaterials, nanogold and nanoPAN, could enhance dramatically the sensitivity of the DNA hybridization recognition.

A simple and sensitive electrochemical DNA biosensor based on in situ DNA amplification with nanosilver as the label and horseradish peroxide (HRP) as the enhancer has been designed by Fu [114].

The good performance of DNA electrochemical biosensors has been proved in several works. For example, the electrochemical detection of harmful algae and other microbial contaminants in coastal waters using hand-held biosensors has been reported by LaGier et al. [115].

3.2.4 Whole-Cell Biosensors

A microbial biosensor is an analytical device that couples microorganisms with a transducer to enable rapid, accurate and sensitive detection of target analytes in fields as diverse as medicine, environmental monitoring, defence, food processing and safety. Earlier microbial biosensors used the respiratory and metabolic functions of the microorganisms to detect a substance that was either a substrate or an inhibitor of these processes.

Recently, genetically engineered microorganisms based on fusing of the *lux*, *gfp* or *lacZ* gene reporters to an inducible gene promoter have been widely applied to assay toxicity and bioavailability.

Whole-cell biosensors in environmental analysis have been applied for the assessment of biological effects such as biological oxygen demand (BOD), toxicity, genotoxicity or estrogenicity, and also for the measurement of target pollutants or groups of pollutants. However, most applications have focussed on whole biological effects. Ecotoxicological assessment using whole-cell biosensors is reviewed in depth in the chapter by J. Adrián et al. (this volume).

Most reported transduction elements in bacterial biosensors are electrochemical and optical. One of the most relevant applications has been the determination of the biochemical oxygen demand [116].

Most BOD sensors rely on the measurement of the bacterial respiration rate in close proximity to a transducer, commonly the Clark type (an amperometric sensor developed by Clark in 1956 for measuring dissolved oxygen [117]). Some BOD sensors have been developed and marketed by various manufacturers in both biofilm and bioreactor-type configurations.

Instrument information about many BOD commercial biosensors was provided by Liu and Mattiason [117]. BOD biosensor systems still present a series of limitations that restrict their applications: the lack of standardization and legislation in most countries, complicated maintenance requirements, and insufficient resistance to various toxic compounds. It is possible to eliminate the toxic effects of heavy-metal ions by using a chelating agent that complexes the ions, for example ethylene diamine tetra-acetate (EDTA) or sodium diethyl dithiocarbamate (DDTC) [118]. Prevention of contamination by other microbes is also important.

However, the conventional standard method for the determination of BOD measures the microorganisms oxygen consumption/respiration over a period of 5 days [116] and is reported as BOD_5. While BOD_5 is a good indicator of the concentration of organic pollutants in water, it is extremely slow and hence not suitable for process control [119]. To address this limitation, several BOD biosensors based on amperometric oxygen electrode transducers modified with microorganisms degrading/metabolizing organic pollutants have been reported. The microbial strains used as the biological sensing element include among others *Torulopsis candida* [120], *Trichosporon cutaneum* [121], *Pseudomonas putida* [122], *Klebsiella oxytoca* AS1 [123], *Bacillus subtilis* [118], *Arxula adeninivorans* LS3 [124] and *Serratia marcescens* LSY4 [125].

However, bacterial biosensors using amperometry as the transduction method have also been relevant in other fields such as in environmental toxicity assessment. Neurotoxic organophosphate (OP) compounds have found wide application as pesticides and insecticides in agriculture and as chemical warfare agents for the military [119]. Amperometric biosensors based on genetically engineered *Moraxella* sp. and *P. putida* with surface-expressed organophosphorus hydrolase (OPH) have been developed by Lei at al. [119] for the detection of organophosphorus pesticides. The biosensor consisted of recombinant PNP-degrading/oxidizing bacteria *P. putida* JS444, and under the optimum operating conditions, the biosensor was able to measure as low as 55 ppb of paraxon and 53 ppb of met.

Recently, a novel single-walled carbon nanotube-based biosensor for real-time detection of organophosphate has been developed by Liu [43]. Horizontally aligned SWNTs were assembled to desirable electrodes using the AC dielectrophoresis technique. Organophosphorus hydrolase immobilized on the SWNTs by non-specific binding triggers enzymatic hydrolysis of organophosphates, such as paraoxon, consequently causing a detectable change in the conductance of the SWNTs. The conductance change is correlated to the concentration of organophosphate.

3.3 Autonomous Biosensor Wireless Networks

A wireless sensor network is a wireless network consisting of spatially distributed autonomous devices using sensors and/or biosensors to cooperatively monitor physical or environmental conditions, such as temperature, pressure, salinity, motion or pollutants, at different locations (Fig. 3). This approach could be in the near future one of the main advances for continuous environmental monitoring, and for the implementation of new EU directives, especially in environments difficult to monitor, such as coastal areas and the open sea.

Pioneers of internet-scale control were the TJ Watson IBM researchers Alex Morrow and Ron Ambrosio. They postulated that the future world would operate on the basis of complex interlocking control loops that range from localized sensor-actuator systems to platforms that aggregate information from multiple heterogeneous sources.

Today, molecular and biological events can be monitorized and translated to digital signals by means of biosensors and this fact opens up a new window in analytical devices. This new approach could be a revolution in the future of environmental monitoring. However, the cost of reliable autonomous biosensing is still far too great for massively scaled-up deployments. Some of the drawbacks that should be first overcome are: the complexity of biosensing processes, the need for regular calibrations, which should be minimized, and the need for more stable biological reagents; energetic costs and residue production should also be minimized.

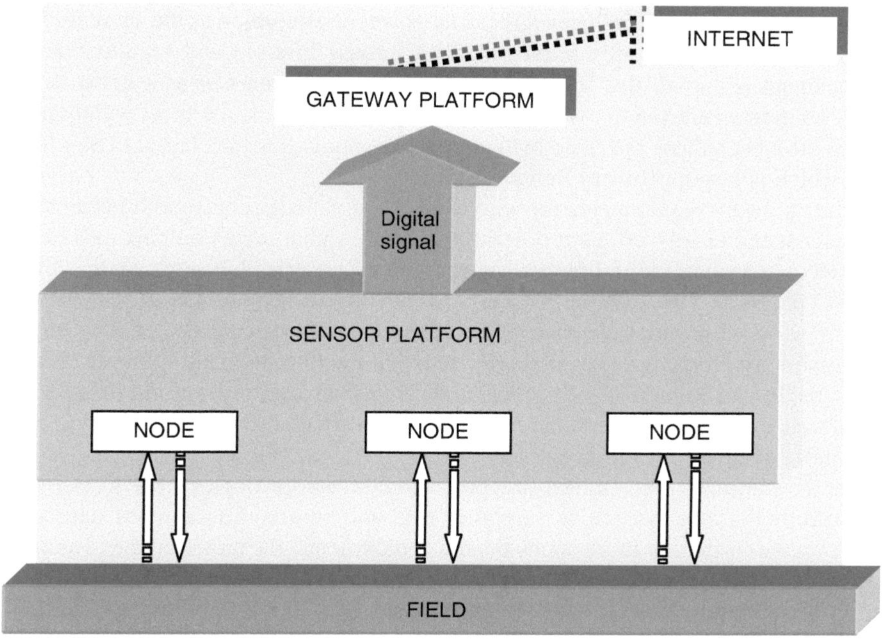

Fig. 3 Wireless sensors network diagram

Considerable advances have been made in recent years in hardware and software for building wireless biosensor networks. However, to ensure effective data gathering by biosensor networks for monitoring remote outdoor environments, the following problems remain:

- *Reactivity:* the ability of the network to react to its environment and provide only relevant data to users;
- *Robustness:* the ability of network nodes to function correctly in harsh outdoor environments;
- *Network lifetime:* maximizing the length of time the network is able to deliver data before nodes' batteries are exhausted.

The fist step in wireless sensor networks is the development of autonomous blocks of sensing platforms. A sensor *node* is the smallest component of a sensor network that has integrated sensing and communication capabilities. It contains basic networking capabilities through wireless communications with other nodes as well as some data storage capacity and a microcontroller that performs basic processing operations. They usually come with a sensor board that usually slots onto the controller board, which allows the user to interface with other sensors, provided the signal is presented in the appropriate form for the controller. They also include a power supply, usually provided by an on-board battery.

In recent years, there has been a focus on power consumption as the small lithium button batteries commonly employed have limited lifetime and regular manual replacement is unrealistic. The sensor nodes within a wireless sensor network are also commonly referred to as "motes". During the last years, the most widely used motes have been those provided by Crossbow Technologies Inc., based in San Jose, CA, which is a spin off from Berkeley.

The hardware requirements for wireless sensors include robust radio technology, a low-cost and energy-efficient processor, flexible signal inputs/outputs for linking a variety of sensors, a long-lifetime energy source, and a flexible, open source development platform. They also outline a number of software requirements for a wireless sensor node which include a small footprint capable of running on low-power processors, small memory requirement, and high modularity to aid software rescue. Thus, the basic components of a sensor node are a microcontroller, radio transceiver, set of transducers, and power source, and the software that runs on these nodes must be small and allow for efficient energy use.

In the simplest case, motes are programmed before deployment to perform measurements at a particular sampling rate and return the captured data in a prearranged format. In more sophisticated deployments, the motes are programmed to facilitate sampling rates that adapt to external events and function cooperatively in terms of finding the optimum route for returning data to remote base stations. However, new transducer interfaces are presently under development.

Because of the limited computing power of sensor motes, they often employ an operating system called TinyOS, although more recently, products that are fully C compliant have become available, and these are generally preferred by experienced programmers. TinyOS is an operating system written in the nesC programming language, which is a dialect of C specifically designed for restricted operating environments as exists on sensor motes where there is limited memory and processor power available [126].

On the basis of those computing, communication, and robotic schemes researchers have investigated biosensors and ancillary technologies for environmental applications, especially those that require continuous monitoring. Biosensors are currently available for monitoring biochemical oxygen demand [127] and bacteria [128]. The Oak Ridge National laboratory USA has developed a 1.5-cm^2 infrared microspectrometer which uses a light source to excite certain types of compounds in gases, liquids and solids which emit infrared light of various wavelengths [129].

However, sensor and biosensor networks are in their infancy, an example is the wireless, remote query ammonia sensor developed by Cai et al. [130]. Recently, the same group has described a wireless magnetoelastic biosensor for the detection of acid phosphatase [131].

Larger projects include an investigation of air quality on the MIR space station [132], and air pollution monitoring in Ferrera [133, 134, 135].

Bendikov et al. [136] performed research on developing sensors on the millimetre scale as well as on sampling. Under the EU FP programmes, the Automated Water Analyser Computer Supported System (AWACSS) project developed an optical sensor to detect river pollution [16]. These projects address the problems of mobility, scale and sensing in water as well as the development of new sensors.

4 Future Trends and Conclusions

Biosensor technology constitutes a rapidly expanding field of research that has been transformed over the past two decades through new discoveries related to novel material technologies, novel means of signal transduction, and powerful computer software to control devices.

However, in order to obtain improved and more reliable devices, future trends in biosensor research activities will be focused on the achievement of continuous monitoring, multi-analyte determinations, using robust transduction elements, and miniaturized devices.

Some systems have been developed for continuous monitoring and can provide easy, rapid and on-site measurements [137, 138]. They may also be useful for mapping of contamination when it is important to obtain rapid results in the field, such as after accidental spills or pollution events [139].

In addition, autonomous sensor platforms and wireless network technologies are very promising for the development of real-time warning systems, revealing dynamic changes in monitored variables in the environment. But to date there are significant problems to be overcome.

As regards advances in sensing technologies, the advancements in optical transduction, which offers robustness, versatility, and facility of miniaturization, must be emphasized. For example, SPR devices appear as some of the most promising technologies to be implemented in environmental warning programmes. Several examples of SPR applications have been reported in recent years. Another important objective is the development of sensors capable of determining several analytes simultaneously. Large-scale biosensor arrays, composed of highly miniaturized signal transducer elements, enable the real-time parallel monitoring of multiple species and are an important driving force in biosensor research [140].

Also relevant advances in electrochemical devices have been achieved recently thanks to the introduction of new nanotechnologic materials, such as nanoparticles and carbon nanotubes, which have been the basis of new systems with improved responses.

On the other hand, the advances in microelectronics and microfluidics have permitted the miniaturization of analytical systems, allowing reduction of energetic requirements, handling of low volumes of samples and therefore a reduction in reagent consumption and waste generation, as well as increasing sample throughput.

New sensing elements that can improve the affinity, specificity and mass production of the molecular recognition components may ultimately dictate the success or failure of detection technologies. In this sense, it should be emphasized that the contribution of gene engineering focuses on two main fields in the area of biosensors: genetically transformed cells and genetically engineered receptor molecules.

Finally, the last limitation to be overcome by biosensors for environmental monitoring is validation. This is an important issue for their integration into pollution control programs.

Acknowledgements The work described in this article was supported by the Spanish Ministerio de Ciencia e Innovación Project CEMAGUA. This work reflects only the author's views and the European Community is not liable for any use that may be made of the information contained therein.

References

1. Rodriguez-Mozaz S, Lopez De Alda MJ, Barcelo D (2006) Biosensors as useful tools for environmental analysis and monitoring. Anal Bioanal Chem 386:1025–1041
2. Gonzalez M, Puchades R, Maquieira A (2007) Optical immunosensors for environmental monitoring: How far have we come? Anal Bioanal Chem 387:205–218
3. Kurosawa S, Park JW, Aizawa H, Wakida SI, Tao H, Ishihara K (2006) Quartz crystal microbalance immunosensors for environmental monitoring. Biosens Bioelectron 22:473–481
4. Farre M, Brix R, Barcelo D (2005) Screening water for pollutants using biological techniques under European Union funding during the last 10 years. Trends Anal Chem 24:532–545
5. Rodriguez-Mozaz S, De Alda ML, Barcelo D (2005) Analysis of bisphenol A in natural waters by means of an optical immunosensor. Water Res 39:5071–5079
6. Hartley HA, Baeumner AJ (2003) Biosensor for the specific detection of a single viable *B. anthracis* spore. Anal Bioanal Chem 376:319–327
7. Rowe-Taitt CA, Golden JP, Feldstein MJ, Cras JJ, Hoffman KE, Ligler FS (2000) Array biosensor for detection of biohazards. Biosens Bioelectron 14:785–794
8. Fan Z, Ho JC, Jacobson ZA, Yerushalmi R, Alley RL, Razavi H, Javey A (2008) Wafer-scale assembly of highly ordered semiconductor nanowire arrays by contact printing. Nano Lett 8:20–25
9. Ho JC, Yerushalmi R, Jacobson ZA, Fan Z, Alley RL, Javey A (2008) Controlled nanoscale doping of semiconductors via molecular monolayers. Nat Mater 7:62–67
10. Tschmelak J, Proll G, Riedt J, Kaiser J, Kraemmer P, rzaga L, Wilkinson JS, Hua P, Patrick Hole J, Nudd R, Jackson M, Abuknesha R, Barcelo D, Rodriguez-Mozaz S, Lopez De Alda MJ, Sacher F, Stien J, Slobodnick J, Oswald P, Kozmenko H, Korenkova E, Tothova L, Krascsenits Z, Gauglitz G (2005c) Biosensors for unattended, cost-effective and continuous monitoring of environmental pollution: automated water analyser computer supported system (AWACSS) and river analyser (RIANA). Int J Environ Anal Chem 85:837–852
11. Rodriguez-Mozaz S, Marco MP, Lopez De Alda MJ, Barcelo D (2004) Biosensors for environmental monitoring of endocrine disruptors: a review article. Anal Bioanal Chem 378:588–598
12. Proll G, Kumpf M, Mehlmann M, Tschmelak J, Griffith H, Abuknesha R, Gauglitz G (2004) Monitoring an antibody affinity chromatography with a label-free optical biosensor technique. J Immunol Methods 292:35–42
13. Prieto F, Sepulveda B, Calle A, Llobera A, Dominguez C, Lechuga LM (2003) Integrated Mach-Zehnder interferometer based on ARROW structures for biosensor applications. Sens Actuators B Chem 92:151–158
14. Luppa PB, Sokoll LJ, Chan DW (2001) Immunosensors – Principles and applications to clinical chemistry. Clin Chim Acta 314:1–26
15. Tschmelak J, Proll G, Riedt J, Kaiser J, Kraemmer P, rzaga L, Wilkinson JS, Hua P, Hole JP, Nudd R, Jackson M, Abuknesha R, Barcelo D, Rodriguez-Mozaz S, Alda MJLD, Sacher F, Stien J, Slobodnik J, Oswald P, Kozmenko H, Korenkova? E, To?thova? L, Krascsenits Z, Gauglitz G (2005b) Automated Water Analyser Computer Supported System (AWACSS) Part I: Project objectives, basic technology, immunoassay development, software design and networking. Biosens Bioelectron 20:1499–1508
16. Tschmelak J, Proll G, Riedt J, Kaiser J, Kraemmer P, rzaga L, Wilkinson JS, Hua P, Hole JP, Nudd R, Jackson M, Abuknesha R, Barcelo D, Rodriguez-Mozaz S, Lopez De Alda MJ, Sacher F, Stien J, Slobodnik J, Oswald P, Kozmenko H, Korenkova E, Tothova L, Krascsenits Z, Gauglitz G (2005a) Automated Water Analyser Computer Supported System (AWACSS) Part II: Intelligent, remote-controlled, cost-effective, on-line, water-monitoring measurement system. Biosens Bioelectron 20:1509–1519
17. Mallat E, Barzen C, Abuknesha R, Gauglitz G, Barcelo D (2001) Part per trillion level determination of isoproturon in certified and estuarine water samples with a direct optical immunosensor. Anal Chim Acta 426:209–216

18. Lazcka O, Campo FJD, oz FX (2007) Pathogen detection: a perspective of traditional methods and biosensors. Biosens Bioelectron 22:1205–1217
19. Schobel U, Barzen C, Gauglitz G (2000) Immunoanalytical techniques for pesticide monitoring based on fluorescence detection. Fresenius' J Anal Chem 366:646–658
20. Butala HD, Sadana A (2003) A fractal analysis of analyte-estrogen receptor binding and dissociation kinetics using biosensors: environmental effects. J Colloid Interf Sci 263:420–431
21. Lim TK, Oyama M, Ikebukuro K, Karube I (2000) Detection of atrazine based on the SPR determination of P450 mRNA levels in *Saccharomyces cerevisiae*. Anal Chem 72:2856–2860
22. Farre M, Martinez E, Ramon J, Navarro A, Radjenovic J, Mauriz E, Lechuga L, Marco MP, Barcelo D (2007) Part per trillion determination of atrazine in natural water samples by a surface plasmon resonance immunosensor. Anal Bioanal Chem 388:207–214
23. Mauriz E, Calle A, Manclus JJ, Montoya A, Lechuga LM (2007b) On-line determination of 3,5,6-trichloro-2-pyridinol in human urine samples by surface plasmon resonance immunosensing. Anal Bioanal Chem 387:2757–2765
24. Minunni M, Mascini M, Carter RM, Jacobs MB, Lubrano GJ, Guilbault GG (1996) A quartz crystal microbalance displacement assay for *Listeria monocytogenes*. Anal Chim Acta 325:169–174
25. Storri S, Santoni T, Minunni M, Mascini M (1998) Surface modifications for the development of piezoimmunosensors. Biosens Bioelectron 13:347–357
26. Cass AEG, Davis G, Francis GD, Allen OH, Aston WJ, John Higgins I, Plotkin EV, Scott LDL, Turner APF (1984) Ferrocene-mediated enzyme electrode for amperometric determination of glucose. Anal Chem 56:667–671
27. Kulys J, Vidziunaite R, Janciene R, Palaima A (2006) Spectroelectrochemical study of N-substituted phenoxazines as electrochemical labels of biomolecules. Electroanalysis 18:1771–1777
28. Andreescu S, Marty JL (2006) Twenty years research in cholinesterase biosensors: From basic research to practical applications. Biomol Eng 23:1–15
29. Ivanov AN, Evtugyn GA, nyi RE, th K, Budnikov HC (2000) Comparative investigation of electrochemical cholinesterase biosensors for pesticide determination. Anal Chim Acta 404:55–65
30. Perez Pita MT, Reviejo AJ, Manuel De Villena FJ, Pingarron JM (1997) Amperometric selective biosensing of dimethyl- and diethyldithiocarbamates based on inhibition processes in a medium of reversed micelles. Anal Chim Acta 340:89–97
31. Moore E, Pravda M, Guilbault GG (2003) Development of a biosensor for the quantitative detection of 2,4,6-trichloroanisole using screen printed electrodes. Anal Chim Acta 484:15–24
32. Wilmer M, Trau D, Renneberg R, Spener F (1997) Amperometric immunosensor for the detection of 2,4-dichlorophenoxyacetic acid (2,4-D) in water. Anal Lett 30:515–525
33. Mita DG, Attanasio A, Arduini F, Diano N, Grano V, Bencivenga U, Rossi S, Amine A, Moscone D (2007) Enzymatic determination of BPA by means of tyrosinase immobilized on different carbon carriers. Biosens Bioelectron 23:60–65
34. Piao MH, Noh HB, Rahman M, Won MS, Shim YB (2008) Label-free detection of bisphenol A using a potentiometric immunosensor. Electroanalysis 20:30–37
35. Notsu H, Tatsuma T, Fujishima A (2002) Tyrosinase-modified boron-doped diamond electrodes for the determination of phenol derivatives. J Electroanal Chem 523:86–92
36. Kochana J, Gala A, Parczewski A, Adamski J (2008) Titania sol-gel-derived tyrosinase-based amperometric biosensor for determination of phenolic compounds in water samples. Examination of interference effects. Anal Bioanal Chem 391:1275–1281
37. Rippeth JJ, Gibson TD, Hart JP, Hartley IC, Nelson G (1997) Flow-injection detector incorporating a screen-printed disposable amperometric biosensor for monitoring organophosphate pesticides. Analyst 122:1425–1429
38. Lei Y, Mulchandani P, Chen W, Mulchandani A (2007) Biosensor for direct determination of fenitrothion and epn using recombinant Pseudomonas putida JS444 with surface-expressed organophosphorous hydrolase. 2. Modified carbon paste electrode. Appl Biochem Biotechnol 136:243–250

39. Lucarelli F, Kicela A, Palchetti I, Marrazza G, Mascini M (2002a) Electrochemical DNA biosensor for analysis of wastewater samples. Bioelectrochemistry 58:113–118
40. Mugweru A, Wang B, Rusling J (2004) Voltammetric sensor for oxidized DNA using ultrathin films of osmium and ruthenium metallopolymers. Anal Chem 76:5557–5563
41. Del Carlo M, Di Marcello M, Perugini M, Ponzielli V, Sergi M, Mascini M, Compagnone D (2008) Electrochemical DNA biosensor for polycyclic aromatic hydrocarbon detection. Microchim Acta 163:163–169
42. Babkina SS, Ulakhovich NA (2005) Complexing of heavy metals with DNA and new bioaffinity method of their determination based on amperometric DNA-based biosensor. Anal Chem 77:5678–5685
43. Liu N, Cai X, Lei Y, Zhang Q, Chan-Park MB, Li C, Chen W, Mulchandani A (2007) Single-walled carbon nanotube based real-time organophosphate detector. Electroanalysis 19:616–619
44. Sacks V, Eshkenazi I, Neufeld T, Dosoretz C, Rishpon J (2000) Immobilized parathion hydrolase: An amperometric sensor for parathion. Anal Chem 72:2055–2058
45. Ma Y, Jiao K, Yang T, Sun D (2008) Sensitive PAT gene sequence detection by nano-SiO2/p-aminothiophenol self-assembled films DNA electrochemical biosensor based on impedance measurement. Sens Actuators B Chem 131:565–571
46. Tsai YC, Chiu CC (2007) Amperometric biosensors based on multiwalled carbon nanotube-Nafion-tyrosinase nanobiocomposites for the determination of phenolic compounds. Sens Actuators B Chem 125:10–16
47. Fahnrich KA, Pravda M, Guilbault GG (2003) Disposable amperometric immunosensor for the detection of polycyclic aromatic hydrocarbons (PAHs) using screen-printed electrodes. Biosens Bioelectron 18:73–82
48. Carralero V, Gonzalez-Cortez A, Yañez-Sedeño P, Pingarron JM (2007) Nanostructured progesterone immunosensor using a tyrosinase-colloidal gold-graphite-Teflon biosensor as amperometric transducer. Anal Chim Acta 596:86–91
49. Taranova LA, Fesay AP, Ivashchenko GV, Reshetilov AN, Winther-Nielsen M, Emneus J (2004) *Comamonas testosteroni* strain TI as a potential base for a microbial sensor detecting surfactants. Appl Biochem Microbiol 40:404–408
50. Farre M, Pasini O, Carmen Alonso M, Castillo M, Barcelo D (2001) Toxicity assessment of organic pollution in wastewaters using a bacterial biosensor. Anal Chim Acta 426:155–165
51. Shiddiky MJA, Rahman M, Cheol CS, Shim YB (2008) Fabrication of disposable sensors for biomolecule detection using hydrazine electrocatalyst. Anal Biochem 379:170–175
52. Kim GY, Shim J, Kang MS, Moon SH (2008a) Optimized coverage of gold nanoparticles at tyrosinase electrode for measurement of a pesticide in various water samples. J Hazard Mater 156:141–147
53. Kim GY, Shim J, Kang MS, Moon SH (2008b) Preparation of a highly sensitive enzyme electrode using gold nanoparticles for measurement of pesticides at the ppt level. J Environ Monitor 10:632–637
54. Dutta K, Bhattacharyay D, Mukherjee A, Setford SJ, Turner APF, Sarkar P (2008) Detection of pesticide by polymeric enzyme electrodes. Ecotoxicol Environ Safe 69:556–561
55. Gooding JJ, Wibowo R, Liu J, Yang W, Losic D, Orbons S, Mearns FJ, Shapter JG, Hibbert DB (2003) Protein electrochemistry using aligned carbon nanotube arrays. J Am Chem Soc 125:9006–9007
56. Yu X, Chattopadhyay D, Galeska I, Papadimitrakopoulos F, and Rusling JF (2003) Peroxidase activity of enzymes bound to the ends of single-wall carbon nanotube forest electrodes. Electrochem Commun 5:408–411
57. Xiao Y, Patolsky F, Katz E, Hainfeld JF, Willner I (2003) Plugging into enzymes: Nanowiring of redox enzymes by a gold nanoparticle. Science 299:1877–1881
58. Wang J, Musameh M, Lin Y (2003b) Solubilization of carbon nanotubes by Nafion toward the preparation of amperometric biosensors. J Am Chem Soc 125:2408–2409
59. Musameh M, Wang J, Merkoci A, Lin Y (2002) Low-potential stable NADH detection at carbon-nanotube-modified glassy carbon electrodes. Electrochem Commun 4:743–746

60. Vaseashta A, Vaclavikova M, Vaseashta S, Gallios G, Roy P, Pummakarnchana O (2007) Nanostructures in environmental pollution detection, monitoring, and remediation. Sci Technol Adv Mater 8:47–59
61. Wang J, Dai J, Yarlagadda T (2005) Carbon nanotube-conducting-polymer composite nanowires. Langmuir 21:9–12
62. You T, Niwa O, Kurita R, Iwasaki Y, Hayashi K, Suzuki K, Hirono S (2004) Reductive H2O2 detection at nanoparticle Iridium/Carbon film electrode and its application as L-glutamate enzyme sensor. Electroanalysis 16:54–59
63. Ramanathan K, Bangar MA, Yun M, Chen W, Mulchandani A, Myung NV (2004) Individually addressable conducting polymer nanowires array. Nano Lett 4:1237–1239
64. Yang M, Wang J, Li H, Zheng JG, Nick Wu N (2008) A lactate electrochemical biosensor with a titanate nanotube as direct electron transfer promoter. Nanotechnology 19:
65. Trau D, Theuerl T, Wilmer M, Meusel M, Spener F (1997) Development of an amperometric flow injection immunoanalysis system for the determination of the herbicide 2,4-dichlorophenoxyacetic acid in water. Biosens Bioelectron 12:499–510
66. Grennan K, Strachan G, Porter AJ, Killard AJ, Smyth MR (2003) Atrazine analysis using an amperometric immunosensor based on single-chain antibody fragments and regeneration-free multi-calibrant measurement. Anal Chim Acta 500:287–298
67. O'Sullivan CK, Guilbault GG (1999) Commercial quartz crystal microbalances – theory and applications. Biosens Bioelectron 14:663–670
68. O'Connell PJ, Guilbault GG (2001) Future trends in biosensor research. Anal Lett 34:1063–1078
69. Schipper EF, Rauchalles S, Kooyman RPH, Hock B, Greve J (1998) The waveguide Mach-Zender interferometer as atrazine sensor. Anal Chem 70:1192–1197
70. Brummel KE, Wright J, Eldefrawi ME (1997) Fiber Optic Biosensor for Cyclodiene Insecticides. J Agric Food Chem 45:3292–3298
71. Kroger D, Liley M, Schiweck W, Skerra A, Vogel H (1999) Immobilization of histidine-tagged proteins on gold surfaces using chelator thioalkanes. Biosens Bioelectron 14:155–161
72. Nabok AV, Tsargorodskaya A, Hassan AK, Starodub NF (2005) Total internal reflection ellipsometry and SPR detection of low molecular weight environmental toxins. Appl Surf Sci 246:381–386
73. Shimomura M, Nomura Y, Lee KH, Ikebukuro K, Karube I (2001a) Dioxin detection based on immunoassay using a polyclonal antibody against octa-chlorinated dibenzo-p-dioxin (OCDD). Analyst 126:1207–1209
74. Shimomura M, Nomura Y, Zhang W, Sakino M, Lee KH, Ikebukuro K, Karube I (2001b) Simple and rapid detection method using surface plasmon resonance for dioxins, polychlorinated biphenylx and atrazine. Anal Chim Acta 434:223–230
75. Mauriz E, Calle A, Lechuga LM, Quintana J, Montoya A, Manclus JJ (2006a) Real-time detection of chlorpyrifos at part per trillion levels in ground, surface and drinking water samples by a portable surface plasmon resonance immunosensor. Anal Chim Acta 561:40–47
76. Mauriz E, Calle A, Manclus JJ, Montoya A, Escuela AM, Sendra JR, Lechuga LM (2006b) Single and multi-analyte surface plasmon resonance assays for simultaneous detection of cholinesterase inhibiting pesticides. Sens Actuators B Chem 118:399–407
77. Tang DP, Yuan R, Chai YQ (2006) Novel immunoassay for carcinoembryonic antigen based on protein A-conjugated immunosensor chip by surface plasmon resonance and cyclic voltammetry. Bioproc Biosyst Eng 28:315–321
78. Aizawa H, Tozuka M, Kurosawa S, Kobayashi K, Rddy SM, Higuchi M (2007) Surface plasmon resonance-based trace detection of small molecules by competitive and signal enhancement immunoreaction. Anal Chim Acta 591:191–194
79. Skladal P (1999) Effect of methanol on the interaction of monoclonal antibody with free and immobilized atrazine studied using the resonant mirror-based biosensor. Biosens Bioelectron 14:257–263

80. Harris RD, Luff BJ, Wilkinson JS, Piehler J, Brecht A, Gauglitz G, Abuknesha RA (1999) Integrated optical surface plasmon resonance immunoprobe for simazine detection. Biosens Bioelectron 14:377–386
81. Bier FF, Jockers R, Schmid RD (1994) Integrated optical immunosensor for s-triazine determination: regeneration, calibration and limitations. Analyst 119:437–441
82. Mauriz E, Calle A, Manclus JJ, Montoya A, Hildebrandt A, Barcelo D, Lechuga LM (2007a) Optical immunosensor for fast and sensitive detection of DDT and related compounds in river water samples. Biosens Bioelectron 22:1410–1418
83. Abdullah J, Ahmad M, Karuppiah N, Heng LY, Sidek H (2006) Immobilization of tyrosinase in chitosan film for an optical detection of phenol. Sens Actuators B Chem 114:604–609
84. Troegl J, Ripp S, Kuncova G, Sayler GS, Churava A, Parik P, Demnerova K, Halova J, Kubicova L (2005) Selectivity of whole cell optical biosensor with immobilized bioreporter Pseudomonas fluorescens HK44. Sens Actuators B Chem 107:98–103
85. Tschmelak J, Kumpf M, ppel N, Proll G, Gauglitz G (2006) Total internal reflectance fluorescence (TIRF) biosensor for environmental monitoring of testosterone with commercially available immunochemistry: antibody characterization, assay development and real sample measurements. Talanta 69:343–350
86. Marchesini GR, Meulenberg E, Haasnoot W, Irth H (2005) Biosensor immunoassays for the detection of bisphenol A. Anal Chim Acta 528:37–45
87. Rose A, Nistor C, us J, Pfeiffer D, Wollenberger U (2002) GDH biosensor based off-line capillary immunoassay for alkylphenols and their ethoxylates. Biosens Bioelectron 17:1033–1043
88. Evtugyn GA, Eremin SA, Shaljamova RP, Ismagilova AR, Budnikov HC (2006) Amperometric immunosensor for nonylphenol determination based on peroxidase indicating reaction. Biosens Bioelectron 22:56–62
89. Drevensek P, Zupancic T, Pihlar B, Jerala R, Kolitsch U, Plaper A, Turel I (2005) Mixed-valence Cu(II)/Cu(I) complex of quinolone ciprofloxacin isolated by a hydrothermal reaction in the presence of l-histidine: comparison of biological activities of various copper-ciprofloxacin compounds. J Inorg Biochem 99:432–442
90. Lopez-Ortal P, Souza V, Bucio L, Gonzalez E, Gutierrez-Ruiz M (1999) DNA damage produced by cadmium in a human fetal hepatic cell line. Mutat Res Genet Toxicol Environ Mutagen 439:301–306
91. Marrazza G, Chianella I, Mascini M (1999b) Disposable DNA electrochemical sensor for hybridization detection. Biosens Bioelectron 14:43–51
92. Marrazza G, Chianella I, Mascini M (1999a) Disposable DNA electrochemical biosensors for environmental monitoring. Anal Chim Acta 387:297–307
93. Doong RA, Shih HM, Lee SH (2005) Sol-gel-derived array DNA biosensor for the detection of polycyclic aromatic hydrocarbons in water and biological samples. Sens Actuators B Chem 111–112:323–330
94. Chiorcea Paquim AM, Diculescu VC, Oretskaya TS, Oliveira Brett AM (2004) AFM and electroanalytical studies of synthetic oligonucleotide hybridization. Biosens Bioelectron 20:933–944
95. Bakker E (2006) Is the DNA sequence the gold standard in genetic testing? Quality of molecular genetic tests assessed. Clin Chem 52:557–558
96. Lazarides AA, Lance Kelly K, Jensen TR, Schatz GC (2000) Optical properties of metal nanoparticles and nanoparticle aggregates important in biosensors. J Mol Struct Theochem 529:59–63
97. Lucarelli F, Palchetti I, Marrazza G, Mascini M (2002b) Electrochemical DNA biosensor as a screening tool for the detection of toxicants in water and wastewater samples. Talanta 56:949–957
98. Lucarelli F, Tombelli S, Minunni M, Marrazza G, Mascini M (2008) Electrochemical and piezoelectric DNA biosensors for hybridisation detection. Anal Chim Acta 609:139–159
99. Gooding JJ (2002) Electrochemical DNA hybridization biosensors. Electroanalysis 14:1149–1156

100. Wang J, Xu D, Kawde AN, Polsky R (2001b) Metal nanoparticle-based electrochemical stripping potentiometric detection of DNA hybridization. Anal Chem 73:5576–5581
101. Dequaire M, Degrand C, Limoges B (2000) An electrochemical metalloimmunoassay based on a colloidal gold label. Anal Chem 72:5521–5528
102. Cai H, Xu Y, Zhu N, He P, Fang Y (2002) An electrochemical DNA hybridization detection assay based on a silver nanoparticle label. Analyst 127:803–808
103. Wang J, Liu G, Merkoci A (2003a) Particle-based detection of DNA hybridization using electrochemical stripping measurements of an iron tracer. Anal Chim Acta 482:149–155
104. Wang J, Xu D, Polsky R (2002) Magnetically-induced solid-state electrochemical detection of DNA hybridization. J Am Chem Soc 124:4208–4209
105. Wang J, Polsky R, Xu D (2001a) Silver-enhanced colloidal gold electrochemical stripping detection of DNA hybridization. Langmuir 17:5739–5741
106. Kawde AN, Wang J (2004) Amplified electrical transduction of DNA hybridization based on polymeric beads loaded with multiple gold nanoparticle tags. Electroanalysis 16:101–107
107. Liu G, Wang J, Kim J, Jan MR, Collins GE (2004) Electrochemical coding for multiplexed immunoassays of proteins. Anal Chem 76:7126–7130
108. Park HJ, Hwang KO, Kim ES (2002) Construction and characterization of a recombinant bioluminescence streptomycetes for potential environmental monitoring. J Microbiol Biotechnol 12:706–709
109. Hahm JI, Lieber CM (2004) Direct ultrasensitive electrical detection of DNA and DNA sequence variations using nanowire nanosensors. Nano Lett 4:51–54
110. Koehne JE, Chen H, Cassell AM, Ye Q, Han J, Meyyappan M, Li J (2004) Miniaturized multiplex label-free electronic chip for rapid nucleic acid analysis based on carbon nanotube nanoelectrode arrays. Clin Chem 50:1886–1893
111. Mascini M, Palchetti I, Marrazza G (2001) DNA electrochemical biosensors. Anal Bioanal Chem 369:15–22
112. Lucarelli F, Authier L, Bagni G, Marrazza G, Baussant T, Aas E, Mascini M (2003) DNA biosensor investigations in fish bile for use as a biomonitoring tool. Anal Lett 36:1887–1901
113. Feng Y, Yang T, Zhang W, Jiang C, Jiao K (2008) Enhanced sensitivity for deoxyribonucleic acid electrochemical impedance sensor: gold nanoparticle/polyaniline nanotube membranes. Anal Chim Acta 616:144–151
114. Fu XH (2008) Electrochemical measurement of DNA hybridization using nanosilver as label and horseradish peroxidase as enhancer. Bioprocess Biosyst Eng 31:69–73
115. LaGier MJ, Fell JW, Goodwin KD (2007) Electrochemical detection of harmful algae and other microbial contaminants in coastal waters using hand-held biosensors. Marine Pollut Bull 54:757–770
116. Chan C, Lehmann M, Chan K, Chan P, Chan C, Gruendig B, Kunze G, Renneberg R (2000) Designing an amperometric thick-film microbial BOD sensor. Biosens Bioelectron 15:343–353
117. Liu J, Mattiasson B (2002) Microbial BOD sensors for wastewater analysis. Water Res 36:3786–3802
118. Tan TC, Qian Z (1997) Dead Bacillus subtilis cells for sensing biochemical oxygen demand of waters and wastewaters. Sens Actuators B Chem 40:65–70
119. Lei Y, Chen W, Mulchandani A (2006) Microbial biosensors. Anal Chim Acta 568:200–210
120. Sangeetha S, Sugandhi G, Murugesan M, Murali Madhav V, Berchmans S, Rajasekar R, Rajasekar S, Jeyakumar D, Prabhakara Rao G (1996) Torulopsis Candida based sensor for the estimation of biochemical oxygen demand and its evaluation. Electroanalysis 8:698–701
121. Marty JL, Olive D, Asano Y (1997) Measurement of BOD: correlation between 5-day BOD and commercial BOD biosensor values. Environ Technol 18:333–337
122. Chee GJ, Nomura Y, Karube I (1999) Biosensor for the estimation of low biochemical oxygen demand. Anal Chim Acta 379:185–191

123. Ohki A, Shinohara K, Ito O, Naka K, Maeda S, Sato T, Akano H, Kato N, Kawamura Y (1994) A BOD sensor using Klebsiella oxytoca AS1. Int J Environ Anal Chem 56:261–269
124. Riedel K, Lehmann M, Tag K, Renneberg R, Kunze G (1998) Arxula adeninivorans based sensor for the estimation of BOD. Anal Lett 31:1–12
125. Tag K, Lehmann M, Chan C, Renneberg R, Riedel K, Kunze G (2000) Measurement of biodegradable substances with a mycelia-sensor based on the salt tolerant yeast Arxula adeninivorans LS3. Sens Actuators B Chem 67:142–148
126. Diamond D, Lau KT, Brady S, Cleary J (2008) Integration of analytical measurements and wireless communications-Current issues and future strategies. Talanta 75:606–612
127. Kim M, Youn SM, Shin SH, Jang JG, Han SH, Hyun MS, Gadd GM, Kim HJ (2003) Practical field application of a novel BOD monitoring system. J Environ Monitor 5:640–643
128. Li J, Wang S, VanDusen WJ, Schultz LD, George HA, Herber WK, Chae HJ, Bentley WE, Rao G (2000) Green fluorescent protein in Saccharomyces cerevisiae: Real-time studies of the GAL1 promoter. Biotechnol Bioeng 70:187–196
129. Datskou I, Rajic S, Datskos PG (1999) Novel magnetic and chemical micro sensors for in-situ, real-time, and unattended use. Proc SPIE 3713:85–93
130. Cai QY, Jain MK, Grimes CA (2001) A wireless, remote query ammonia sensor. Sens Actuators B Chem 77:614–619
131. Wu S, Gao X, Cai Q, and Grimes CA (2007) A wireless magnetoelastic biosensor for convenient and sensitive detection of acid phosphatase. Sens Actuators B Chem 123:856–859
132. Persaud KC, Pisanelli AM, Szyszko S, Reichl M, Horner G, Rakow W, Keding HJ, Wessels H (1999) Smart gas sensor for monitoring environmental changes in closed systems: results from the MIR space station. Sens Actuators B Chem 55:118–126
133. Carotta MC, Martinelli G, Crema L, Gallana M, Merli M, Ghiotti G, Traversa E (2000) Array of thick film sensors for atmospheric pollutant monitoring. Sens Actuators B Chem 68:1–8
134. Carotta MC, Martinelli G, Crema L, Malagu C, Merli M, Ghiotti G, Traversa E (2001) Nanostructured thick-film gas sensors for atmospheric pollutant monitoring: quantitative analysis on field tests. Sens Actuators B Chem 76:336–342
135. Zen M, Butturi MA, Carotta MC, Cavicchi B, Ferroni M, Malagu C, Martinelli G, Vincenzi D (2002) Gas sensing through thick film technology. Sens Actuators B Chem 84:78–82
136. Bendikov TA, Kim J, Harmon TC (2005) Development and environmental application of a nitrate selective microsensor based on doped polypyrrole films. Sens Actuators B Chem 106:512–517
137. Bhattacharyya J, Read D, Amos S, Dooley S, Killham K, Paton GI (2005) Biosensor-based diagnostics of contaminated groundwater: assessment and remediation strategy. Environ Pollut 134:485–492
138. Han TS, Sasaki S, Yano K, Ikebukuro K, Kitayama A, Nagamune T, Karube I (2002) Flow injection microbial trichloroethylene sensor. Talanta 57:271–276
139. Allan IJ, Vrana B, Greenwood R, Mills GA, Knutsson J, Holmberg A, Guigues N, Fouillac AM, Laschi S (2006) Strategic monitoring for the European water framework directive. Trends Anal Chem 25:704–715
140. Haes AJ, Van Duyne RP (2002) A nanoscale optical biosensor: Sensitivity and selectivity of an approach based on the localized surface plasmon resonance spectroscopy of triangular silver nanoparticles. J Am Chem Soc 124:10596–10604

Hdb Env Chem (2009) 5J: 33–46
DOI:10.1007/698_2008_15

Achievements of the RIANA and AWACSS EU Projects: Immunosensors for the Determination of Pesticides, Endocrine Disrupting Chemicals and Pharmaceuticals

Sara Rodriguez-Mozaz, Maria J. López de Alda, and Damià Barceló

Abstract In the last 15 years the research projects supported by the European Commission River Analyser (RIANA) and Automated Water Analyser Computer Supported System (AWACSS) have developed quick, intelligent and cost-effective biosensors that allow the monitoring of different organic pollutants, ranging from regulated ones such as pesticides to emerging contaminants such as pharmaceuticals and endocrine disrupting compounds. Herein we give an overview of the aims, scope and main achievements of the RIANA and AWACSS projects as well as details about basic technology, immunoassays, software and networking developed within the research project. We also report on the systems performance, real water sample measurements and validation of the biosensors with conventional analytical methods. The biosensors developed were able to measure several organic pollutants at low nanogram per litre level in an analysis taking only a few minutes without any prior sample pre-concentration or pre-treatment steps. This work is a proof-of-concept that biosensors are a practical alternative or complementary methodology to traditional chromatographic techniques.

Keywords Automated Water Analyser Computer Supported System, Environmental monitoring, Emerging pollutants, Immunoassay, Network system, Optical immunosensor, River Analyser, Validation

S. Rodriguez-Mozaz (✉), M.J. López de Alda, and D. Barceló
Department of Environmental Chemistry, IIQAB-CSIC, C/Jordi Girona Salgado 18-26, 08034, Barcelona, Spain
e-mail: sarqam@cid.csic.es

D. Barceló
Catalan Institute for Water Research (ICRA), Parc Científic i Tecnològic de la Universitat de Girona, Edifici Jaume Casademont, C/Pic de Peguera, 15, 17003, Girona, Spain

Contents

1 Introduction

Environmental and water pollution is a global problem which has led to increasing water quality demands and to more stringent legislation measures in the European Union (EU). Standardised methods for water monitoring need to be developed and implemented throughout Europe to ensure effective enforcement of EU water quality directives. Until now, environmental monitoring has been accomplished by the use of classical analytical tools such as liquid and gas chromatography, which are often lengthy and limit the number of analytes that can be determined in a single run to specific chemical families [1]. Therefore, new tools are demanded for the simultaneous monitoring of organic micro-pollutants in water in real-time and at reasonable cost. In the last 15 years two research projects supported by the European Commission – River Analyser (RIANA) and Automated Water Analyser Computer Supported System (AWACSS) – have developed quick, intelligent and cost-effective instruments that allow the simultaneous detection of different organic pollutants, ranging from regulated ones such as pesticides to emerging contaminants such as pharmaceuticals and endocrine disrupting compounds (EDCs). Herein we intend to provide an overview of the main practical achievements of these research projects. Two different biosensor prototypes were developed for a different set of trace organic contaminants and applied in different water matrices during the projects course. The biosensor RIANA (1995–1999; ENV4-CT95-0066) was developed within the project with the same name, funded by the European Commission under the Environment and Climate Program (Fourth Framework Program). Three RIANA biosensor prototypes were constructed by Perkin Elmer (Ueberlingen, Germany). The system was optimised toward ultra-sensitive detection of pesticides and EDCs in water samples. A new more advanced prototype was constructed by

Central Research Laboratories (CRL, Middlesex, UK) within the AWACSS project (2001–2004) (EVK1-CT-2000-00045) under the Fifth Framework Programme. The AWACSS consortium included nine European partners from four different European countries. Four partners out of the nine were also part of the former RIANA consortium. This new and versatile, multi-sensor system was designed based on reliable and low-cost and instrumentation for a quick multi-analyte analysis. Another objective of the AWACSS project was the design of an intelligent warning system with a number of remote measurement stations linked by a communications network.

2 Achievements of the RIANA and AWACSS Projects

The main achievements attained during the RIANA and AWACSS projects are summarised in the following sections. Key features of the two biosensors developed in the corresponding projects are also reviewed.

2.1 RIANA System: Key Features

The RIANA immunosensor was described in detail by Klotz et al. [2]. Basically, the system consists of a flow injection system, a transducer mounted in a flow cell and an optical excitation and detection system. The flow injection system, FIAS 3000, from Perkin Elmer (Ueberlingen, Germany), consists of a six-way distribution valve equipped with a 1-mL syringe pump and connected through a Teflon tubing to the flow cell (dimensions of 40 × 1.7 × 0.06 mm). The transducer is mounted in this flow cell, which includes a flow channel of 1.7 mm width and 0.1 mm depth, and is sealed by an O-ring. Light from a collimated and modulated He–Ne laser source (633 nm, 7 mW) is directly coupled into a glass-slide via a bevelled end-face and guided through the length of the transducer by total internal reflection. An evanescent field is produced at each reflection spot and penetrates a few hundreds of nanometres into the external medium. This configuration avoids interferences from bulk fluorescence and permits the excitation of fluorescently labelled antibodies that are locally and specifically bound to the analyte derivatives. Fluorescence light is subsequently collected through optic fibres, filtered, and detected by photodiodes using lock-in detection. An autosampler AS90/91 from Perkin Elmer is used to deliver the samples to the FIA system. Fluid handling and data acquisition is fully automated and computer controlled.

Immunochemical techniques can be particularly suited for the measurement of organic pollutants because of their specificity, high sensitivity, adaptability for field use and ability to recognise a wide range of substances [1]. Both in the RIANA and AWACSS systems, a solid phase fluoroimmunoassay was selected as an appropriate option to meet the requirements of a useful monitoring immunoprobe. In this immunoassay the immobilisation of an analyte derivate was preferred to the immobilisation of labile antibodies in order to increase the stability and shelf life of the transducer. The configuration also enables its regeneration for re-use without loss of activity, thus allowing semi-continuous water monitoring.

Polyclonal antibodies were prepared against target analytes by immunising sheep with the corresponding hapten–protein conjugates. After about 8 months pure antibodies were isolated from the polyclonal antisera of the sheep by affinity chromatography. After purification, the antibodies were labelled with a fluorescent marker. The immunochemistry utilised in the project takes advantage of a binding inhibition test that requires that either the antibodies directed against specific analytes or analyte derivatives can be covalently bound to a transducer surface. The immobilisation of the immunoassay component in the surface of the transducer was achieved by binding aminodextran derivatives of the target analytes on separated areas on the transducer glass slide following the procedure described by Barzen et al. [3]. The objective of achieving spatially resolved excitation and collection of fluorescence from fluorescently labelled antibodies locally bound at a planar interface can be met by the evanescent field excitation of the fluorophore. Excitation light is guided by total internal reflection within the transducer structure resulting in an evanescent wave which allows the excitation of the fluorophores bound to the transducer surface (Fig. 1). The total internal reflection fluorescence (TIRF) principle has important advantages compared to direct illumination of the active area of the transducer: it allows selective detection of surface bound fluorophores and, therefore, on-line monitoring of binding events [1].

2.2 AWACSS System: Advanced Features

On the basis of the same immunoassay and optical detection principle of the biosensor RIANA, the new device AWACSS was designed to overcome RIANA drawbacks, with major improvements in three critical areas: (1) novel design approaches to the optical detection and fluidics including miniaturised integrated optics and micro-fluidics; (2) expanded multi-analyte analysis capability allowing for simultaneous measurements of up to 32 analytes; and (3) intelligent remote surveillance and control for unattended continuous monitoring [5].

With the RIANA design, up to six different analytes can be determined simultaneously, whereas the AWACSS instrument allows the performance of up to 32 different immunoassays simultaneously. The AWACSS biosensor is based on the same fluorescence detection principle but the fibre-pigtailed chip, driven by a semiconductor laser, consists of a waveguide circuit that distributes excitation light to 32 separate sensing patches on the chip surface. A fibre-coupled detection array is used to monitor the 32 separate fluorescence signals. The optics with 32 channels and 32 photodiodes was a major part of the AWACSS instrument design [5]. In addition, both the optical bench and the photodiode block were fitted with a semiconductor temperature detector. Algorithms to correct for temperature variation

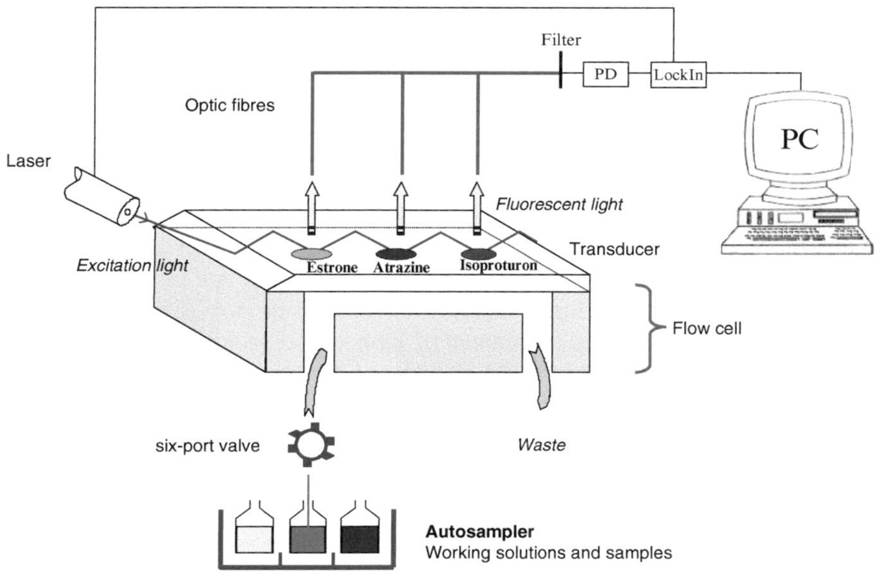

Fig. 1 Scheme of the immunosensor RIANA for a particular application, the simultaneous analysis of atrazine, isoproturon and estrone. The FIA system, equipped with a six-port valve and a 1-mL syringe pump, delivers the buffer solution, the sample or the regeneration solutions to the flow-cell. The transducer – a 1.5-mm thick surface polished sheet glass –is mounted on the flow cell and is sealed with an O-ring. The excitation light (He–Ne laser) is totally internally reflected at the transducer surface. The collected fluorescent light is filtered and detected by photodiodes. Reprinted from [4] with permission of Elsevier

(positive or negative trends) can therefore be programmed into the measurement software. The electronics with low noise, linear amplifier and filter behaviour, external data conversion, and RS232 communications with an external data acquisition unit was the second major part. The electronics design required the assembly of 32 independent linear amplifier circuits. The fluidics box with fully automated sample handling, micro-fluidic flow channel, and interfacing to autosampler was the third major part of the instrument design. The enclosure part with high-quality RF shielding and three separate sections for optics/electronics/fluidics brought the three major parts together. The casing for the instrument was designed as three compartments, separating the main electrical components from the optical bench and amplifier circuits. In this way it was possible to screen the outputs from the 32 channels from digital electronic interference. The optical bench and electronic amplifier compartments were also made light tight to prevent stray light from impinging on the sensor chip or photodiodes. The inside surfaces of the light tight compartments were also blackened to eliminate the possibility of reflected light interfering with the signal [5].

2.3 Limit of Detection

Immunoanalytical methods with very low limits of detection (LOD) and low limits of quantification (LOQ) are becoming increasingly important for environmental analysis and, especially, for monitoring drinking-water quality. Tschmelak et al. devoted some efforts to improve the limit of detection of the biosensor assay. The basis of a sensitive assay is an antibody with a high affinity constant toward the analyte [6, 7]. Reducing the amount of antibody per sample results in better validation parameters and lower limits of detection [7]. During the optimisation process, Tschmelak et al. compared different surface modifications (four different immobilised derivatives) and reduced the amount of antibody per sample. The optimisation of the immunoassay resulted in very low limits of detection in water: 0.6, 0.37 and 2 ng L^{-1} for propanil, [8] progesterone [6] and estrone [7], respectively, without any sample pre-treatment or pre-concentration. These low LOD achieved, in the nanogram per litre range, show that biosensors can compete with common analytical methods in the field of water analysis [6].

2.4 Multi-Analyte Determination

Sensors capable of determining several analytes simultaneously allow shortening of the time of analysis and the volume of sample and other reagents, and thus constitute a valuable tool for environmental monitoring. The development of multi-analyte immunoassays was one of the aims of the RIANA project. The spatially resolved immobilisation of the biomolecular recognition elements in transducer surfaces enabled the real-time parallel monitoring of multiple organic contaminants by means of separate immunoreactions. Aminodextran derivatives of the target analytes were covalently bound on separated areas on the transducer glass slide following the procedure described by Barzen et al. [3]. Briefly, a glass interference layer of 58 × 10 × 1.5 mm with a 45° bevel on the short side was cleaned by immersion in a freshly prepared hot mixture of concentrated H_2SO_4/H_2O_2 2:1 for 30 min, rinsed with water afterwards and dried at room temperature. Silanisation of the surface was achieved by treatment with 3-glycidoxypropyltrimethoxysilanxysilane (GOPTs) for not more than 1 h. Thereafter, it was rinsed with acetone and dried in a nitrogen stream. Conjugates of the analyte derivatives with aminodextran were then placed in separated areas of the activated transducer surface allowing the reaction to take place. Such a multi-analyte transducer with predefined detection spots permits the performance of simultaneous multi-analyte measurements.

The first attempts at multi-analyte determination were carried out during the RIANA project [3]. Feasibility studies on multi-analyte measurements were performed with a mixture of atrazine, isoproturon and estrone [9]. The method developed was used to investigate the occurrence and removal of these three compounds throughout the treatment process (sand filtration, ozonation, activated carbon filtration and chlorination) in a waterworks [4]. In the frame of the AWACSS project further efforts were devoted to this multi-analyte aspect; different studies and assays were developed for the determination of pharmaceuticals, antibiotics, hormones, endocrine disrupting chemicals and pesticides. In the final stage of the project, the multi-analyte determination capability of the AWACSS was successfully tested with a mixture of six analytes. The AWACSS chip was modified with derivatives of atrazine, bisphenol A, estrone, isoproturon, sulphamethizole and propanil according to the previously described immobilisation protocol. Subsequently, a simultaneous calibration in Milli-Q water with all six analytes mixed with an antibody stock solution containing the six corresponding polyclonal antibodies (anti-atrazine, anti-bisphenol A, anti-estrone, anti-isoproturon, anti-mixed sulphonamides and anti-propanil) was performed. For all compounds, the calculated LOD was below 0.020 μg L^{-1} [10]. This multi-analyte calibration demonstrated the possibility to quantify pesticides from three different classes (triazines, phenylureas and anilides), EDCs (bisphenol A), steroid hormones (estrone), and pharmaceuticals (sulphamethizole) within one single measurement cycle, which only takes approximately 18 min. No cross-reactivity effect was observed for any of the tested analytes.

2.5 *Automatisation*

Automation and minimal or no sample manipulation are among the most important features in any analytical technique to enhance the precision and accuracy of the measurements and to help diminish potential contamination problems in the analysis process. An autosampler AS90/91 from Perkin Elmer was used in the RIANA project, and initially during the AWACSS project, to deliver the samples to the FIA system of the biosensor. Fluid handling and data acquisition was fully automated and computer controlled. Later, this autosampler was replaced with an autosampler from CTC Analytics, the HTC PAS, which was coupled both to the RIANA and to the new AWACSS device. With this new system there was no need to prepare and incubate samples and standards before analysis since the new autosampler was able to select, mix and inject the samples following a programmed sequence. The new autosampler also permitted improvement in reproducibility, by lowering the standard deviations (RSD), and as a consequence, the LODs.

2.6 *Network System*

Contaminant concentrations in water courses are dynamic, changing both as a result of receiving inputs and changes in water flow. With monthly sampling and analysis it is extremely unlikely that mapping of contamination profiles or obtaining rapid results will occur when it is important, such as after accidental spills or pollution events [11]. The establishment of network analytical systems generates huge interest. These network systems consist of multiple autonomous analytical stations that can control extensively the sites of interest in rivers, lakes, wells or even water treatment plants. In this respect, biosensors can be used as automated continuous monitoring systems that can provide easy, rapid and on-site measurement. One of the main achievements of the AWACSS project was the establishment of an early warning system by means of a network of measurement and control stations [5].

Figure 2 illustrates the capabilities of the AWACSS monitoring network. The network concept is based on remote measurement stations and an internet-accessible database on the server station. The measurement station uploads the measurement data to the server and downloads new parameters for the next measurement. For each analyte, the incoming data on the server side is checked regarding the limit value concentration. Different limit value concentrations are held in the database depending on analyte type, sampling site location and additional factors. A comparison of these values with the incoming analyte concentration values originating from the different sampling stations is performed by a watch dog module. Automatic alert messages and e-mails are generated following the early-warning concept defined in the database and watch dog module [5].

2.7 AWACSS–Marketing Aspects: Water Quality Survey

A market survey covering 12 European countries and the USA was performed to check the feasibility of the AWACSS system in the water quality market [5]. A standardised questionnaire was completed by institutions dealing with water quality matters in the selected countries to help identify end-user requirements. The data from this survey represented the situation in analytical practice in the different

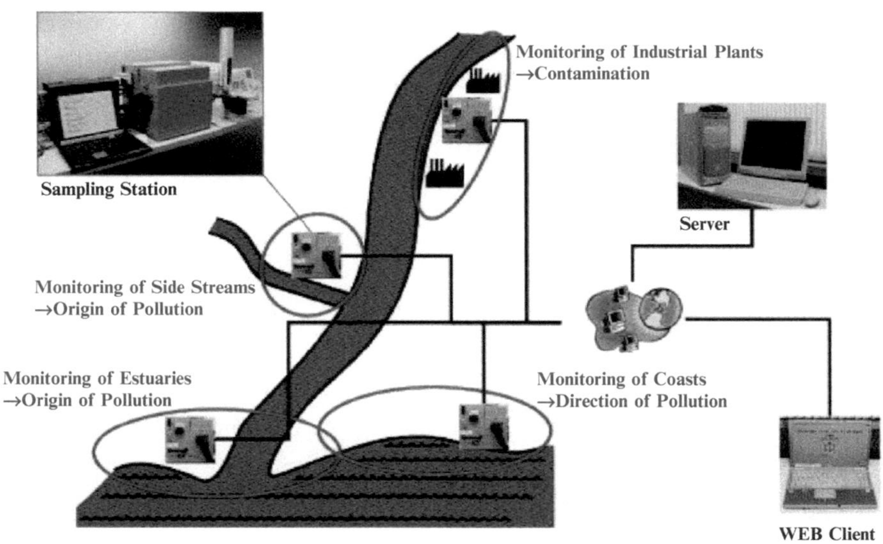

Fig. 2 AWACSS monitoring network. Reprinted from [5] with permission of Elsevier

countries based on the present legislative needs and available laboratory equipment and assays. The collected data covered lists of substances, number of sampling sites, sampling frequencies, number of samples, detection limits, methods applied, number of positive findings, determinations below 0.1 $\mu g\ L^{-1}$, and analytical costs. Real-time sensing, improved regulatory and public acceptance and less expensive monitoring were found to be prime end-user requirements. The AWACSS system not only meets the market demand but exceeds the current competitive detection methods in terms of meeting end-user requirements for current and future expenditures and expectations: A system like AWACSS, equipped to measure up to 32 top compounds, could be well placed to the actual market with increasing acceptance in the future due to the flexibility of adaptation of new compounds as well as the capability of online, unattended and centrally controlled monitoring and surveillance.

3 Environmental Applications

In the previous section, biosensor features were described in terms of analytical achievements and contribution to biosensor field development. Practical aspects related to biosensor performance in the analysis of environmental samples as well as different aspects and achievements of the RIANA and AWACSS projects are reviewed in this section.

3.1 Target Pollutants

A set of immunochemical reagents was developed and generated for selected organic micro-pollutants of interest to be analysed in environmental water samples. Major criteria for selection of target analytes were their presence in the environment, the existing environmental legislation, the increasing interest in certain emerging contaminants and the technical possibilities to prepare sufficiently selective polyclonal antibodies and their corresponding analyte. Analytical methods for the analysis of pesticides, EDCs, WFD priority substances, industrial pollutants and pharmaceuticals have been developed using the biosensor devices from the RIANA and AWACSS projects. Table 1 gathers together the organic pollutants and the water matrices studied and the limits of detection achieved.

3.2 Real-World Water Analysis

Different assays using the RIANA and AWACSS biosensors have been developed in Milli-Q water in the process of developing and optimising the appropriate analytical method for the target compound analysis. This step is necessary before application

Table 1 List of target organic pollutants analysed by RIANA or AWACSS using either single or multi-analyte configurations

Analytes	LOD [ng L^{-1}]	Environmental matrix	References
Paraquat	10	Milli-Q and river water	[12]
Isoproturon	10	Milli-Q and river water	[13]
Triazines	60–200	River water	[14]
Propanil	0.6	Milli-Q water	[6]
Bisphenol A	14	Waterworks	[15]
Progesterone	0.37	Spiked Milli-Q, drinking and river water	[8]
Testosterone	0.2	Surface water	[16]
Estrone	0.2	Milli-Q water	[7]
	1		[17]
Sulphonamides	2.7–6.6	Drinking, ground, and surface water	[18]
Multi-analyte determination			
Atrazine	30		
Simazine	30		
Isoproturon	110	Milli-Q Water	[3]
Alachlor	70		
2,4-D	70		
PCP	4,230		
Isoproturon	50	Natural water samples	[4, 9]
Atrazine	160		
Estrone	80		
Bisphenol A	5		
Atrazine	2	Milli-Q, surface water	
Estrone	19	Groundwater	[19]
Isoproturon	16		

2,4-D: 2,4-dichlorophenoxy acetic acid, PCP: Pentachlorophenol

to real-world analysis regardless of whether single or multi-analyte measurements are to be performed. The application of new biodevices to real-world environmental samples has to be accomplished in the final steps of development. However, despite the large number of newly developed biosensors, most of the biosensors described in the literature overlook this step and only report applications in distilled water or buffer solutions [20]. One reason is the matrix effect in the measurements accuracy. Sample matrix can affect dramatically the application of biosensors to real samples. RIANA and AWACSS are based on competitive interactions between antibodies, analytes and analyte derivatives. These interactions can be affected by the pH or the ionic strength of natural water matrices. It has also been reported that the presence of other natural substances, such as dissolved organic carbon that interact weakly with the antibody, can induce an overestimation of the immunoassay response [21]. The influence of ground and river water matrices on the immunosensor response was normally evaluated for the different assays developed for RIANA and AWACSS by spiking these matrices with analyte standard solutions and observing the displacement of the calibration curves.

On the other hand, several strategies have been proposed and tested to prevent matrix interferences of the water samples, such as diluting and buffering the sample, adjusting the sample pH and conductivity, sample clean-up, enrichment of the sample by solid phase extraction and subsequent reconstitution of the extract, and adding a background protein. The addition of detergents or immolation proteins reduces the probability of the antibody to adhere to passive particles in water and has become a current routine to overcome matrix interferences in immunoassays [4, 22, 23]. This is the strategy adopted in the case of the RIANA and AWACSS assays where the background protein ovoalbumine is added to a real sample as an immolation protein.

As part of the AWACSS project, Tschmelak et al. proposed and demonstrated the efficacy of a new method to overcome the partly occurring matrix problems in real-world immunosensor monitoring [24]. They developed an easy matrix referencing method using a synthetic organic carbon standard that could be adapted to other applications depending on a similar test-format. The organic carbon level of monitored waters should be determined in parallel or in frequent intervals. With this referencing method, very good recovery rates for the ultra-sensitive estrone assay were possible.

3.3 Validation

Despite the practical advantages of biosensors, they need to be comparable to conventional analytical systems in terms of reliability, sensitivity, selectivity, specificity and robustness. Biosensor measurements need therefore to be verified and validated before being accepted. The RIANA and AWACSS biosensors were validated by comparing the biosensor performance with that of conventional methods. The overall performance of the new system in comparison to the conventional analytical and immunosensor techniques was tested in an inter-laboratory collaborative trial among the AWACSS partners at the late stages of the project. Different analytical set-ups were used in six laboratories; among them four automated online and offline solid-phase extraction (SPE) systems coupled to liquid chromatography and either mass spectrometry or diode array detection (SPE–LC–MS, SPE–LC–DAD), a large volume injection-gas chromatography-mass spectrometry system (LVI–GC–MS), two RIANA biosensors, one AWACSS device and one ELISA assay. The tested matrices were Milli–Q water and freeze-dried river sediments from the Nitra River in the Slovak Republic. Each of them was spiked with the analytes atrazine, bisphenol A and estrone at 0.1 and 1.0 $\mu g\ L^{-1}$ in water matrices and 50 and 500 ng g^{-1} in sediments. All samples, including blanks, were prepared in triplicate and distributed among the partners. For more details on the procedures, the reader is addressed to reference [10]. The results showed that, in terms of accuracy, the AWACSS performance in Milli-Q water and sediment samples is fully comparable to conventional chromatography-based techniques. In general, the AWACSS results were less biased towards higher values than the ELISA and RIANA immunosensor techniques.

3.4 *Water Monitoring*

The performance of both, the RIANA and the AWACSS biosensors has been tested for a variety of organic pollutants and water matrices. However, only two works have reported their use in environmental monitoring studies. The immunosensor RIANA was evaluated to monitor total chlorotriazine pesticides in river water samples in the Ebre area (Tarragona, Spain) from April to June 1998 [14]. Simazine and atrazine were detected (levels ranging from 0.18 to 1 μg L^{-1}) in only eight samples out of 18 measurements carried out. The results obtained with the immunosensor were validated by online solid-phase extraction followed by liquid chromatography-atmospheric pressure chemical ionisation-mass spectrometry (LC–APCI–MS). Although some overestimation was observed with the biosensor no false-positive was detected. Thus, the method developed with the biosensor allows the unequivocal determination of chlorotriazines in natural waters and can be applied to comply both, with the EPA legislation for drinking waters, which requires the monitoring of atrazine at the 2 μg L^{-1} level and also with the European Community (EEC) Drinking Water Directive (DWD), which establishes limit values for individual and total pesticide concentrations at 0.1 and 0.5 μg L^{-1}, respectively.

The RIANA system was also applied to determine the occurrence of the organic pollutants bisphenol A, atrazine, isoproturon and estrone in the waterworks Sant Joan Despí (Barcelona, Spain) between May and August 2002 [4, 15]. The removal efficiency of the target pollutants throughout the different purification stages at the waterworks was assessed using two different biosensor configurations: single analyte mode for bisphenol analysis [15] and multi-analyte configuration for the determination of atrazine, isoproturon and estrone [4]. Chromatographic methods were also applied in parallel for the analysis of the samples. On the basis of the comparative performance of the biosensor and liquid chromatography based methods, the biosensor emerged as a suitable tool for fast, simple and automated screening of water pollutants without sample pre-treatment and was proved to be a useful tool for water resources protection; in most water production companies the quality of the raw water and the removal efficiency of target contaminants during drinking water production needs to be assessed.

4 Other Applications

The significant progress made with RIANA in environmental monitoring and water analysis was successfully adapted to milk analysis for use in the field of reproduction management, in particular for the sensitive determination of progesterone in cow milk, as an indicator of ovulation [25]. The sensitivity and robustness of the existing progesterone assay for water analysis was improved, resulting in a LOD of only 0.2 ng L^{-1}. Progesterone was then analysed in three different types of milk (UHT milk, fresh milk and raw milk) (levels typically in the range 5–15 μg L^{-1}). LOD achieved for added progesterone (i.e. spiked samples) were between 45.5 and 56.1 ng L^{-1} depending on milk type. In this work, a commercially available antibody was incorporated for the first time into the immunosensor. The biosensor operation was fully automated and the LOD achieved for progesterone in bovine milk was below 1 $\mu g\ L^{-1}$. The assay was further optimised by reducing the time per measurement to about 5 min and tested by measuring the progesterone level in daily milk samples for 25 days, covering a whole oestrus cycle [26].

5 Conclusions

Herein we have reported the main achievements attained during the EU research projects RIANA and AWACSS, and the successful application of biosensors for environmental monitoring. Two different biosensor prototypes have been designed and constructed during the projects and different software modules for data storage, communication, data treatment, networking and data evaluation have been successfully developed and combined into a software package. Feasibility studies have been conducted on multi-analyte analysis and on matrix effects. Detection limits of most developed immunoassays have been in the few nanogram or even sub-nanogram per litre range.

The biosensors capability for analysing a variety of environmental organic micropollutants, such as pesticides, EDCs and pharmaceuticals in surface, ground, drinking and waste water has been compared with conventional analytical methods for validation purposes. The biosensors described in this report have been demonstrated not only to be cost-effective alternatives to chromatographic techniques for environmental screening, but also very convenient techniques when the information is needed immediately (measurements are performed very rapidly; maximum 20 min).

Most of the compounds analysed with the RIANA and AWACSS biosensors meet the end-users demands, their monitoring is required by the present EU environmental legislation (98/83/EC, 1998; 2000/60/EC, 2000) and all of them are frequently detected in real water samples all over Europe. On the basis of the features reported, good perspectives are given to the biosensor system to be placed among the current state-of-the-art analytical instruments for water monitoring.

Acknowledgements This work was funded by the "Automated Water Analyser Computer Supported System" (AWACSS) (EVK1-CT-2000-00045) research project supported by the European Commission under the Fifth Framework Programme and contributing to the implementation of the Key Action "Sustainable Management and Quality of Water" within Energy, Environment and Sustainable Development. This work reflects only the authors' views and the EU is not liable for any use that may be made of the information contained therein.

References

1. Brecht A, Klotz A, Barzen C, Gauglitz G, Harris RD, Quigley GR, Wilkinson JS, Sztajnbok P, Abuknesha R, Gascon J (1998) Anal Chim Acta 362:69–79
2. Klotz A, Brecht A, Barzen C, Gauglitz G, Harris RD, Quigley GR, Wilkinson JS, Abuknesha RA (1998) Sens Actuators B Chem 51:181–187

3. Barzen C, Brecht A, Gauglitz G (2002) Biosens Bioelectron 17:289–295
4. Rodriguez-Mozaz S, Lopez de Alda M, Barceló D (2006) Talanta 69:343–350
5. Tschmelak J, Proll G, Riedt J, Kaiser J, Kraemmer P, Barzaga L, Wilkinson JS, Hole JP, Nudd R, Jackson M, Abuknesha R, Barcelo D, Rodriguez-Mozaz S, Lopez de Alda MJ, Sacher F, Stien J, Slobodnik J, Oswald P, Kozmenko H, Korenkova E, Tothova L, Krascsenits Z, Gauglitz G (2005) Biosens Bioelectron 20:1499–1508
6. Tschmelak J, Proll G, Gauglitz G (2004) Anal Bioanal Chem 379:1004–1012
7. Tschmelak J, Proll G, Gauglitz G (2004) Anal Bioanaly Chem 378:744–745
8. Tschmelak J, Proll G, Gauglitz G (2004) Anal Chim Acta 519:143–146
9. Rodriguez-Mozaz S, Reder S, Lopez de Alda M, Gauglitz G, Barceló D (2004) Biosens Bioelectron 19:633–640
10. Tschmelak J, Proll G, Riedt J, Kaiser J, Kraemmer P, Barzaga L, Wilkinson JS, Hole JP, Nudd R, Jackson M, Abuknesha R, Barcelo D, Rodriguez-Mozaz S, Lopez de Alda MJ, Sacher F, Stien J, Slobodnik J, Oswald P, Kozmenko H, Korenkova E, Tothova L, Krascsenits Z, Gauglitz G (2005) Biosens Bioelectron 20:1509–1519
11. Allan IJ, Vrana B, Greenwood R, Mills DW, Roig B, Gonzalez C (2006) Talanta 69:302–322
12. Mallat E, Barzen C, Abuknesha R, Gauglitz G, Barcelo D (2001) Anal Chim Acta 427:165–171
13. Mallat E, Barzen C, Abuknesha R, Gauglitz G, Barcelo D (2001) Anal Chim Acta 426:209–216
14. Mallat E, Barzen C, Klotz A, Brecht A, Gauglitz G, Barcelo D (1999) Environ Sci Technol 33:965–971
15. Rodriguez-Mozaz S, Lopez de Alda M, Barceló D (2005) Water Res 39:20
16. Tschmelak J, Kumpf M, Kappel N, Proll G, Gauglitz G (2006) Talanta 69:343–350
17. Hua P, Hole JP, Wilkinson JS, Proll G, Tschmelak J, Gauglitz G, Jackson M, Nudd R, Griffith H, Abuknesha R, Kaiser J, Kraemmer P (2005) Opt Express 13:1124–1130
18. Tschmelak J, Kumpf M, Proll G, Gauglitz G, (2004) Anal Lett 37:1701–1718
19. Tschmelak J, Proll G, Gauglitz G (2004) Biosens Bioelectron 20:743–752
20. Kroger S, Law RJ (2005) Trends Biotechnol 23:250–256
21. Hennion M-C, Barcelo D (1998) Anal Chim Acta 362:3–34
22. Coille I, Reder S, Bucher S, Gauglitz G (2002) Biomol Eng 18:273–280
23. Gonzalez-Martinez MA, Puchades R, Maquieira A (1999) TrAC Trends Anal Chem 18:204–218
24. Tschmelak J, Proll G, Gauglitz G (2005) Biosens Bioelectron 21:979–983
25. Tschmelak J, Kappel N, Gauglitz G (2005) Anal Bioanal Chem 382:1618–2650
26. Käppel ND, Pröll F, Gauglitz G (2007) Biosens Bioelectron 22:2295–2300

Hdb Env Chem (2009) 5J: 47–68
DOI:10.1007/698_2008_10

Biosensors for Pharmaceuticals and Emerging Contaminants Based on Novel Micro and Nanotechnology Approaches

Javier Adrián, Fátima Fernández, Alejandro Muriano, Raquel Obregon, Javier Ramón-Azcon, Nuria Tort, and M.-Pilar Marco

Abstract The investigation of new sensing principles and technologies for the detection of molecular binding events has created great expectations on numerous major industrial sectors, such as healthcare, food, water and agriculture. Combining many of these advances with the potential of the immunochemical systems has allowed developing novel biosensors that provide interesting advantages against the traditional strategies for analysis, such as the possibility of multianalysis, development of field analytical methods and fabrication of easy end-user devices. Specifically, many efforts have been lately invested to control residues of pharmaceuticals in food and environmental samples, as an indication of the impact of the human activity in the media. Human and veterinary drugs, such as antibiotics, hormones, analgesics, cytostatics or β-blockers, show a high potential risk of negative effects in the environment and public health. Thus, there is a great need for low-cost and highly efficient tools for quick, reliable, and accurate detection of these contaminating bioactive agents. In particular, the scope of the present chapter is addressed to provide an overview of the potential of novel micro(nano)technology approaches to develop biosensors useful for the analysis of emerging pollutants.

Keywords Amperometric biosensor, Antibody, Antibiotics, Biosensor, Evanescent wave, Hormones, Impedimetric biosensor, Microcantilevers, Nanobiotechnology, Pesticides, Piezoelectric crystal, Plasmon resonance

---

J. Adrián, F. Fernández, A. Muriano, R. Obregon, J. Ramón-Azcon, N. Tort, and M.-P. Marco (✉)
Applied Molecular Receptors Group (AMRg), IQAC-CSIC, Networking Research Center on Bioengineering, Biomaterials and Nanomedicine (CIBER-BBN), Jordi Girona, 18-26, 08034, Barcelona, Spain
e-mail: mpmqob@iiqab.csic.es

Contents

1 Introduction

Biomolecular analytical devices have evolved rapidly over the last years. Novel molecular diagnostic approaches to refine and extend the limits of detection have emerged as a consequence of the combination of micro(nano)technological and biotechnological advances. Biosensors are among the potential applications of new materials and devices and of novel nanobiotechnological approaches. Thus, subtle changes in properties such as the dielectric field or the refractive index produced after biomolecular recognition events can be detected if they are taking place at the surface of appropriately developed transducers. New knowledge of the unique properties of nanostructured materials has opened up the possibility to investigate the influence that biorecognition phenomena have on the new optical and/or electrical properties of these systems, possibly leading to the development of more sensitive and flexible biosensing systems.

Among biosensors, immunosensors using antibodies as biorecognition elements have fascinating features such as the possibility to respond selectively to biological or bioactive substances and the capability to respond in a physiological manner. The unique properties of certain nanomaterials combined with the excellent features of the antibodies allow envisaging novel exquisitely sensitive chemical and biological sensors. Thus, antibodies are natural molecules with inherent capabilities to specifically react with their counter antigen. Antibodies can be produced, in principle, against all kinds of substances and their features can be tailored according to the requirements of each application. Their homogeneity regarding chemical structure and properties allow standardization of several procedures. Moreover, antibodies show greater improved stability, when compared to other biomolecules.

A very important aspect in biosensor development is the construction of biofunctional surfaces. Sensor solid surfaces are in general solid inorganic matter and per se not suitable for immobilizing biomolecules. Hence, further modification is required to adapt them for the immobilization of biomolecules. In addition, functional sensor surfaces place several demands such as biocompatibility, homogeneity, stability, specificity, and functionality. Thus, a challenge in biosensor development is to construct adequate surfaces as well as to design molecules suitable for site-directed immobilization. Surface architecture depends on the nature of the transducer and on the features of the biomolecule, as well as the type of measurements needed [1–4]. The surface has to be activated appropriately for further tethering of the proteins with a particular immobilization method. Subsequent layers can be generated in place, and textured following specific demands.

Environmental monitoring of chemical and biological hazard is a potential application of biosensors, and therefore several developments can be found in the literature trying to provide better analytical tools to assess any risk derived from contamination. This interest in developing more efficient bioanalytical devices responds to the increasing production of these chemicals, including so-called emerging contaminants such as pharmaceutical active ingredients and personal care products (PPCPs) [5] and the potential toxicological risks derived from them. Pharmaceuticals and biogenic hormones are normally discharged to the environment through wastewater treatment plants (WWTP), which are often not designed to remove them from the effluent. From there on, transport, fate, or possible adverse consequences of these pollutants on human health and on the ecosystem, are frequently unknown or not clearly understood [6]. Potential concerns include reproductive impairment [7–9], increased incidence of cancer [10], development of antibiotic-resistant bacteria [11], or the potential increased toxicity of chemical mixtures due to synergistic effects [12].

In Europe, regulations and regulatory methods to assess and control the impact of these substances in the aquatic environment, aim to protect the ecosystem and public health while monitoring contamination levels and any potential negative effects. In addition to the REACH (Registration, Evaluation, Authorization, and Restriction of Chemicals) law (EC 1907/2006), regarding chemicals and their safe use, which entered into force on 1st June 2007, there are specific regulations addressed to protect health and ensure quality of all water resources such as the Drinking Water Directive (DWD, Council Directive 98/83/EC), the Bathing Water Directive (2006/7/EC), and the Urban Waste Water Directives (91/271/EEC and 98/15/EC). Moreover, the Water Framework Directive (WFD, 2000/60/EC) intends to provide an overall framework for a cleaner and safer water ecosystem, particularly regarding surface freshwater and ground water bodies (i.e., lakes, streams, rivers, estuaries, coastal waters etc.). Thus, consistent with this directive, the Marine Strategy Framework Directive (2008/56/EC) aims to achieve good environmental status of the EU's marine waters by 2021. Similarly, a River Basin Management Plan (RBMP) is being developed to be implemented for each river basin district which must include knowledge of the particular pressures and impacts of human activity, protection programs, controls, and remediation measures. The first RBPM is scheduled to be published at the end of 2009 and according to it the new Groundwater Directive (2006/118/EC) establishes a regime which sets underground water quality standards and introduces measures to prevent or limit input of pollutants into the groundwater.

On the other hand, pharmaceutical residues can also contaminate food through animal treatment with veterinary drugs. During their lifetime animals may have to be treated with medicines for prevention or cure of diseases. In food-producing animals such as cattle, pigs, poultry, and fish this may lead to residues of the substances used for the treatment in the food products derived from these animals (e.g., meat, milk, eggs). The residues should, however, not be harmful to the consumer, for which reason EC legislation requires that the toxicity of potential residues is evaluated before use of a medicinal substance in food-producing animals is authorized. If considered necessary, maximum residue limits (MRLs) are established and in some cases the use of the relevant substance is prohibited. Thus, Directive 96/23/EC establishes the frequencies and level of sampling and the groups of substances to be controlled for each food commodity. Moreover, Commission Decision 2002/657/EC (and the following amending directives) establishes criteria and procedures for the validation of analytical methods to ensure the quality and comparability of analytical results generated by official laboratories. Additionally, the Decision introduces a procedure to progressively establish minimum required performance limits (MRPL) for analytical methods employed to detect substances for which no permitted limit (maximum limit) has been established, such as hormones.

Pharmaceuticals are by definition very active substances and therefore there is an increased demand for new and more efficient analytical methods in order to protect ecosystems and public health. Herein we aim to illustrate this with examples of recent developments of new biosensor devices reported to detect certain emerging pollutants, antibiotics, and androgenic anabolic steroids. The physical principles and micro/nanotechnological approaches behind the examples presented herein should be contemplated as universal and with general applicability to the analysis of other chemical or biological hazards as soon as a bioreceptor is available to specifically detect these substances.

2 Electrochemical Biosensors

Because of its simplicity electrochemical transduction has given rise to a great variety of low-cost devices based on different formats and configurations (for recent reviews see [13–16, 17]). From all possible electrochemical transduction systems those based on recording the amperometric signal generated after an enzymatic reaction have seen the greatest development [16]. However, new approaches have been described aimed at directly detecting biomolecular recognition reactions without using enzyme labels, such as the impedimetric devices, or incorporating nanoparticles to generate or increase the electrochemical signal. As an example, gold nanoparticles that provide a suitable microenvironment for biomolecule immobilization whilst retaining biological activity, and to facilitate electron transfer between the immobilized proteins and electrode surfaces, have led to an intensive use of this nanomaterial for the construction of electrochemical biosensors with enhanced analytical performance with respect to other biosensor designs [13, 15]. Similarly, the particular structure of carbon nanotubes and their unique properties make them a very attractive material for the design of electrochemical biosensors. Among the potential advantages are the promotion of different electron transfer reactions, especially those related to biomolecules and the increase of the active surfaces which has led to different strategies for constructing carbon nanotube-based electrochemical sensors [14].

2.1 Amperometric Biosensors

Amperometric biosensors are based on the measurement of the current generated by oxidation/reduction of redox species at the electrode surface, which is maintained at an appropriate electric potential. The current observed has a linear relationship with the concentration of the electroactive species. The electrode is usually constructed of platinum, gold, or carbon. Adjacent to the electrode, entrapped by a membrane or directly immobilized, the bioreceptor is placed. Label enzymes used in electrochemical immunoassays are usually oxidoreductases such as horseradish peroxidase (HRP), or hydrolytic enzymes, such as alkaline phosphatase (AP), which yield electroactive species as a product of the enzymatic reaction (see Fig. 2). Other enzymes used as sensing elements for environmental monitoring are tyrosinase, laccase, aldehyde dehydrogenase, etc. Sometimes the substrate or the product of the enzymatic reaction can be monitored amperometrically, without the need for a mediator. These electrodes are called *unmediated amperometric enzyme biosensors*. However, a number of factors must be taken into account when assessing the suitability of an enzyme substrate to be used in an electrochemical detection system: the electrochemistry of the substrate, the electrochemistry of the product of the enzymatic reaction, the medium in which the measurements will be performed, and the electrochemistry of endogenous materials in the test sample. A problem often encountered with unmediated sensors is that other species present in the samples being analyzed are also electroactive at the potential applied. For example ascorbic acid and uric acid, present in many biological samples, are oxidized at an anodic potential of +0.35 V. AP combined with *p*-aminophenyl phosphate (PAPP) as a substrate has been shown to be a good alternative when measuring with such a system. Although *p*-aminophenyl phosphate has an irreversible wave in cyclic voltammetry at around 0.45 V vs. Ag/AgCl, its hydrolysis product *p*-aminophenol (PAP) shows a reversible electrochemistry with a half wave potential of –0.065 V vs. Ag/AgCl. Consequently, measurements on biological matrices can take place at lower potentials avoiding interference of endogenous compounds. Choosing an alternative electron transfer acceptor can also circumvent these problems. Usually the *mediator* is a species of low molecular weight (MW) that shuttles electrons between the redox center of the enzyme and the working electrode. These sensors are called *mediated amperometric enzyme biosensors*. A mediator should react rapidly with the enzyme, exhibit reversible heterogeneous kinetics, possess a low overpotential for regeneration, and be stable at a certain range of pH, temperature, redox state, and dioxygen. Some mediators frequently used are I^-, $[Fe(CN)_6]^{-4}$, *o*-phenylenediamine, diaminobenzidine, hydroquinone, and 5-aminosalicylic acid.

A recent example has been reported to detect xenoestrogens using an amperometric immunosensor based on a quinoprotein glucose dehydrogenase (GDH)-modified thick-film sensor. The immunosensor is used to detect the trazer used in a capillary immunoassay (CIA) [18]. With this configuration, detection of the alkylphenols and their ethoxylates in μg L^{-1} range is possible using a β-galactosidase tracer and anti alkylphenol ethoxylate antibodies. The assay is performed off-line in small disposable PVC capillaries coated with immobilized antibodies, which allows the combination of the assay with a small portable device potentially useful for on-site environmental monitoring. The bioelectrocatalytic properties of this biosensor offer an additional amplification and allow a very sensitive quantification of the 4-aminophenol, generated by the β-galactosidase. However, in the last few years, the advances and possibilities presented by easily fabricated screen-printed electrodes (SPEs) has given rise to a significant number of examples demonstrating the potential and the flexibility of this technology to develop low-cost biosensor devices for a variety of targets including microorganisms, pesticides, and also pharmaceuticals. As an example, disposable carbon SPE immunosensors have been described for the detection of anabolic androgenic steroids (AAS) using horseradish peroxidase as the enzyme label and tetramethylbenzidine/hydrogen peroxide (TMB/H_2O_2) as the substrate for chronoamperometric detection at +100 mV [19–21]. The immunosensors were fabricated by immobilizing hapten-bovine serum albumin conjugates on the surface of the SPEs and followed by the competition between the free analyte and coating conjugate with corresponding antibodies. The signal is recorded after an anti IgG-HRP conjugate determining the degree of competition. By using different antibodies with different affinities and cross-reactivity patterns the immunosensor detects different AAS with limits of detection (LOD) below 200 pg mL^{-1}. Thus, LODs of 30.9 ± 4.3 pg mL^{-1} for boldenone and 120.2 ± 8.2 pg mL^{-1} for methylboldenone, respectively, have been reported. Testosterone can be detected in buffer with a LOD of 26 and 1.8 pg mL^{-1} in a complex matrix such as bovine urine. Similarly, nortestosterone can be detected in the same matrix with a LOD of 15 pg mL^{-1}.

The possibility to analyze progesterone using amperometric immunosensors has been recently reported using a colloidal gold-graphite-Teflon-tyrosinase composite biosensor as the amperometric transducer [22]. A sequential competitive configuration between the analyte and progesterone labeled with alkaline phosphatase was selected using in this case phenyl phosphate as the AP-substrate and the enzyme reaction product, phenol, was oxidized by tyrosinase to o-quinone, which was subsequently reduced at −0.1 V at the biocomposite electrode. A LOD of 0.43 ng mL^{-1} was accomplished while the time needed to reach the steady-state current from the addition of phenyl phosphate was 30–40 s. A lifetime of 14 days with no need to apply any regeneration procedure was reported. The immunosensor was applied to the detection of progesterone in milk spiked after just sample dilution.

Recently, several papers have been published combining magnetic particles with electrochemical amperometric detection. Magnetic beads are known to be a powerful and versatile tool in a variety of analytical and biotechnology applications. The use of magnetic beads greatly improves the performance of the immunological reaction, minimizing the matrix effect due to improved washing and separation steps. Additionally, the magnetic beads can be easy magnetically manipulated by using permanent magnets or electromagnets. Therefore, the analysis of samples performed on magnetic beads can be easily achieved without any pre-enrichment, purification, or pretreatment steps, which are normally necessary for standard methods. In this context, an electrochemical immunosensing strategy based on magnetic sensors is presented as a sensitive, simple, cheap, and user-friendly analytical alternative to classical methods for the detection of residues. The strategy combines the advantages taken from immunological assays, magnetic bead separation, and electrochemical transduction and is based on the immunochemical competitive immunoreactions between the contaminant and the enzyme tracer with the specific antibody immobilized on magnetic beads followed by its separation from the biological matrix by the use of a magnetic field. Thus, after the immunological reaction, the magnetic beads can be easily captured by a novel electrochemical and magnetic transducer, based on a magneto graphite-epoxy composite [12, 13] (m-GEC), by merely dipping the magneto electrode in the solution where the beads are suspended (Fig. 1). This strategy has been used for the detection of sulfonamide antibiotics in milk samples [23]. The antibodies raised to provide an appropriate recognition pattern of this antibiotic family [24] are immobilized on the surface of magnetic beads through covalent attachment using tosyl-activated commercially available magnetic beads. Other immobilization strategies tested before also gave suitable results, but this strategy was demonstrated to be the one providing the best performance [25]. The immunological reaction for the detection of sulfonamide antibiotics is based on a direct competitive assay using a peroxidase (HRP) tracer as the enzymatic label. Excellent results were accomplished detecting sulfamethazine directly in a diluted raw full cream sample with detectability values far below the MRLs established by the EC in those types of samples (IC50 of 9.8 $\mu g\ L^{-1}$ and LOD of 0.36 $\mu g\ L^{-1}$). The same results were obtained with UHT (ultra high temperature) milk, such as full cream (about 3.25% fat), semi-skimmed (about 1.5–1.8% fat), and

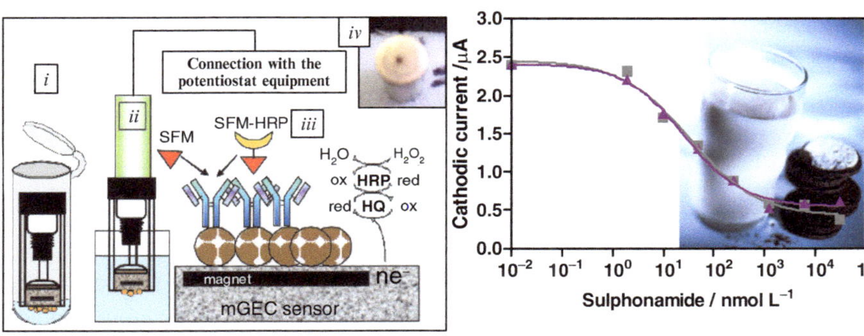

Fig. 1 *Left:* Schematic representation of the electrochemical magnetoimmunosensing strategy for the detection of sulfonamides. After the immunoreaction, the antibody-modified magnetic beads were captured for the m-GEC electrode (i). Chemical reactions occurring at the m-GEC surface polarized at –0.150 V (vs. Ag/AgCl) upon the addition of H_2O_2 in the presence of mediator (HQ) (ii and iii). Appearance of the m-GEC electrode covered with the magnetic beads (iv). *Right:* Calibration curves for sulfapyridine run in PBS and in 1:5 whole milk. Parts of this figure are reprinted from Zacco et al. [23] with permission from Elsevier

skimmed (0.1% fat). This strategy has been proven to be useful for the analysis of other residues in complex matrices such as atrazine in orange juice [25] or fluoroquinolone antibiotics in milk (unpublished results).

The same strategy has also been applied to the detection of polychlorinated biphenyls (PCBs) though in this case using SPEs [26, 27]. The sensor was based on a direct competitive immunoassay scheme in which antibody-coated magnetic beads were used on SPEs to detect the product of the enzymatic reaction of alkaline phosphatase used as the tracer. α-Naphthyl phosphate was used as the substrate and the α-naphthol produced during the enzymatic reaction was detected using differential pulse voltametry (DPV). Aroclor 1248 PCB mixture could be measured with a detection limit of 0.4 ng mL^{-1}. The performance of the sensor was demonstrated analyzing PCBs on marine sediment extracts and spiked soil samples.

2.2 *Impedimetric Biosensors*

Different sensor approaches have been accomplished to avoid the use of labels when the target analyte is a biomacromolecule, bacteria, or virus, but only on a few occasions has the performance of these transduction principles been demonstrated to detect small organic molecules at low concentration levels. In this context, *electrochemical impedance spectroscopy* (EIS) is being rapidly developed because of the possibility to record directly information on biorecognition events occurring at the electrode surfaces and inducing capacitance and resistance changes [28, 29]. Thus, when a biological receptor binds to its counterpart, there is a change in the impedance of the system that enables direct measurement of an electrical signal, allowing the development of label-free biosensing devices. A major drawback of early sensors was that impedance changes due to biorecognition are very small. However, new electrode configurations have significantly increased the detectability of these devices. Thus, impedimetric label-free devices have been described to detect DNA hybridization [30] or biomolecules such as interferon-gamma [31], but direct detection of contaminants and residues in complex matrices is still a challenge. Label-free detection of bisphenol A (BPA) based on the impedance measurement has also been achieved with an impedimetric immunosensor [87]. The immunosensor was fabricated by the covalent immobilization of a polyclonal antibody on carboxylic acid group-functionalized nano-particle comprised conducting polymer and bisphenol A hapte, (4,4-bis(4-hydroxyphenyl) valeric acid, BHPVA) BSA conjugate. The immunosensor showed specific recognition of BPA with a LOD of 0.3 ± 0.07 ng mL^{-1}. The proposed immunosensor was applied to the analysis of BPA in human serum samples.

In this context, interdigitated microelectrodes (IDμE) have recently drawn great attention since their sensitivity is higher than conventional electrodes [32–35]. Using thin Au/Cr (~200 nm thickness) IDμE's (3.85 μm thick and with an electrode gap of 6.8 μm) on a Pyrex 7740 glass substrate, our group has recently reported detection of atrazine without the use of any label with a limit of detection of 0.04 μg L^{-1}[36]. The sensor has been evaluated to assess its potential to analyze pesticide residues in a complex sample matrix, such as red wine. An atrazine hapten-BSA conjugate was covalently immobilized within the μ-electrodes on the glass substrate. An appropriate functionalization of the transducer surfaces was crucial in this case. A key problem is the nonspecific binding of molecules present in the sample matrix to the surface of the sensor. This concerns any surface from a lower to higher extent, but particularly, gold is very suitable to capture nonspecifically organic molecules and components from the media. For this reason, a two-step activation procedure based on the distinct properties of the gold (electrodes) and the silicon oxide surfaces (between the electrodes) was used. The whole activation procedure was monitored by impedance spectroscopy using IDμE as is shown in Fig. 2 In a first step, modification of the gold surfaces was accomplished making use of thiol chemistry using N-acetylcysteamine to cover the gold electrodes and to protect the sensor from undesired nonspecific absorptions (Step I). Following this, immobilization of the glass material between the electrodes was activated making use of silane chemistry, using (3-glycidoxypropyl)trimethoxysilane (GOPS) (Step II). The epoxy group provided the necessary reactivity for further attachment of the immunoreagents through a nucleophylic attack of the amino groups of the lysine

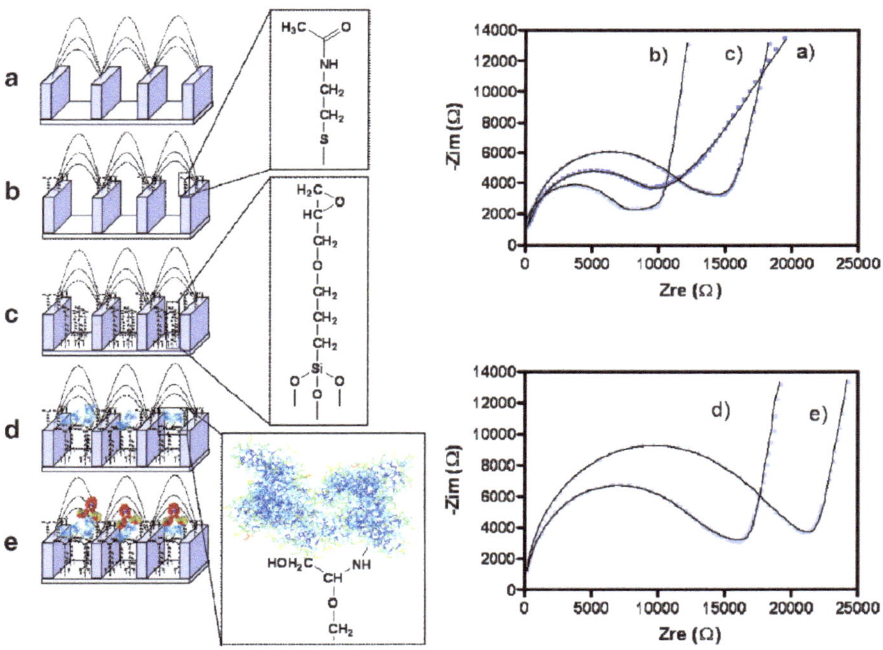

Fig. 2 *Left:* Scheme showing steps used to prepare the immunosensor surfaces and antibody binding. *Right:* Nyquist plots of impedance spectra corresponding to: (**a**) IDμE, (**b**) Step I: N-acetylcysteamine, gold protection, (**c**) Step II: functionalization of Pyrex substrate with (3-glycidoxypropyl)trimethoxysilane, (**d**) Step III: coating antigen 2d-BSA, covalent immobilization (1 μg mL^{-1}), and (**e**) Step IV: Antibody Ab11, incubation step (1 μg mL^{-1}). *Symbols* represent the experimental data. *Solid curves* represent the computer fitting data with the parameters calculated by the commercially available software Zplot/Zview (Scibner Associates). Parts of this figure are reprinted from [36] with permission from Elsevier

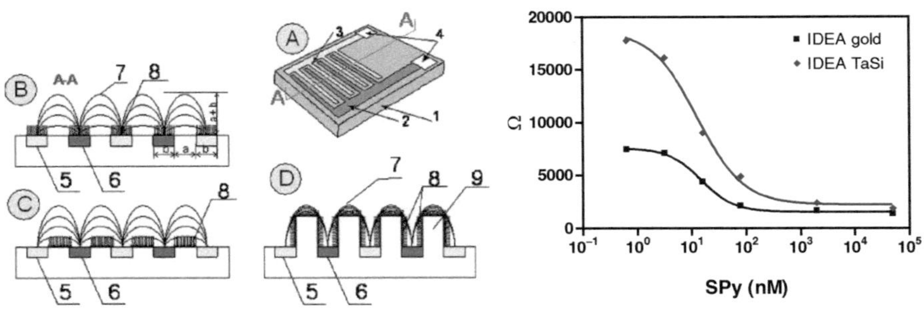

Fig. 3 *Left:* Schemes of the planar and barrier electrodes: (A) Planar IDEA device; (B), (C) cross-section A-A of the planar IDEA device; (D) IDEA device with insulating barriers between electrode digits. The numbers represent: 1-Insulating substrate; 2, 3-electrode collector bars; 4-contact pads; 5, 6-electrode "digits"; 7-electric field lines; 8-immobilized biomolecules; 9-insulating barrier. *Right:* Calibration curves obtained for sulfapyridine using either the planar electrodes or the electrodes with barriers. Parts of this figure are reprinted from Bratov et al. [37] with permission from Elsevier

residues of the antigen. As can be observed, the covalent immobilization of the atrazine antigen 2d-BSA (Step III) produced a significant change in the Nyquist plot. Finally, exposure of the immunosensor to a solution of the specific antibody (Step IV) produced a shift in the impedance spectra that is dependent on the concentration of atrazine present in the sample.

Also based on IDμE an improved immunosensor for sulfonamide antibiotic residue analysis has been reported [37, 38]. This new transducer used a three-dimensional interdigitated electrode in which the different electrode digits are separated by insulating barriers. Using this strategy it was possible to use most of the electric field instead of using just a small percentage (Fig. 3, left). Moreover, there is a significant increase of the available area to immobilize the immunoreagents. On the other hand, the electordes in this case are made of TaSi instead of gold, diminishing the nonspecific adsorptions on the surface. The immunosensor allows direct detection of the antibiotic sulfapyridine (SPY) and shows an IC_{50} value of 5.3 μg L^{-1} in buffer. As can be observed in Fig. 3 (right), the three-dimensional sensor shows considerable improvement compared with a standard planar InterDigitated Electrode Array (IDEA) design.

3 Optical Biosensors

Optical biosensors show great promise for residue analysis. Although first developments took advantage of the flexibility and low cost of the optical fibers measuring the absorption or emission of light of one of the components of the bioreaction [39], later developments were mainly based on micro/nanotechnology and on the possibility to fabricate complex micro/nanostructures to guide light which has opened up the possibility to investigate novel optical phenomena and the optical properties of materials at the nanometric level (for reviews on new optical biosensing systems see [40–44]). Some of these developments have found application in the detection of emerging pollutants and also pharmaceuticals (i.e., [45–47]) and herein we will describe some of these examples.

3.1 Biosensors Based on the Evanescent Wave Optical Principle

An evanescent wave (EW) is produced in the external media (refractive index, n_2) of a waveguide (n_1) by the electromagnetic field associated to the light guided by total internal refection (TIR). The electromagnetic field does not abruptly switch to zero at the interface between the two media ($n_1 > n_2$), but decays exponentially with distance from the interface. The penetration depth of the evanescent field is defined as the distance where its strength is reduced to 1/e of its value at the interface and generally has a value around a hundred nanometers. The penetration depth is dependent on the incidence angle at the interface and is proportional to the wavelength of the excitation light. When molecules with an absorption spectrum including the excitation wavelength are located in the evanescent field, they absorb energy leading to attenuation (attenuated total reflection, ATN) in the reflected light of the waveguide. One of the advantages of the biosensors based on this principle is that possible interferences from the bulk media are avoided since only directly absorbed substances interfere with the electromagnetic field.

With the aim to improve detectability, many reported immunosensors combine this principle with the use of labeled molecules that are able to re-emit the absorbed evanescent photons at a longer wavelength as fluorescence. Part of this emission is coupled back to the waveguide and in this way is transmitted to the receptor. This phenomenon is known as total internal reflection fluorescence (TIRF). As an example a fully automated TIRF-based biosensor study for progesterone in bovine milk has recently been reported measuring progesterone levels in daily milk samples for 25 days, covering the whole estrus cycle [48]. The immunochemical assay was designed as a binding-inhibition test with a progesterone derivative covalently immobilized on the sensor surface and a monoclonal antiprogesterone antibody as the biological recognition element. Progesterone levels can be measured in about 5 min with a LOD of 0.04 ng mL^{-1}. The biosensor can be used in-line in the milking parlor and thus could be an important tool for reproductive management of dairy cattle detecting heat and predicting pregnancy, which are critical parameters in milk production. Similarly, the River Analyzer (RIANA) is also a highly sensitive fully automated biosensor able to detect multiple organic targets rapidly and simultaneously [49–51]. Thus, the system has been used to analyze a variety of analytes in environmental samples with very good limits of detection. Hormones, pesticides, antibiotics, and endocrine-disrupting chemicals are among the organic pollutants that have been detected with this biosensor system [47, 49, 52, 53].

More recently, other EW immunosensor approaches such as *Grating Couplers* (i.e., [54–56]) or *Mach-Zehnder Interferometers* (MZI) [57–60] have been investigated in order to make possible direct measurements of small analytes without the use of fluorescent labels. Thus, planar optical waveguides with an input grating coupler, or input/output gratings, have been reported with the aim to be used as label-free biodetection devices. Nanoimprint lithography is used to integrate the grating patterns with thin-film waveguides [55]. An optical waveguide lightmode spectroscopy (OWLS)-based biosensor is a recently developed device in the field of integrated optics, and exploits the science of light guided in structures smaller than the wavelength of the light. The system OWLS has been commercialized by MicroVacuum Ltd (Budapest, Hungary). The immobilization of a bioligand to an OWLS-based biosensor occurs by the silanization of the waveguide layer and the coupling of it to the silanized sensor surface. The structure of the chip consists of an input grating and a waveguide layer. The light of a He-Ne laser ($\lambda = 632.8$ nm) is diffracted by the grating at the interface and starts propagating via TIR inside the waveguide. At a well-defined incident angle (α), the phase shift during one internal reflection equals zero (constructive interference) and a guided mode is excited, which generates an evanescent field penetrating into the covering medium. A change in refractive index at the surface as a consequence of a biorecognition reaction is monitored on-line by precise measurement of incoupling angle as a function of time with the operating software of the immunosensor system. This principle has been used to analyze sulfamethazine, by immobilizing the antibody on the surface of the chip and reaches a LOD of 10^{-8} M.

A similar system has been developed by the Centre Suisse d'Electronique et de Microtechnique (CSEM, Neuchâtel, Switzerland) [61–63]. The wavelength interrogated optical immunosensor (WIOS) also measures changes in the refractive index of the evanescent wave. In this case, the chip has an input and an output grating (Fig. 4). Light from a VCSEL laser emitting at around 763 nm is incident on a first grating and when coupling takes place, the waveguide mode is excited and propagates into the waveguide layer until the second grating where the light couples out

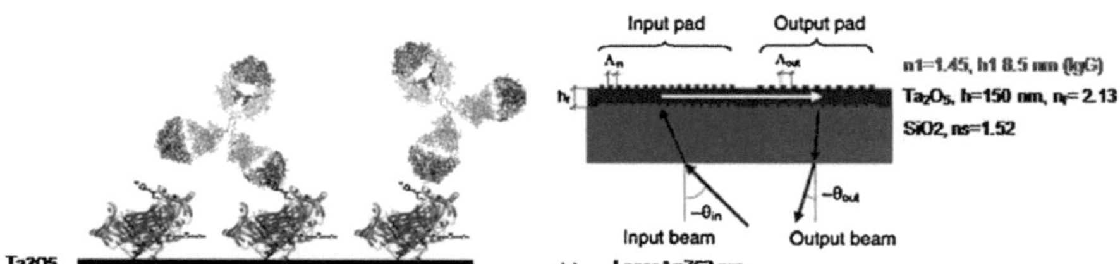

Fig. 4 Scheme of the WIOS system. The chip is a customized platform that contains 24 sensing sites. Each site is waveguide grating. Light from a VCSEL laser emitting at around 763 nm is incident on a first grating. The waveguide mode is excited and propagates into the waveguiding layer. The second grating couples out the guided light (at a different angle), which is collected with large plastic optical fibers. Electronics controls the laser wavelength modulation and amplifies the detected signals. A computer does the data acquisition and processing

of the waveguide (at a different angle) and is collected with large plastic optical fibers. Since the resonant coupling occurring at the first grating is also dependent on the wavelength (λr) of the incident light, an optical configuration has been envisaged that allows working at a fixed incidence angle (θ) while the wavelength at which coupling takes place is monitored through the output grating. This allows recording of effective index variations of the waveguide by continuously monitoring this parameter. On the basis of this principle, an immunosensor to detect sulfonamide antibiotic residues in milk samples has been developed. Immunoreagents appropriately produced to detect a wide range of sulfonamide antibiotic congeners [64] have been incorporated on the surface of the chip making use of a photopolymerizable dextran layer. The WIOS immunosensor is able to reach the necessary detectability to determine the most frequently used sulfonamide antibiotics used in the veterinary field, in compliance with the EC regulations. Thus, sulfapyridine can be detected in milk with a LOD value of 0.50 ± 0.12 and 0.24 ± 0.06 μg L^{-1} and buffer samples, respectively. Other sulfonamides tested showed LODs below the MRLs values [65].

On a *Mach-Zehnder interferometer* the propagating light is split into two arms, one with the appropriate sensing layer and the other acting as a reference. The evanescent field of the measuring arm collects information regarding the bioreaction via the change produced in the refractive index. Consequently, the velocity of the wave in this arm varies. At the end recombination of the waves from both arms allows observation of a constructive or destructive interference, which is related to the extent of the bioreaction that has occurred on the sensing arm. Although there is intensive research in this field to our knowledge no applications to the development of biosensors for pharmaceuticals or emerging pollutants have been reported [57–60].

3.2 Plasmon Resonance Biosensors

The plasmon resonance is an evanescent electromagnetic field generated at the surface of a metal conductor (usually Ag or Au) when excited by the impact of light of an appropriate wavelength at a particular angle (θ_p). Surface plasmons are generated by electrons at the metal surfaces that behave differently from those in the bulk of the metal. These electrons are excited by the incident light, producing an oscillation (resonance) at different frequency from that in the bulk of the metal film. The absorption of light energy by the surface plasmons during resonance is observed as a sharp minimum in light reflectance when the varying angle of incidence reaches the critical value. The critical angle depends on the wavelength and polarization state of the incident light, but also on the dielectric properties of the medium adjacent to the metal surface and therefore is affected by analytes binding to that surface. This principle has been extensively investigated for monitoring of biological interactions in different types of configuration. We will provide herein some examples of late developments regarding detection of pharmaceuticals using this principle, however for additional information the reader is directed to some recent reviews (i.e., [39, 43, 66]).

Since distinct surface plasmon resonance (SPR) prototypes (Biacore, IASys, etc.) have appeared in the market, a significant number of applications of this principle have been reported during the last year. For example, in a recent paper surface plasmon resonance has been used to analyze 17β-estradiol (E2) in sewage in the coastal marine environment [67]. Using the commercial Biacore system, the small molecule E2 is immobilized on a CM5 sensor chip for an indirect competitive immunoassay. The LOD achieved was 0.445 $\mu g\ L^{-1}$ although to analyze environmental samples, the SPR was couplet to a solid-phase extraction system [67]. Similarly, several examples of the use of SPR commercialized by Sensia to analyze environmental organic pollutants (DDT, atrazine, chlorpyrifos, and carbaryl) in natural water samples have been reported [46, 68–70]. BPA has also been analyzed using a SPR biosensor immobilizing a BPA-ovalbumin (BPA-OVA) conjugate on the Au thin film of the SPR sensor chip by just physical adsorption [71]. The lowest detection limit for BPA by SPR was around 1 ppb, almost the same as the LOD reached by ELISA (enzyme-linked immunosorbent assay) using the same immunoreagents.

An interesting example also reports the development of a novel surface plasmon resonance-based biosensor assay for the bioeffect-related screening of chemicals with thyroid-disrupting activity [72]. For this purpose two thyroid transport proteins (TPs), thyroxine-binding globulin (TBG) and recombinant transthyretin (rTTR), were applied in an inhibition assay format in a Biacore 3000 using CM5 biosensor chips coated with l-thyroxine (T4), the main hormone of the thyroid system. Known thyroid disruptors and structurally related compounds such as halogenated phenols, halogenated bisphenols, bisphenol A, 3,5-dichlorobiphenyl, and its hydroxylated metabolite 4-hydroxy-3,5-dichlorobiphenyl (4-OH PCB 14) were tested in this system to their relative potency (RP) with T4. The TBG-based assay was highly specific for T4, and the rTTR-based assay was sensitive toward several compounds, the highest sensitivity (RP = 4.4) being obtained with 4-OH PCB 14, followed by tetrabromobisphenol A (RP = 1.5) and tetrachlorobisphenol A (RP = 0.75).

Regarding antibiotic residues, a rapid and sensitive screening qualitative method using a surface plasmon resonance biosensor has been reported that detects all fenicol antibiotic residues in shrimps from a single sample extract [88]. This method requires ethyl acetate extraction followed by a single wash with isooctane/chloroform. Each sample extract is injected over the surfaces of two biosensor chip flow cells, one surface having the capability to detect florefenicol amine (FF amine), florefenicol (FF), and thiamphenicol (TAP) and the second surface for chloramphenicol (CAP) detection. The estimated detection capabilities (CCβ) were 0.1, 0.2, 250, and 0.5 ppb for CAP, FF, FF amine, and TAP, respectively. This quick, simple test allowed the detection of CAP residues in shrimps at the minimum required performance limit of 0.1 $\mu g\ kg^{-1}$ for this compound and of FF, FF amine, and TAP below their maximum residue limits. Previously, detection of chloramphenicol and the mammalian metabolite chloramphenicol glucuronide (CAP-Glu) with the Biacore Q biosensor has also been reported. Honey, porcine kidney, and prawn extracts were analyzed after solvent extraction and solid-phase clean-up. Similarly, the determination of β-lactams using

Biacore has been reported, by immobilizing either an enzyme receptor [73] or a specific antibody [74]. In the first case, the specific recognition is based on the enzymatic activity of a carboxypeptidase converting a 3-peptide into a 2-peptide, a reaction that is inhibited in the presence of β-lactams. Antibodies were used to measure either the amount of formed enzymatic product or the amount of remaining enzymatic substrate. Different β-lactams were detected at or below European maximum residue limits. Thus, the LOD for penicillin G was 1.2 and 1.5 g kg^{-1}, depending on whether the 2- or 3-peptide was detected, respectively. The Biacore biosensor was also employed to detect β-lactams using a commercial antibody against ampicillin which had much higher affinity for the open β-lactam ring than for the closed ring. Two different pretreatments were tested prior to the biosensor assay to open the ring in order to increase the assay sensitivity. The LODs achieved for ampicillin in milk were 33 and 12.5 μg L^{-1} after enzymatic and chemical pretreatment, respectively. Conversely, penicillins G and M could be detected at or below their respective MRLs (4 and 30 μg L^{-1} in milk, respectively) and ampicillin and amoxicillin could be detected only at about the MRLs after chemical pretreatment.

A step forward in the use of plasmon resonant (PR) properties of materials is the use of particles for biological detection. Thus, colloidal gold or silver particles, typically 10–120 nm in diameter, are plasmon resonant. The particle geometric confinement of the surface plasmon gives rise to a spectral resonance in the light scattering that does not occur in thin films such as those used in the Biacore. The light scattered by nanometer-sized colloidal metal particles is dominated by the collective oscillation of the conduction electrons induced by the incident electric field (light), as mentioned above for the plasmon resonance of flat surfaces. However, in this case, the frequency band (specific color) of the scattered light is a function of the size, shape, and material properties of the particle. Thus, silver or gold nanoparticles, with diameters of 30–120 nm, efficiently scatter light in the visible spectrum. Variations in the refractive index around the nanoparticle surface also affect the resonance peak (frequency band of the scattered light), for which reason biomolecular interactions can be detected using nanoparticles (Fig. 5). This physical phenomenon is known as localized surface plasmon resonance (LSPR). An example of the application of this physical transducer principle to the analysis of an anabolic androgenic steroid such as stanozolol, has been recently reported [75, 76]. The chip could be fabricated in house by first activating the glass substrate with a silanizing agent possessing either a NH_2 or a SH group to bind the gold nanoparticles on the top of a flat SiO_2 (glass) surface without the observation of clusters. Following this, a stanozolol hapten-BSA conjugate was used to coat the gold nanoparticles and the detection of the hormone was made under competitive conditions by just mixing the sample with the antibody and adding a drop on top of the LSPR chip. The extinction spectra of the gold colloids were measured by conventional dark-field spectroscopy, illuminating the sample with a halogen lamp (100 W). The light scattered by the particles was analyzed with a microspectrometer and a CCD camera. Using this simple set-up stanozolol could be measured to a level below the minimum required performance level.

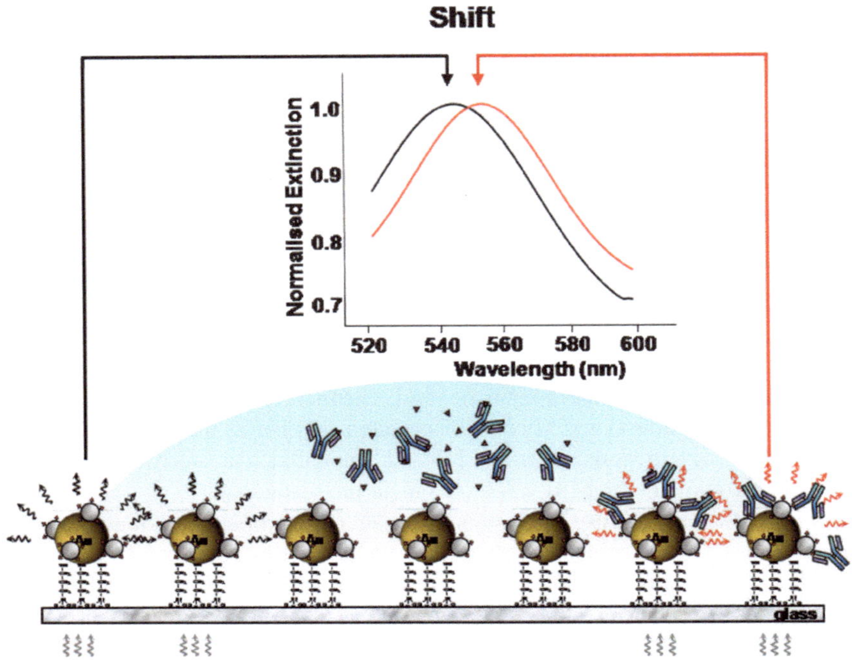

Fig. 5 Schematic of the chip using Au nanoparticles on top of a glass substrate used to detect stanozolol, and androgenic steroid used illegally to increase meat production or athletic efficiency in sport competitions. Specific binding of the antibody to stanozolol hapen-BSA coated on top of the nanoparticles, produces a shift in the resonance peak of the nanoparticles that has been used to detect this anabolic androgenic steroid. Reproduced from [75]

3.3 *Nanomechanical Biosensors*

Finally, biosensors based on microcantilevers (MCLs) have become a promising tool for directly detecting biomolecular interactions. Microcantilevers translate molecular recognition of biomolecules into nanomechanical motion that is commonly coupled to an optical or piezo-resistive read-out detector system. Biosensors based on cantilevers are a good example of how nanotechnology and biotechnology can go together. Microcantilever sensors rely on their deflection to indicate sensing. Thus, molecular adsorption onto the sensing element shifts the resonance frequency and changes its surface forces (surface stress). Surface stress due to conformation change of proteins and other polymers has been a recent focus of MCL research. Since conformational changes in proteins can be produced through binding of analytes at specific receptor sites, MCLs that respond to conformational change-induced surface stress are promising as transducers of chemical information and are ideal for developing microcantilever-based biosensors. On the other hand, high-throughput platforms using arrays of cantilevers have been developed for simultaneous measurement and read-out of hundreds of samples. Although many interesting applications in the environmental and clinical diagnostic field have been reported (for recent reviews see [42, 77–80]), examples of their use to detect pharmaceuticals are still scarce.

4 Piezoelectric Immunosensors

Piezoelectrics are materials that may be brought into resonance by the application of an external alternating electric field. The frequency of the resulting oscillation is determined by the mass of the crystal. By coating a piezoelectric with an appropriate biomolecule such as an antibody, these kinds of immunosensors can, in principle, directly detect the binding of the corresponding analyte (for reviews see [81, 82]). Piezoelectric immunosensors may adopt two modes: (1) *Bulk acoustic (BA) devices* where adsorption of the analytes occurs on the coated surface of a piezoelectric crystal connected to an oscillator circuit. Resonance occurs on the entire mass of the crystal. If for example an antibody-coated crystal is placed in an atmosphere containing the selected analyte the immunoreaction will produce an increase in the mass of the crystal. The resonant frequency will therefore decrease according to the Sauerbrey equation: $\Delta f = -2.3 \times 10^6 f^2(\Delta m/A)$, where f is the oscillation frequency in Hz, Δm is the adsorbed mass in g, and A is the sensing area in cm^2. (2) *Surface acoustic wave (SAW) devices* where an acoustic wave moves just at the surface of the crystal. Mass loading on the acoustic path between two sets of electrodes will alter the phase wave velocity and cause a shift in the frequency. This technology can, in principle, detect binding events produced at the surface of the piezoelectric material.

Several examples can be found in the literature regarding the use of this type of sensor for environmental, food safety, and clinical applications. Liu et al. reported the detection of various polycyclic aromatic hydrocarbons (PAHs) by immobilizing the antigen (benzo[a]pyrene-BSA conjugate, BaP-BSA) on gold-coated quartz crystals, with a basic resonant frequency of 10 MHz, through thioctic acid [83]. The monoclonal antibody 10c10 (mAb10c10) allowed one to determine benzo[a]pyrene (BaP) in the flow injection system by using a competitive pattern, down to the nanomolar level. Similarly, a piezoelectric immunosensor system was developed for the rapid detection of polychlorinated dibenzo-p-dioxins (PCDDs) using a direct enzyme immunoassay format [84]. The purpose of using the enzyme was just to increase the mass of the binding molecule. The mouse monoclonal antibody against 2,3,7,8-tetrachlorodibenzo-p-dioxin (2,3,7,8-TCDD) was deposited on a 10-MHz AT-cut quartz crystal resonator modified with a self-assembly monolayer of dithiobis-N-succinimidyl propionate. Competition took place between the analyte and a conjugate of a dioxin-like competitor coupled to the enzyme horseradish peroxidase. Measurement of the frequency responses allowed detection of 2,3,7,8-TCDD in the concentration range of 0.01–1.3 ng mL^{-1}. In another example BPA was detected down to 10 ng mL^{-1} using antibodies coupled to nanoparticles, approximately 200 nm in diameter, with the aim to increase the mass change on the surface of the immunosensor and thereby increase the frequency shift detected [85]. PCBs have also been detected using a piezoelectric immunosensor, by immobilizing a polyclonal antibody on a gold electrode via protein A. Detection in the range of 1–50 ppm was achieved using either enzyme or polystyrene particles to increase the response of the sensor [86]. As a representative example of the use of this principle to analyze pharmaceutical residues, Park et al. reported the direct detection of chloramphenicol by covalently immobilizing specific antibodies onto the gold electrode surface of piezoelectric crystals using a self-assembled monolayer, obtaining a significant shift of the resonance frequency in about 10 min [85]. However, the detectability achieved was much worse than in those formats where a competitive configuration was used.

5 Conclusions

Future advances in biosensor development will need scientists from different disciplines combining their research efforts. First, improvement of transducer technology is necessary in order to allow direct detection on environmental samples and to confer the necessary flexibility to develop field analytical methods. Frequent new micro/nanotechnological advances are opening up new and exciting possibilities that should be investigated in order to find novel ways to convert biomolecular recognition events into electrical or optical signals. Second, increasing the number of analytes that can be measured using the biosensor technology is necessary. Especially interesting are those compounds that are difficult to measure using currently accepted methods. Antibody-based biosensors continue to show great potential due to the excellent features of this type of biomolecule (stability, chemical homogeneity etc.) and the capability to produce them against almost any type of analyte, which may broaden the number of compounds that can be analyzed. This versatility of the immune system combined with the progress that recombinant DNA technology is making in this field may increase the potential of these bioreceptors. A third trend should be developing integrated units able to perform multianalyte measurements. And finally the extensive research that is taking place in this area will not give the expected fruit without performing rigorous validation studies with environmental samples. All these facts will contribute to improving investment by private companies to finally deliver reliable devices into the market.

Acknowledgments This work was supported by the Ministry of Science and Technology (Contract numbers AGL2005-07700-C06-01, NAN2004-09415-C05-02, NAN2004-09195-C04-04, and DEP2007-73224-C03-01) and by the European Community (Conffidence, Proposal No.: 211,326 - CP Collaborative Project). The AMR group is a Group de Recerca de la Generalitat de Catalunya and has support from the Departament d'Universitats, Recerca i Societat de la Informació la Generalitat de Catalunya (expedient 2005SGR 00207). We would also like to give our thanks for the contribution from research groups from the CNM-CSIC, CSEM, Unisensors, and UAB to some of the results presented herein.

References

1. Siow KS et al (2006) Plasma methods for the generation of chemically reactive surfaces for biomolecule immobilization and cell colonization – A review. Plasma Process Polym 3(6–7):392–418

2. Kandimalla VB, Tripathi VS, Ju HX (2006) Immobilization of biomolecules in sol-gels: Biological and analytical applications. Crit Rev Anal Chem 36(2):73–106
3. Meyer-Plath AA et al (2003) Current trends in biomaterial surface functionalization-nitrogen-containing plasma assisted processes with enhanced selectivity. Vacuum 71(3 Suppl):391–406
4. Cosnier S (1999) Biomolecule immobilization on electrode surfaces by entrapment or attachment to electrochemically polymerized films. A review. Biosens Bioelectron 14(5):443–456
5. Xia K et al (2005) Occurrence and fate of pharmaceuticals and personal care products (PPCPs) in biosolids. J Environ Qual 34(1):91–104
6. Rodriguez-Mozaz S, de Alda MJL, Barcelo D (2006) Biosensors as useful tools for environmental analysis and monitoring. Anal Bioanal Chem 386(4):1025–1041
7. Keren Barel-Cohen LSS, Shemesh M, Wenzel A, Mueller J, Kronfeld-Schor N (2006) Monitoring of natural and synthetic hormone in a polluted river. J Environ Manage 78:16–23
8. Nozaki O (2001) Steroid analysis for medical diagnosis. J Chromatogr A 935(1–2):267–278
9. Ying G-G, Kookana RS, Ru Y-J (2002) Occurrence and fate of hormone steroids in the environment. Environ Int 28(6):545–551
10. IARC (1985) I.A.f.r.o.c., Polynuclear aromatic compounds: bituminous, coal tars and derived products, shale oils and soots. IARC monographs on the evaluation of the carcinogenic risks of chemicals to humans 35:4
11. Wegener HC (2003) Antibiotics in animal feed and their role in resistance development. Curr Opin Microbiol 6(5):439–445
12. Sumpter JP, Jobling S (1995) Vitellogenesis as a biomarker for estrogenic contamination of the aquatic environment. Environ Health Perspect 103:174–178
13. Pumera M et al (2007) Electrochemical nanobiosensors. Sens Actuators B Chem 123(2):1195–1205
14. Rivas GA et al (2007) Carbon nanotubes for electrochemical biosensing. Talanta 74(3):291–307
15. Pingarrón JM, Yáñez-Sedeño P, González-Cortés A (2008) Gold nanoparticle-based electrochemical biosensors. Electrochim Acta 53(19):5848–5866
16. Li X-M, Yang XY, Zhang S-S (2008) Electrochemical enzyme immunoassay using model labels. Trends Anal Chem 27(6):543–553
17. Badihi-Mossberg M, Buchner V, Rishpon J (2007) Electrochemical biosensors for pollutants in the environment. Electroanalysis 19(19–20):2015–2028
18. Rose A et al (2002) GDH biosensor based off-line capillary immunoassay for alkylphenols and their ethoxylates. Biosensors Bioelectron 17(11–12):1033–1043
19. Lu H et al (2006) Screening of boldenone and methylboldenone in bovine urine using disposable electrochemical immunosensors. Steroids 71(9):760–767
20. Conneely G et al (2007) Development of an immunosensor for the detection of testosterone in bovine urine. Anal Chim Acta 583(1):153–160
21. Conneely G et al (2007) Electrochemical immunosensors for the detection of 19-nortestosterone and methyltestosterone in bovine urine. Sens Actuators B Chem 121(1):103–112
22. Carralero V et al (2007) Nanostructured progesterone immunosensor using a tyrosinase-colloidal gold-graphite-Teflon biosensor as amperometric transducer. Anal Chim Acta 596(1):86–91
23. Zacco E et al (2007) Electrochemical magneto immunosensing of antibiotic residues in milk. Biosens Bioelectron 22(9–10):2184–2191
24. Font H et al (2008) Immunochemical assays for direct sulfonamide antibiotic detection in milk and hair samples using antibody derivatized magnetic nanoparticles. J Agric Food Chem 56(3):736–743
25. Zacco E, Pividori MI, Alegret S (2006) Electrochemical magnetoimmunosensing strategy for the detection of pesticide residues. Anal Chem 78(6):1780–1788
26. Centi S et al (2005) A disposable immunomagnetic electrochemical sensor based on functionalised magnetic beads and carbon-based screen-printed electrodes (SPCEs) for the detection of polychlorinated biphenyls (PCBs). Analytica Chimica Acta 538(1–2):205–212

27. Centi S, Laschi S, Mascini M (2007) Improvement of analytical performances of a disposable electrochemical immunosensor by using magnetic beads. Talanta 73(2):394–399
28. Katz E, Willner I (2003) Probing biomolecular interactions at conductive and semiconductive surfaces by impedance spectroscopy: routes to impedimetric immunosensors, DNA-sensors, and enzyme biosensors. Electroanalysis 15(11):913–947
29. Guan JG, Miao YQ, Zhang QJ (2004) Impedimetric biosensors. J Biosci Bioeng 97(4):219–226
30. Ma KS et al (2006) DNA hybridization detection by label free versus impedance amplifying label with impedance spectroscopy. Sens Actuators B 114(1):58–64
31. Bart M et al (2005) On the response of a label-free interferon-gamma immunosensor utilizing electrochemical impedance spectroscopy. Biosens Bioelectron 21(1):49–59
32. Moreno-Hagelsieb L et al (2007) Electrical detection of DNA hybridization: three extraction techniques based on interdigitated Al/Al2O3 capacitors. Biosens Bioelectron 22(9–10):2199–2207
33. Laschi S, Mascini M (2006) Planar electrochemical sensors for biomedical applications. Med Eng Phys 28(10):934–943
34. Navratilova I, Skladal P (2004) The immunosensors for measurement of 2,4-dichlorophenoxyacetic acid based on electrochemical impedance spectroscopy. Bioelectrochemistry 62(1):11–18
35. Berggren C, Bjarnason B, Johansson G (2001) Capacitive biosensors. Electroanalysis 13(3):173–180
36. Ramón-Azcón J et al (2008) An impedimetric immunosensor based on interdigitated microelectrodes (ID[μ]E) for the determination of atrazine residues in food samples. Biosens Bioelectron 23(9):1367–1373
37. Bratov A et al (2008) Three-dimensional interdigitated electrode array as a transducer for label-free biosensors. Biosens Bioelectron 24(4):729–735
38. Bratov A et al (2008) Characterisation of the interdigitated electrode array with tantalum silicide electrodes separated by insulating barriers. Electrochem Commum 10(10):1621–1624
39. Leung A, Shankar PM, Mutharasan R (2007) A review of fiber-optic biosensors. Sens Actuators B Chem 125(2):688–703
40. McDonagh C, Burke CS, MacCraith BD (2008) Optical Chemical Sensors. Chem Rev 108(2):400–422
41. Borisov SM, Wolfbeis OS (2008) Optical biosensors. Chem Rev 108:423–461
42. Carrascosa LG et al (2006) Nanomechanical biosensors: a new sensing tool. Trends Anal Chem 25(3):196–206
43. Bally M et al (2006) Optical microarray biosensing techniques. Surf Interf Anal 38(11):1442–1458
44. Vaseashta A, Irudayaraj J (2005) Nanostructured and nanoscale devices and sensors. J Optoelectron Adv Mater 7(1):35–42
45. Gonzalez-Martinez MA, Puchades R, Maquieira A (2007) Optical immunosensors for environmental monitoring: How far have we come? Anal Bioanal Chem 387(1):205–218
46. Mauriz E et al (2006) Determination of environmental organic pollutants with a portable optical immunosensor. Talanta 69(2):359–364
47. Tschmelak J, Proll G, Gauglitz G (2005) Optical biosensor for pharmaceuticals, antibiotics, hormones, endocrine disrupting chemicals and pesticides in water: Assay optimization process for estrone as example. Talanta 65(2):313–323
48. Kappel ND, Proll F, Gauglitz G (2007) Development of a TIRF-based biosensor for sensitive detection of progesterone in bovine milk. Biosens Bioelectron 22(9–10):2295–2300
49. Tschmelak J et al (2006) Total internal reflectance fluorescence (TIRF) biosensor for environmental monitoring of testosterone with commercially available immunochemistry: Antibody characterization, assay development and real sample measurements. Talanta 69(2):343–350
50. Tschmelak J et al (2005) Automated water analyser computer supported system (AWACSS): Part II: Intelligent, remote-controlled, cost-effective, on-line, water-monitoring measurement system. Biosens Bioelectron 20(8):1509–1519

51. Tschmelak J et al (2005) Automated water analyser computer supported system (AWACSS) Part I: Project objectives, basic technology, immunoassay development, software design and networking. Biosens Bioelectron 20(8):1499–1508
52. Tschmelak J, Proll G, Gauglitz G (2004) Verification of performance with the automated direct optical TIRF immunosensor (River analyser) in single and multi-analyte assays with real water samples. Biosens Bioelectron Microarrays Biodefense Environ Appl 20(4):743–752
53. Rodriguez-Mozaz S et al (2004) Simultaneous multi-analyte determination of estrone, isoproturon and atrazine in natural waters by the RIver ANAlyser (RIANA), an optical immunosensor. Biosens Bioelectron 19(7):633–640
54. Szekacs A et al (2003) Development of a non-labeled immunosensor for the herbicide trifluralin via optical waveguide lightmode spectroscopic detection. Anal Chim Acta 487(1):31–42
55. Grego S, McDaniel JR, Stoner BR (2008) Wavelength interrogation of grating-based optical biosensors in the input coupler configuration. Sens Actuators B Chem 131(2):347–355
56. Kim N, Park IS, Kim WY (2007) Salmonella detection with a direct-binding optical grating coupler immunosensor. Sens Actuators B Chem 121(2):606–615
57. Hsu SH, Huang YT (2005) Design and analysis of Mach-Zehnder interferometer sensors based on dual strip antiresonant reflecting optical waveguide structures. Optics Lett 30(21): 2897–2899
58. Hsu SH, Huang YT (2005) A novel Mach-Zehnder interferometer based on dual-ARROW structures for sensing applications. J Lightwave Technol 23(12):4200–4206
59. Kinrot N (2004) Analysis of bulk material sensing using a periodically segmented waveguide Mach-Zehnder interferometer for biosensing. J Lightwave Technol 22(10):2296–2301
60. Prieto F et al (2003) Integrated Mach-Zehnder interferometer based on ARROW structures for biosensor applications. Sens Actuators B Chem 92(1–2):151–158
61. Cottier K et al (2003) Label-free highly sensitive detection of (small) molecules by wavelength interrogation of integrated optical chips. Sens Actuators B Chem 91(1–3):241–251
62. Kunz RE, Cottier K (2006) Optimizing integrated optical chips for label-free (bio-) chemical sensing. Anal Bioanal Chem 384(1):180–190
63. Cottier K, Kunz RE, Herzig HP (2004) Efficient and practical modeling of finite waveguide grating couplers. Jpn J Appl Phys Part 1 Reg Pap Short Notes Rev Pap 43(8B):5742–5746
64. Adrian J et al (2009) Generation of broad specificity antibodies for sulfonamide antibiotics and development of an enzyme-linked immunosorbent assay (ELISA) for the analysis of milk samples. J Agric Food Chem 57(2):385–394
65. Adrian J et al (2009) Waveguide interrogated optical immunoSensor (WIOS) for detection of sulfonamide antibiotics in milk. Biosens Bioelectron, submitted
66. Shankaran DR, Gobi KV, Miura N (2007) Recent advancements in surface plasmon resonance immunosensors for detection of small molecules of biomedical, food and environmental interest. Sens Actuators B Chem 121(1):158–177
67. Zhang W-W et al (2007) Analysis of 17[β]-Estadiol from sewage in coastal marine environment by surface plasmon resonance technique. Chem Res Chin Univ 23(4):404–407
68. Mauriz E et al (2007) Optical immunosensor for fast and sensitive detection of DDT and related compounds in river water samples. Biosens Bioelectron 22(7):1410–1418
69. Mauriz E et al (2006) Determination of carbaryl in natural water samples by a surface plasmon resonance flow-through immunosensor. Biosens Bioelectron 21(11):2129–2136
70. Farre M et al (2007) Part per trillion determination of atrazine in natural water samples by a surface plasmon resonance immunosensor. Anal Bioanal Chem 388(1):207–214
71. Matsumoto K et al (2005) A surface plasmon resonance-based immunosensor for sensitive detection of bisphenol A. J Facul Agric Kyushu Univ 50(2):625–634
72. Marchesini GR et al (2006) Biosensor recognition of thyroid-disrupting chemicals using transport proteins. Anal Chem 78(4):1107–1114
73. Gustavsson E et al (2004) Determination of B-lactams in milk using a surface plasmon resonance-based biosensor. J Agric Food Chem 52(10):2791–2796
74. Gaudin V, Fontaine J, Maris P (2001) Screening of penicillin residues in milk by a surface plasmon resonance-based biosensor assay: comparison of chemical and enzymatic sample pre-treatment. Anal Chim Acta 436(2):191–198

75. Kreuzer M et al (2008) Colloidal-based localized surface plasmon resonance (LSPR) biosensor for the quantitative determination of stanozolol. Anal Bioanal Chem 391:1813–1820
76. Kreuzer MP et al (2006) Quantitative detection of doping substances by a localised surface plasmon sensor. Biosens Bioelectron 21(7):1345–1349
77. Raiteri R et al (2001) Micromechanical cantilever-based biosensors. Sens Actuators B Chem 79(2–3):115–126
78. Fortina P et al (2005) Nanobiotechnology: the promise and reality of new approaches to molecular recognition. Trends Biotechnol 23(4):168–173
79. Goeders KM, Colton JS, Bottomley LA (2008) Microcantilevers: sensing chemical interactions via mechanical motion. Chem Rev 108(2):522–542
80. Ji HF et al (2008) Microcantilever biosensors based on conformational change of proteins. Analyst 133(4):434–443
81. O'Sullivan CK, Vaughan R, Guilbault GG (1999) Piezoelectric immunosensors - theory and applications. Anal Lett 32(12):2353–2377
82. Bunde RL, Jarvi EJ, Rosentreter JJ (1998) Piezoelectric quartz crystal biosensors. Talanta 46(6):1223–1236
83. Liu M, Li QX, Rechnitz GA (1999) Flow injection immunosensing of polycyclic aromatic hydrocarbons with a quartz crystal microbalance. Anal Chim Acta 387(1):29–38
84. Zhou XC, Cao L (2001) High sensitivity microgravimetric biosensor for qualitative and quantitative diagnostic detection of polychlorinated dibenzo-p-dioxins. The Analyst 126(1):71–78
85. Park IS et al (2004) Development of a direct-binding chloramphenicol sensor based on thiol or sulfide mediated self-assembled antibody monolayers. Biosens Bioelectron 19(7):667–674
86. Pribyl J, Hepel M, Skladal P (2006) Piezoelectric immunosensors for polychlorinated biphenyls operating in aqueous and organic phases. Sens Actuators B Chem (Special Issue - In honour of Professor Karl Cammann) 113(2):900–910
87. Rahman MA et al. (2007) An impedimetric immunosensor for the label-free detection of bisphenol A. Biosens Bioelectron 22(11):2464–2470
88. Dumont V et al (2006) A surface plasmon resonance biosensor assay for the simultaneous determination of thiamphenicol, florefenicol, florefenicol amine and chloramphenicol residues in shrimps. Analytica Chimica Acta 567(2):179–183

Hdb Env Chem (2009) 5J: 69–113
DOI:10.1007/698_2008_14

Recombinant Yeast Assays and Gene Expression Assays for the Analysis of Endocrine Disruption

Benjamin Piña, Susanna Boronat, Marta Casado, and Alba Olivares

Abstract Endocrine disruption is a particularly pervasive form of pollution, as it affects the reproductive and metabolic systems at very low concentrations. In its most usual form, endocrine disruption is mediated by the binding of exogenous substances to certain natural hormone receptors that regulate the endocrine system. Here we describe two tools to study potential or actual impacts of endocrine disruptors in wildlife and humans. The Recombinant Yeast Assay, RYA, is an in vitro method to evaluate the ability of a given compound or a mixture of substances to bind a specific receptor and to elicit, at last partially, the physiological responses associated to this binding in vivo. The second tool, the analysis the expression of hormone-responsive genes by quantitative real-time PCR, measures directly endocrine disruption effects in exposed organisms in vivo, either in the field or in controlled experiments. Both strategies are based on the concept that hormone receptors are ligand-dependent transcription factors and that the principal primary responses to the endocrine regulation (and also to endocrine disruption) are changes on the expression of hormone-responsive genes. These methods are aimed to the rapid screening for potential ligands for hormone receptors and to the identification of endocrine disruption signals at lower concentrations and shorter times of exposure than those required to develop the associated pathologies.

Keywords Biomarkers, Bioassays, Endocrine disruption, Gene expression, Hormone receptors, Recombinant Yeast Assay, Quantitative Real-Time PCR

B. Piña (✉), S. Boronat, M. Casado, and A. Olivares
Institute of Environmental Assessment and Water Research (IDAEA-CSIC), Jordi Girona 18, 08034, Barcelona, Spain
e-mail: bpcbmc@cid.csic.es

Contents

> Till, piercing each the other's flesh, they run/Together, and incorporate in one:/[...] Shoot up the same, and wear a common rind:/Both bodies in a single body mix,/A single body with a double sex.
>
> The story of Salamacis and Hermaphroditus, Metamorphoses/Book IV, Ovid

Separation of sexual roles has been a constant for most human cultures for as long as we have historical records. Deviations from the standard male/female dichotomy have been historically regarded as a perversion of the natural order or considered a sophisticated way of divine punishment, like in the Hermaphroditus myth. This cultural background fuelled the almost universal upheaval provoked by revelations in the early 1960s indicating that chemical pollutants may perturb reproductive functions of wildlife and humans, including sterility and sex reversal. This was the origin of the concept of endocrine disruption [1, 2]

The international definition for endocrine disruptors, agreed in the 1996 Weybridge workshop, defines them as "exogenous substances which cause adverse effects in an organism, or its progeny, subsequent to changes in the endocrine system" [3]. Endocrine disrupting compounds include a variety of chemical classes, including hormones, plant metabolites, pesticides, compounds used in the plastics industry and in consumer products, and other industrial by-products and pollutants. Some of them are persistent organic pollutants (POP's) that can be transported long distances across national boundaries and have been found in virtually all regions of the world. Others are rapidly degraded either by physical agents or microbiological metabolism in the environment, remaining active for only short periods of time. Finally, many of these compounds may be metabolised in the exposed individuals, with subsequent changes in their chemical and biological properties.

Since the seminal publication of Rachel Carson's Silent Spring, endocrine disruption effects have been recorded in different wildlife populations: adverse effects include abnormal thyroid function and development in fish and birds; decreased fertility in shellfish, fish, birds and mammals; decreased hatching success in fish, birds and reptiles; demasculinisation and feminisation of fish, birds, reptiles and mammals; defeminisation and masculinisation of gastropods, fish and birds; decreased offspring survival; and alteration of immune and behavioural function in birds and mammals [4, 5].

Epidemiological effects of endocrine disruption in humans are (luckily) much scarcer, although this has not mitigated the general concern about them, under the general understanding that human health is inextricably linked to the health of the environment. Many reports of endocrine disruption in humans came not from environmental risks, but rather from side effects of hormonal and obstetric medication. A paradigmatic case was the anti-abortive drug diethylstilbestrol (DES), which was discovered to affect the development of the reproductive system and cause vaginal cancer in the offspring of treated mothers [4]. Nevertheless, many health effects have been attributed to endocrine disrupting compounds, including a range of reproductive problems: reduced fertility, male and female reproductive tract abnormalities, skewed male/female sex ratios, loss of foetus, menstrual problems, changes in hormone levels; early puberty; brain and behaviour problems; impaired immune functions; and different cancers [6].

Different toxicity mechanisms may result in endocrine disruption, and their effects can be as complicated as the endocrine system is. Endocrine systems display a delicate conjunction of production of small molecules that are released into the bloodstream (the ligands) by specific secretory glands, and the presence of equally specific proteins in the target tissues, the hormone receptors, which recognise the ligands and that have the capacity to modify the cell metabolism according to their presence or absence. Generally speaking, there are two versions of this mechanism, depending on the nature of both the receptors and the ligands (Fig. 1). Membrane receptors are anchored in the cell membrane and are structured in two well-defined functional domains. The outer domain interacts with the cell media and recognises the ligand, whereas the inner domain originates a secondary signal as a response to this interaction. This secondary signal is transmitted from the inner plasma membrane into the

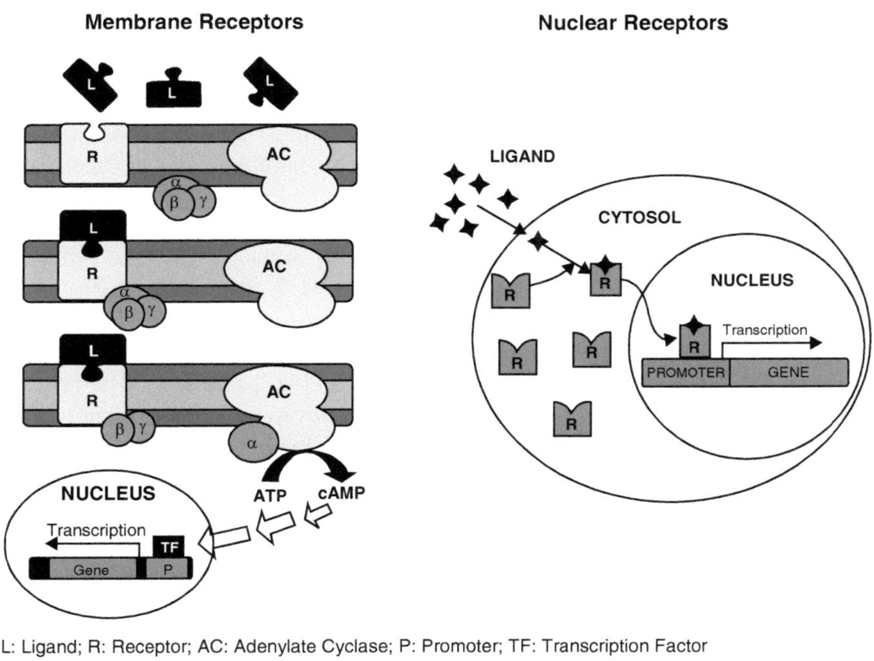

Fig. 1 Model of mechanisms of action of membrane-bound (*left*) and nuclear hormone receptors (*right*)

cell nucleus through a cascade of protein kinases, which result in a modification of the functionality of transcription factors and a subsequent change in the expression pattern of the target cells. In contrast, nuclear receptors are soluble proteins present in the cytoplasm or in the nucleus of target cells waiting for their ligands to cross the plasma membrane. In the presence of the ligand, the resulting receptor–ligand complex acts as a transcription factor, binding to DNA and modifying the expression of hormone-responsive genes. The physiological ligands for membrane receptors are typically small peptides, like insulin or neuropeptides, whereas nuclear receptors bind to small organic molecules, in many cases related to the steroid pathway. In consequence, nuclear receptors are sometimes dubbed as “steroid hormone receptors”, although in some cases their ligands are definitively not steroids; the best known example is the thyroid hormone receptor.

Cell receptors vary in their ability to bind exogenous substances. Because of the proteinaceous nature of their ligand, membrane receptors are in general less prone to be fooled by non-endogenous substances, and therefore, they are more resilient to endocrine disruption. There are notorious exceptions to this rule, opioids being an excellent example. Nevertheless, most current research on endocrine disruptors focuses on nuclear receptors, for anthropogenic compounds may easily share critical structural features with their natural ligands.

The so-called nuclear hormone receptor superfamily includes receptors for thyroid and steroid hormones (estrogens, progestins, androgens, mineralocorticoids and glucocorticoids), retinoids and vitamin D, ecdysone in invertebrates, as well as different orphan receptors of unknown ligands [7]. New ligands for some of these receptors have been identified, such as fatty acids [8], prostaglandin [9], or cholesterol derivatives [10]. All these receptors share a common ancestor; representatives of the superfamily are found in all metazoans but, as far as we know, they are absent in fungi, plants and other eukaryotes [11]. Up to date, a single representative of this family has been reported in Porifera [12]. The different receptor families and their ligands are shown in Table 1 (modified from [7]). Many potential exogenous ligands have been reported for several of these receptors, particularly for hormone receptors related to sex steroids (ER, AR and PR, Table 1, [4]), and many of them may cause endocrine disruption in exposed fauna and/or human populations. A deeper knowledge of the mechanisms of control of the endocrine system resulted in a notable increase in the number of receptors for which putative exogenous ligands have been reported, including TR, RAR, PPAR, PXR, CAR and RXR [13–17] (Table 1). The possible effects of these exogenous ligands in exposed populations are largely unknown, but it has been suggested that molecules interacting with the PPAR/RXR complex may promote obesity in humans [16]; in any case, it is clear that these ligands may affect sexual development and other physiological processes in molluscs [18, 19].

An unexpected addition to the list of endocrine disruptors occurred with the discovery that toxic effects of dioxines and benzopyrenes were mediated through their interaction with the Aryl hydrocarbon receptor (AhR); mice devoid of this receptor become refractory to the effects of TCDD [20] and of B[a]P [21]. The AhR belongs to a completely different family of transcriptional regulators, the bHLH-PAS family [22], with representatives in both Prokaryotes and Eukaryotes, although their ability to interact with ligands originated apparently with Chordates [22]. The

Table 1 Subfamilies of mammalian nuclear receptors

	Receptor	Denomination	Ligand	Binding Pattern[a]	Reported disrupters[b]
Class I	TR	Thyroid hormone receptor	Thyroid hormone (T3)	H	X
	RAR	Retinoic acid receptor	Retinoic acid	H	X
	VDR	Vitamin D receptor	1-25(OH)2 vitamin D3	H	
	PPAR	Peroxisome proliferator activated receptor	Benzotriene B4; Wy 14.643 Eicosanoids; thiazolidinediones (TZDS); 15-deoxy-12,41 - polyunsaturated fatty acids	H	X
	PXR	Pregnane X receptor	Pregnanes; C21 steroids	H	X
	CAR/MB67	Constitutive androstane Receptor	Androstanes; 1,4-bis[2-(3,5- dichloropyridyloxy)]benzene	H	X
	LXR	Liver X receptor	Oxysterols	H	
	FXR	Farnesoid X receptor	Bile acids	H	
	RevErb	Reverse ErbA	Unknown	M, D	
	RZR/ROR	Retinoid Z receptor/ retinoic acid-related orphan receptor	Unknown	M	
	UR	Ubiquitous receptor	Unknown	H	
Class II	RXR	Retinoid X receptor	9-Cis-retinoic acid	D	X
	COUP-TF	Chicken ovalbumin upstream promoter transcription factor	Unknown	D, H	
	HNF-4	Hepatocyte nuclear factor 4	Fatty acyl-CoA thioesters	D	
	TLX	Tailles-related receptor	Unknown	M, D	
	PNR	Photoreceptor-specific nuclear receptor	Unknown	M, D	
	TR2	Testis receptor	Unknown	D, H	
Class III	GR	Glucocorticoid receptor	Glucocorticoids	D	X
	AR	Androgen receptor	Androgens	D	X
	PR	Progesterone receptor	Progestins	D	X
	ER	Estrogen receptor	Estradiol	D	X
	MR	Mineralocorticoid receptor	Mineralocorticoids	M, D	X
	ERR	Estrogen-related receptor	Unknown		

[a] M, monomer; D, homodimer; H, heterodimer.

[b] "X" indicates receptors for which that putative or bona fide exogenous ligands have been reported (see text)

physiological role of AhR, as well as the precise nature of its presumptive physiological ligands has not been completely elucidated yet, but it is known to become activated by a numerous family of aromatic compounds and to regulate specific phase I and II metabolic enzymes, among others [23, 24]. Exogenous ligands of AhR are known to elicit a broad range of physiological effects, including immune dysfunction, endocrine disruption, reproductive toxicity, developmental defects, and cancers in vertebrates [25–27].

There are essentially two ways to study potential or actual impacts of endocrine disruptors in wildlife and humans. First, there is an array of in vitro methodologies to evaluate the endocrine-disruption potential of a given compound or of an environmental sample [28–33], methodologies that can be coupled to surveillance protocols or to quantitative structure-activity relationship (QSAR) analysis [34]. A more direct approach is to analyse endocrine disruption effects in exposed organisms, either in the field or in controlled experiments. Different end-points for endocrine disruption have been described for both human populations and wildlife (reviewed in [6]). Here we describe two tools covering both strategies, both aimed at the rapid screening for potential ligands for hormone receptors and the identification of endocrine disruption signals at lower concentrations and shorter times of exposure than those required to develop the associated pathologies: yeast-based receptor binding assays and the quantification of gene expression biomarkers.

1 The Recombinant Yeast Assay

Genetically modified yeast strains have become common tools to detect exogenous ligands for different receptors. These tools have been applied to detect the presence of receptor-binding activity in the environment and to screen for new putative receptor binding substances among the ever-increasing number of new chemical compounds appearing in the market every year [29, 35–37] (Table 2). Yeast bioassays have been used to correlate the presence of suspected or bona-fide endocrine disruptors and estrogenic activity in environmental samples [38–40], and to establish relationships between chemical structures and affinity for vertebrate hormone receptors [37, 41–44]. These systems, also known as recombinant yeast assays (RYA) do not provide a precise chemical characterisation of the hormone-receptor ligands, but their low cost and easiness of handling makes them a first choice when testing large numbers of ill-characterised samples or compounds. A very detailed protocol to perform RYA has been recently published [33].

Yeast-based bioassays consist of an engineered yeast strain harbouring at least two foreign genetic elements, one of them allowing the expression of a given vertebrate nuclear receptor in the yeast cell and the second one devised to allow monitoring of the activity of the expressed receptor. In this second genetic element, a convenient gene whose expression is easy to quantify (the reporter gene) is under transcriptional control of the expressed nuclear receptor. The setup ensures that the genetic product of the reporter gene (an enzyme or a fluorescent protein, for example) will only be expressed in the presence of the receptor–ligand complex.

Table 2 Examples of use of RYA to monitor exogenous ligands for vertebrate hormone receptors

Receptor	Species	Receptor structure	Receptor expression/ Promoter	Reporter	Applications	References
PR	chicken	Ub-PR	Inducible/CUP1	PRE-LacZ	Test strain	[130]
	human	Full length	Inducible/CUP1	PRE-LacZ	Test ligands	[41]
	human	GAL4DBD-hPRDN23	Constitutive/ADH1	GAL1-LacZ	Test ligands	[131]
	human	Full length	n.a.	PRE-LacZ	Environment	[132]
	human	Full length	n.a.	PRE-LacZ	Environment	[133]
	human	Full length	n.a.	PRE-yEGFP	Test ligands	[134]
	chicken	Ub-PR	Inducible/2 × ERE-CYC1-ub-cPRA	PRE-LacZ	Test strain	[135]
	human	lexADBD-hPRLBD	Constitutive/ADH1	lexA-LacZ	Test strain	[60]
	rat	Full length	Constitutive/ADH1	PRE-LacZ	Environment	[29]
AR	human	Full length	Inducible/CUP1	ARE-LacZ	Test ligands	[41]
	human	Full length	Constitutive/PGK	ARE-lux A/B + ARE-lux CDE+frp	Test ligands	[113]
	rat	Full length	Constitutive/GPD	GRE-LacZ	Test ligands	[136]
	human	Full length	Constitutive/GPD	ARE-LUC-slk	Environment	[137]
	human	Full length	Constitutive/GPD	ARE-yEGFP	Test ligands	[44]
	human	Ub-FLAG-AR	Inducible/CUP1	ARE-LacZ and ARE-GFP	Test ligands	[138]
	mouse	lexADBD-ARLBD	Constitutive/ADH1	lexA-LacZ	Test ligands	[139]
	human	Full length	Constitutive/PGK	ARE-LacZ	Test ligands	[140]
	human	Full length	n.a.	ARE-LacZ	Environment	[141]
	human	Full length	n.a.	ARE-LacZ	Environment	[142]
PPAR	human	GAL4DBD-PPAR+GAL4AD-CBPID	Constitutive/ADH1	MEL1-LacZ	Test ligands	[63]
	rat	Ub-PPAR + Ub-RXR	Inducible/CUP1	PPARRE-LacZ	Test strain	[143]
RXR	human	RXR-LBD + TIF2-ID	Constitutive/ADH1	GAL1-LacZ	Environment	[144]

(continued)

Table 2 (continued)

Receptor	Species	Receptor structure	Receptor expression/ Promoter	Reporter	Applications	References
AhR	AhR	AhR + ARNT + SRCia	Inducible/GAL1	MELIDUAS-DRE-Luc	Environment	[145]
	human	AhR + ARNT	Inducible/GAL1	DRE-LacZ	Test ligands	[146]
	human	AhR + ARNT	Inducible/GAL1	DRE-LacZ	ligands human	[147]
	human	AhR + ARNT	Inducible/GAL1	DRE-LacZ	Environment	[148]
	human	AhR + ARNT	Inducible/GAL1	DRE-LacZ	Environment	[149]
	human	AhR + ARNT	Inducible/GAL1	DRE-LacZ	Environment	[150]
	human	AhR + ARNT	Inducible/GAL1	DRE-LacZ	Environment	[151]
	human	AhR + ARNT	Inducible/GAL1	DRE-LacZ	Test ligands	[152]
	human	AhR + ARNT	Inducible/GAL1	DRE-LacZ	Test ligands	[153]
	human	AhR + ARNT	Inducible/GAL1	DRE-LacZ	Test strain	[29]
	human	AhR + ARNT	Inducible/GAL1	DRE-LacZ	Environment	[74]
	human	AhR + ARNT	Inducible/GAL1	DRE-LacZ	Test ligands	[66]
GR	human	DN	Inducible/GAL1	GRE-LacZ	Test ligands	[154]
PXR	mouse	GAL4-DBD-mPXR+GAL4AD-SRC1 or GAL4AD-RIP140	Constitutive/ADH1	GAL4UAS-LacZ	Environment	[155]
TR	human	GAL4DBD-TRa1LBD + GAL4AD-RXRaLBD	Constitutive/ADH1	GAL1-LacZ	Test ligands	[156]
VDR	human	GAL4DBD-VDR + GAL4AD-RXRβ	Constitutive/ADH1	GAL1-LacZ	Test ligands	[157]
	human	GAL4DBD-VDR + GAL4AD-hRXRa	Constitutive/ADH1	GAL4UAS-LacZ	Test ligands	[158]

Abbreviations: n.a. not available: DBD, DNA Binding Domain; LBD, Ligand binding Domain; AD, Activation Domain; ID, Interaction Domain; ΔN, truncated N-terminal domain; UAS, Upstream Activating Sequence

This imitates the mechanism by which many hormone-responsive genes are regulated in vertebrates; the fundamental similarity of all eukaryotes ensures that the same mechanism also works in yeast. A comparison between the physiological pathway in vertebrate cells and its engineered yeast version is shown in Fig. 2.

The expression of the reporter gene is monitored by its enzymatic or fluorescent activity by using convenient quantitative techniques. This activity is directly related to the effective concentration of estrogenic compound(s) through a typical dose–response sigmoidal curve. The system does not provide a direct measurement of the molar (or mass) concentration of endocrine disruptors, but of their estrogenic activity. For simplification, results are given in "hormone equivalents", defined as the amount of a standard ligand (estrogen for the estrogen receptors, progesterone for the progesterone receptors, testosterone or di-hidroxytestosterone for the androgen receptor, etc.) that should be present in a given sample to account for all the observed response.

1.1 The Genetic Setup

1.1.1 Receptors

Yeast lack nuclear receptor homologues; therefore, they have to be introduced in the strain's genetic make-up. This apparently complicated step is much facilitated by the power of Molecular Biology tools available for *Saccharomyces cerevisiae*,

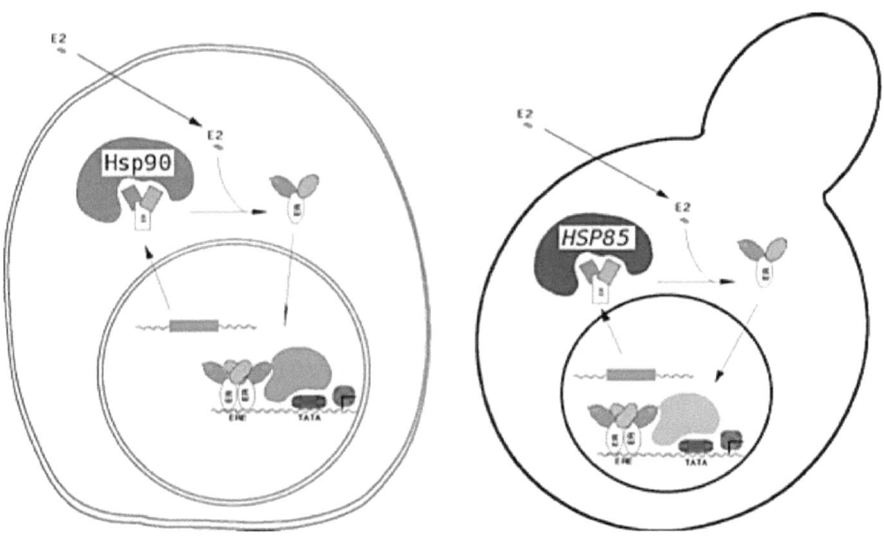

Fig. 2 Schematic representation of the transfer of the mechanism of hormone response from a vertebrate cell (*left*) to yeast (*right*). Note that the receptor itself and the reporter gene are the only two elements that have to be added to the yeast genetic set-up; the rest of the components are provided by the yeast genome

the yeast species almost universally used for RYA assay. Cloning and expressing non-homologous proteins in yeast are standard procedures. In the design of a RYA system, there are at least three choices to make: the receptor to clone, the species to clone it from, and the yeast expression system.

In principle, DNA sequences coding for any desired receptor could be obtained from appropriate tissues of any species. In practice, the methodology (extraction of RNA from the tissue, its retrotranscription into cDNA, and amplification of the specific cDNA molecule encompassing the desired gene) requires not only physiological information of the tissue to start from and, some times, the appropriate conditions of age and maturation of the specimen, but also DNA sequence information to design the appropriate primers and probes. Because of the multi-family nature of nuclear receptors, there are different genes with similar sequences but different functions, the so-called orphan receptors being good exponents of this. A number of different receptors for mammals and teleostei has already been cloned and tested; there is a much shorter list for birds, reptiles and amphibians, and an even shorter one for the rest of animal Phyla. To have a hint of how imperfect our knowledge of the endocrine system in non-vertebrates is, it is illustrative of the fact that we do not have a good understanding of the mechanism of action of the juvenile hormone in insects, despite the massive knowledge on development and molecular biology of *Drosophila* [45, 46].

1.1.2 Yeast Expression Vectors

Expression of any heterologous gene in yeast can be brought about by a handful of well-studied genetic mechanisms (see [47–50] for detailed protocols and methods). The basic strategy is to place the desired DNA coding sequence under the control of a yeast promoter; the addition of a yeast terminator 3' of the inserted gene normally facilitates the post-transcriptional processing of the heterologous transcript and, many times, increases its stability in the cell. The conjunction of promoter-coding sequence-terminator is sometimes described as an "expression cassette", for it constitutes a genetic unit that directs the expression of the cloned protein (the receptor, in our case) when introduced into the yeast cell. Two main types of yeast promoters are used to express foreign proteins. Constitutive promoters are those that induce expression of contiguous genes permanently, which results in a stable level of transcript. On the contrary, inducible promoters are silent unless a given inductor is present in the cell. Expression vectors activated upon addition of galactose or copper are the most used inducible systems to express foreign proteins in yeast. The use of one or another system is essentially a trial and error decision: some proteins inhibit yeast growth and, therefore, it is better to use an inducible system to propagate yeast cells in their absence and express it only when is needed. However, this introduces a necessary activation step that may increase both variations among experiments and the time required to complete the process.

The combination of promoter-coding sequence-terminator has to be afterwards introduced into a recipient yeast cell, and maintained in a form as stable as possible.

There are essentially three ways of introducing and maintaining foreign DNA sequences in yeast: by integration into the yeast genome, by creating artificial yeast chromosomes, either linear or circular, or by using the naturally occurring yeast plasmid 2 μ [50].

Yeast integrates into its genome linear sequences by homologous recombination, provided their flanking sequences coincide with DNA sequences already present in the genome. As a result, the cell receives a single copy of the introduced sequence. Yeast genome has minimal positioning effects and they are restricted to perfectly documented regions, such as telomeres and subtelomeric sequences [51].

Unlike most eukaryotes, yeast chromosomes have very small and well-defined centromers, which span less than 1 Kb and that are fully functional when cloned into any DNA fragment. This property can be used to construct artificial DNA elements that segregate into daughter cells with efficiencies similar to the natural chromosomes. In the simplest version, circular molecules encompassing a yeast centromer and an origin of replication (ARS) can be used to carry any desired DNA sequence. These constructs, called ARS/CEN plasmids, are very useful for maintaining DNA fragments up to 10–20 Kb in yeast. Longer sequences require the use of linear constructs that encompass, in addition to centromer and ARS, yeast telomeres at both ends. These constructs are called YAC (for yeast artificial chromosomes), but they are not usually used for the purposes described here. ARS/CEN plasmids are usually maintained in one to three copies in the yeast cell [52–54].

The 2 μ (two-micron) plasmid is a genetic autoreplicative element found in many, but not all, yeast strains, and for which no specific function has been described yet. Its replication is uncoupled from the yeast genomic DNA and it maintains some 200 copies in the cell, being distributed passively into mother and daughter cells. The wild plasmid encodes different enzymes required for its own replication; however, a defective form only encompassing the origin of replication (about 1 Kb) is stable in yeast provided the complete 2 μ version supplies necessary trans elements. The defective 2 μ has been widely used for introducing and maintaining a stable number of copies (from 10 to 20) of chimeric bacterial/yeast constructs, able to propagate in both systems and that can be engineered to carry different sequences, including gene expression cassettes [54].

Expression of foreign proteins confers no selective advantage to yeast strains; on the contrary, it can reduce their fitness. Therefore, receptor-expressing cells would be rapidly overgrown by non-expressing counterparts. For this reason, it is necessary to associate the receptor-expressing cassette with some genetic element that provides the necessary selective advantage. These elements, called markers, exist in two main forms. In the simplest one, the marker provides the recipient cell with enzymes required for its growth in selective media. This is usually achieved by using yeast strains defective for some crucial step in some catabolic process and introducing the missing gene along with (usually, physically linked to) the construct [52]. A second type of marker includes those conferring resistance to antibiotics, like gentamicin or hygromycin. These so-called dominant markers have the advantage of not requiring any specific yeast genetic background, but they do require the continuous presence of the antibiotic to maintain the selective pressure [55].

1.1.3 Reporter Genes

Once the receptor is expressed in the yeast cell, it is necessary to design a method to monitor its binding to any ligand that may exist in the medium. Nuclear receptors are facultative transcription factors, able to recognise specific DNA sequences and to activate neighbouring promoters when the ligand is present. RYA takes profit of this fundamental property by devising the so-called reporter genes, that is, genes encoding proteins easy to detect under the control of a silent promoter complemented with DNA recognition sites for the expressed receptor (HRE, hormone response element). Upon the presence of an appropriate ligand, the receptor-ligand complex would bind to the HRE and activate transcription of the reporter gene; the concentration of the resulting protein would reflect the amount of ligand present (see below).

The most used reporter genes are bacterial enzymes with activities completely absent in natural yeast strains, like β-galactosidase (lacZ) or β-glucuronidase (GUS) [56–58]. The resulting enzymatic activities may be easily monitored by using chromogenic (ONPG, X-gal blue, etc.) [59] or fluorogenic substrates (MUG, MUGal, etc.) [60]. An alternative possibility is to use autofluorescent proteins, like GFP. In this version no substrate is required, as GFP fluorescence can be detected and quantified by irradiating GFP-expressing cells with appropriate wavelengths [61]. Another approach comes from the use of bioluminescent bacteria operons that recapitulate in yeast the bioluminescence phenomenon observed in bacteria upon activation of its expression. The feasibility of these approaches depends on the capabilities of sophisticated fluorescence readers able to differentiate the natural yeast fluorescence from the induced one. Whatever the technique used to measure the expression levels of the reporter gene, it reflects the concentration of the ligand following a sigmoidal dose-response curve. The mathematics underlying this kind of response will be discussed below

1.1.4 Requirement for Cofactors

Nuclear receptors differ in their capability to trigger transcriptional activation of reporter genes in yeast. Mammalian ER is a very good activator in yeast, indicating that the yeast transcriptional machinery readily recognises activation domains in the ER as such, in an amazing demonstration of evolutionary conservation of cell regulatory signals. However, this is not always the case, and many nuclear receptors are very poor activators in yeast. In this case, the habitual solution is the use of the so-called "two hybrid strategy", in which both the receptor and recognised targeted factors are cloned in yeast, in a way that the receptor bears the DNA recognition domain (homologous or heterologous, see below) and the cofactor is supplied with a bona fide yeast activation domain. The presence of the ligand favours the binding of the receptor to DNA and its interaction with the homologous counterpart. The presence of the yeast activation domain transforms this ternary complex into a true yeast activator [62–64].

A particular case occurs when the receptor needs some coactivator to bind DNA. This is the case, for example, of VDR (Table 1), which requires RXR to bind to the corresponding HRE. The unrelated AhR also requires a cofactor, known as AhR Nuclear Translocator, or ARNT, to bind DNA. In this case, there are at least two possible strategies. An obvious one is to supply the yeast strain with all cofactors required for allowing the receptor to bind DNA and promote transcription. A second possibility is to replace the homologous DNA binding domain (DBD) of the receptor by a heterologous DBD. In many cases, these chimeric constructs maintain most pharmacological features of the homologous system [65], although some fundamental differences on response to different ligands have been reported recently [66].

1.2 Mathematical Modelling

The simplest and most popular model for dose-response curves assumes equilibrium between hormone-free and hormone-loaded hormone receptor molecules in solution,

$$R + h \overset{K_d}{\rightleftarrows} hR,$$

in which R represents the concentration of hormone-free receptor molecules, h the hormone concentration, hR the concentration of the hormone-loaded receptor molecule, and K_d is the dissociation constant. This particular model assumes a single agonist molecule binding to each receptor molecule, as has been demonstrated by numerous physical and crystallographic data for nuclear receptors [67]. Assuming a hormone concentration much larger than the receptor concentration, the fraction of receptor bound to the hormone Φ_r can be described by the Hill equation:

$$\Phi_r = \frac{[hR]}{R_t} = \frac{1}{1 + \frac{K_d}{[h]}}. \tag{1}$$

In which R_t is the total receptor concentration (bound and free) [68].

The Hill equation only describes the equilibrium for the hormone–receptor interaction. It is frequently assumed that this is the rate-limiting step in the cascade of events going from the presence of a given concentration of hormone and the final physiological effects – expression of β-galactosidase, GUS or GFP fluorescence. K_d represents the dissociation constant of the ligand-hormone complex; its value coincides with EC_{50}, the ligand concentration giving 50% of the maximal response. Apparent K_d values can be calculated for each sample or tested compound by measu-ring the observed response at different concentrations or dilution factors. There are different ways to calculate apparent K_d values from an experimental dose–response curve. A simple way is to determine the LOD (lower observable dilution) and estimate that it corresponds to EC_{10}, approximately [60]. More rigorously, the mid-point of the resulting dose–response curve (by definition, EC_{50}) can be calculated from the experimental data by non-linear regression methods, assuming the Hill's equation [60]. Some linear approaches to this method have also been published [69, 70]. Figure 3 shows a typical example of a dose-response graph for the estrogen receptor.

Different K_d values obtained by RYA can be used to estimate the relative ligand potency of a given substance or environmental sample, when compared to a standard ligand we use as reference. The nature of this reference ligand depends on the receptor. In some cases, it is easy to determine which molecule has to be selected, like estrogen for the estrogen receptors, progesterone for the progesterone receptors, testosterone or di-hidroxytestosterone for the androgen receptor, etc. For some receptors, like the PPAR or the AhR, there is no such "natural standard", and some well-known and easily manipulable heterologous chemical (rosiglitazone, β-naphthoflavone) has to be used instead [63, 71–73]. Once the standard ligand has been selected and its EC_{50} values have been established for each receptor, the relative ligand potency of a substance can be then calculated as:

$$\frac{EC_{50}(\text{compound})}{EC_{50}(\text{reference})}. \tag{2}$$

A very similar equation applies to complex samples:

$$\frac{EC_{50}(\text{dilution factor})}{EC_{50}(\text{reference})}. \tag{3}$$

In this case, the relative potency of the sample can be interpreted as the concentration of standard ligand it should contain to elicit the observed response. This approach can be used in the inverse direction; if we know the concentration of different ligands in a sample and their relative potencies, we can predict the expected potency of the sample, following the equation:

$$\text{Total potency} = \sum X_i \varepsilon_i. \tag{4}$$

In which X_i represents the concentrations for each compound calculated from chemical analysis and ε_i the corresponding relative potency factors. The adjustment of this prediction to the observed potency of the sample would indicate that the chemical analysis is able to identify and quantify all relevant ligands for a particular receptor. Significant differences between the two values indicate either the presence of an unchecked ligand (Calculated < Observed) or interferences between different ligands (Observed < Calculated). Examples of such analysis and relative potency values have been already published (Table 3, data from [73, 74]).

1.2.1 Inhibitory Effects

All biological assays are more or less prone to suffer from inhibitory effects that reduce its sensitivity. In the case of RYA, an obvious possibility is that the sample or compound to analyse may be toxic to the yeast cell and therefore it prevents the

Table 3 EC_{50} values for different chemicals tested with ER-RYA and AhR-RYA systems

COMPOUND	AhR EC_{50} (mg/L)	ER EC_{50} (mg/L)
17-b-Estradiol	n.d	$(7.3\pm0.4)\times10^{-5}$
b-Naphtoflavone	0.014 ± 0.02	n.d
PCB9	0.4 ± 0.2	b.d.l
PCB29	1.1 ± 0.3	8 ± 5
PCB30	b.d.l	63 ± 23
PCB50	b.d.l	7.5 ± 2.3
PCB58	b.d.l	b.d.l
PCB77	0.014 ± 0.005	b.d.l
PCB104	b.d.l	5.3 ± 3.7
PCB126	0.9 ± 0.3	b.d.l
PCB188	b.d.l	b.d.l
HCB	7.7 ± 0.5	b.d.l
pp -DDT	b.d.l	b.d.l
Diuron	0.26 ± 0.10	>200
N-Demethyl diuron	2.7 ± 1.2	b.d.l
1-(3,5-Dichlorophenyl)urea	40.6 ± 0.7	b.d.l
Irgarol	b.d.l	b.d.l
Sea-Nine	b.d.l	b.d.l
BZT-OH	6.9 ± 0.8	>10
BZT	10.2 ± 0.6	5.5 ± 1.3
MetBZT	11.0 ± 0.6	8 ± 3
DEHP	n.d.	2.3 ± 1.5
OP	n.d.	0.19 ± 0.08
Estriol	n.d.	$1.2\times10^{-4}\pm6.4\times10^{-5}$
estrone	n.d.	$2.1\times10^{-3}\pm1.2\times10^{-3}$
Et-estradiol	n.d.	$2.3\times10^{-4}\pm1.3\times10^{-4}$
DES	n.d.	$8.7\times10^{-4}\pm4.4\times10^{-4}$
BPA	n.d.	1.6 ± 0.9
BBP	n.d.	5.1 ± 2.2
DEP	n.d.	7.0 ± 2.3
DMP	n.d.	4.1 ± 1.2
BZT	n.d.	5.5 ± 1.3
MeTBZT	n.d.	7.8 ± 3.2
DBP	n.d.	4.0 ± 1.5
HOBT	n.d.	6.3 ± 1.7
NP	n.d.	0.08 ± 0.02
NP1EO	n.d.	3.0 ± 0.4

bdl. below detection limit; nd. not done

Abbreviations: BBP: butyl benzyl phthalate; BPA: bisphenol A; BZT-OH: 3′-hydroxy benzothiazole; BZT: benzothiazole; DBP: dibutyl phthalate; DEHP: bis(2-ethylhexyl) phthalate; DEP: diethyl phthalate; DES: dietilestilbestrol; DMP: dimethyl phthalate; HCB: hexachlorobenzene; HOBt: 1-Hydroxybenzotriazole; MetBZT: 2-methyl benzothiazole; NP: nonylphenol; NPIEO: nonylphenol mono etoxylated; OP: octylphenol; p,p'-DDT: 1,1,1-trichloro-2,2-bis(4-chlorophenyl)ethane; PCB: polychlorobiphenyl.

required biological response to the presence of ligands. There are several procedures to assess whether or not a negative value is due to toxicity or to the absence of ligand activity. An obvious one is to spike the samples or compounds with a saturating concentration of the standard ligand; toxicity would prevent response whatever the ligand would be. A more sophisticated method is to measure endogenous yeast

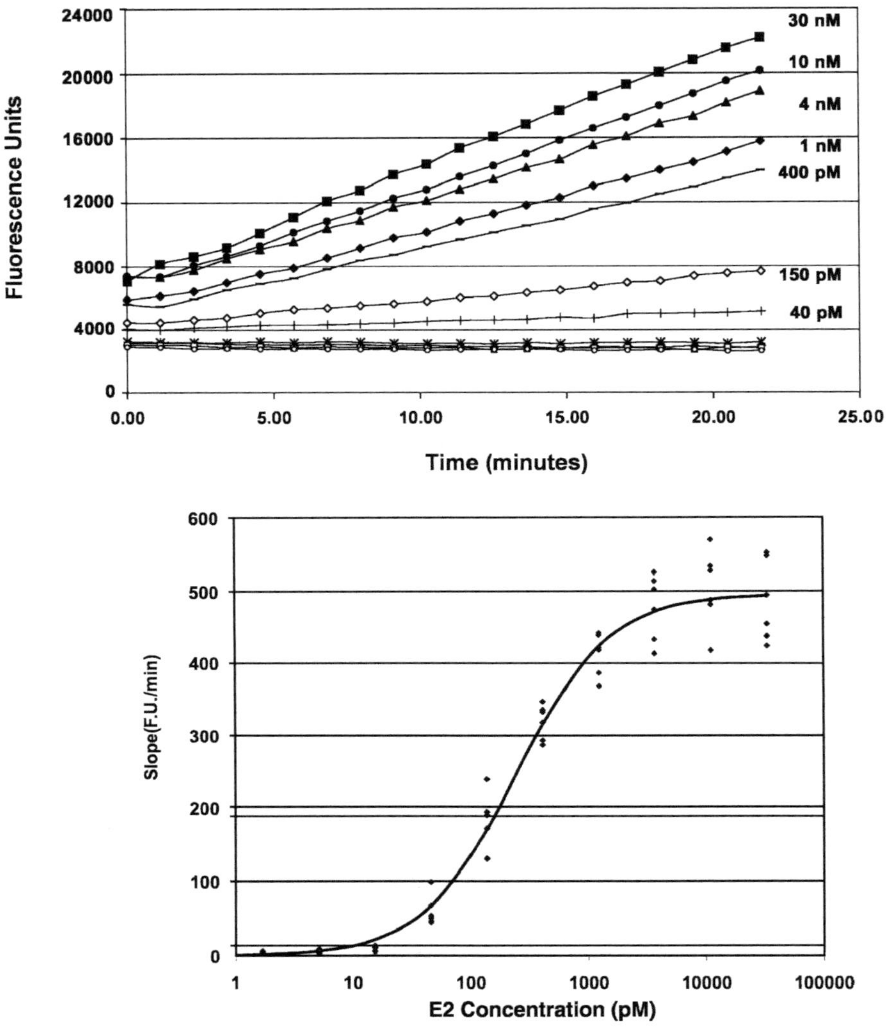

Fig. 3 Response ER–RYA to different concentrations of estradiol (fluorogenic substrate). *Top*, linear plots of the increase of fluorescence (in fluorescence units, FU) with time for different estradiol concentrations in the ER-RYA system. Concentrations of estradiol are indicated on the right margin. *Bottom*, Dose–response curve for estradiol (E2) in the ER-RYA system. The *y*-axis represents β-galactosidase activity in FU/min, *diamonds* represent values from six independent determinations for each estradiol concentration. Graphs indicate theoretical curves as defined by Hill's equation. The corresponding parameters were calculated by non-linear regression methods. Data from [60]

activity (for example, expression of a reporter gene under a yeast promotor) in the presence and absence of the sample; in this case a quantitative estimation of the toxicity can be obtained [42]. However, there is a radically different mechanism preventing a bona fide ligand to elicit biological response. There are ligands, the binding of which not only does not elicit response, but it prevents the system from responding to the presence of other ligands present in the same solution. Ligands that elicit the physiological response are known as receptor agonists, whereas those preventing the same response are called antagonists. The term "partial agonists" is used to define ligands able to elicit a limited subset of the physiological responses of pure agonists.

The interplay between agonists, antagonists and partial agonists is central in the physiology and the pharmacology of many endocrine-acting compounds. However, most of this delicate balance is simplified in RYA, as yeast lacks most of the cofactors and interactions intervening in the vertebrate hormonal regulation. As a consequence, most ligands able to bind to a given receptor would be classified as agonists in the yeast assay. However, there are substances acting as antagonists in yeast, and this peculiarity is important in the characterisation of the potential toxicity of a new substance. The antagonistic effect can be precisely studied in yeast, producing data on the inhibitory potential, in a similar way as the ligand potential described before (Fig. 3, [42, 73]). In this regard, one has to consider three mechanisms of mutual interaction among a pure agonistic ligand (h_1) and an inhibitor (h_2).

The competitive inhibition model proposes an equilibrium between free receptor, R, and two ligands that bind alternatively to a single site of the receptor molecule, with dissociation constants K_{d1} and K_{d2}:

$$R_O + h_2 \rightarrow Rh_2 + R; R + h_1 \overset{Kd_1}{\leftrightharpoons} Rh_1 \tag{5}$$

$$K_{d_1} = \frac{[R][h_1]}{[Rh_1]};\ K_{d_2} = \frac{[R][h_2]}{[Rh_2]}. \tag{6}$$

The total amount of receptor molecules, free and occupied (R_o), can be expressed as

$$R_o = [Rh_1]\left[1 + \frac{K_{d_1}}{[h_1]} + \frac{K_{d_1}[h_2]}{K_{d_2}[h_1]}\right]. \tag{7}$$

At any given concentrations of h_1 and h_2, any target gene would show a fraction of its maximal activation at a saturating concentration of h_1 (A/A_{max}), which that could be expressed as:

$$\frac{A}{A_{max}} = \frac{[Rh_1]}{R_o} = \frac{1}{K_{d_1} + \frac{\frac{K_{d_1}[h_2]}{K_{d_2}}}{1 + \frac{K_{d_2}}{[h_1]}}} \tag{8}$$

In this variant of the Hill equation, A_{max} is independent from h_2, whereas the apparent EC_{50} for h_1 (EC_{50app}) equals K_{d1} only when $h_2 = 0$. At any other concentration of h_2, EC_{50app} would equal

$$\mathrm{EC}_{50\,\mathrm{app.}} = K_{\mathrm{d}_1} + [h_2]\frac{K_{\mathrm{d}_1}}{K_{\mathrm{d}_2}} \tag{9}$$

And therefore,

$$\frac{\mathrm{EC}_{50(h_2)}}{\mathrm{EC}_{50(h_2=0)}} = 1 + \frac{[h_2]}{K_{\mathrm{d}_1}} \tag{10}$$

in which $\mathrm{EC}_{50(h2)}$ and $\mathrm{EC}_{50(h2=0)}$ correspond to the EC_{50} for h_1 in the presence and in the absence of a given concentration of h_2. K_{d2} is then identical to IC_{50}; its value can be calculated as the inverse of the slope of the plot corresponding to (10).

A second possible method of inhibition implies that the binding of h_2 to the receptor irreversibly inactivates it, reducing the amount of available receptor molecules for binding to h_1:

$$R + h_1 \overset{Kd_1}{\leftrightharpoons} Rh_1\ ;\ R + h_2 \overset{Kd_2}{\leftrightharpoons} Rh_2 \tag{11}$$

If the proportion of receptor molecules becoming inactivated follows a typical logistic function with an inhibitory constant K_i,

$$\frac{Rh_2}{R_o} = \frac{1}{1 + \frac{K_i}{[h_2]}};\ R = R_o - Rh_2. \tag{12}$$

In this model, the maximal activity at saturating concentrations of h_1 depends on the concentration of h_2, as follows:

$$h_1 \to \infty, \frac{A_{\max,\,h_2\,0}}{A_{\max,\,h_2}} = \frac{R_o}{R} = 1 + \frac{[h_2]}{K_i}. \tag{13}$$

K_i is therefore equivalent to the concentration of h_2 reducing to 50% the maximal activity at saturating concentrations of h_1, IC_{50}.by definition. This model is indistinguishable from a general toxicity of the tested compound for the yeast cells, with identical parameters.

1.3 Practical Examples

Table 2 shows some examples of how RYA can be used in different ways to monitor putative endocrine disruptors. This kind of study can be divided into at least three groups, relating to their intended purpose. The largest group is constituted by experiments aimed at assessing the potential affinity of a known substance to a given receptor, as a method to evaluate environmental risks [44, 72, 73, 75]. A derivative of this group is the use of RYA to monitor changes of the affinity upon chemical modification of the ligand, either to evaluate the risk of ligand derivatives appearing in the environment, or to monitor the degradation or inactivation of the ligand by wastewater processing schemes [42, 76]. Finally, RYA can be used to evaluate endocrine disruption activities in environmental or industrial samples, without previous knowledge of the active compounds. Combined with chemical analysis, this system may help in the identification of the possible contaminants with potential endocrine disruption activities and monitor the appropriate remediation programs. As an example of this, Fig. 5 shows the results of a survey on EDC loads in different points along a complete basin in the Spanish Northern Mediterranean area to estimate the contribution of STPs to these loads [77]. In this particular case, chemical analysis data and RYA data showed a quasi-linear correlation, demonstrating the complementariness of both methods. Moreover, the

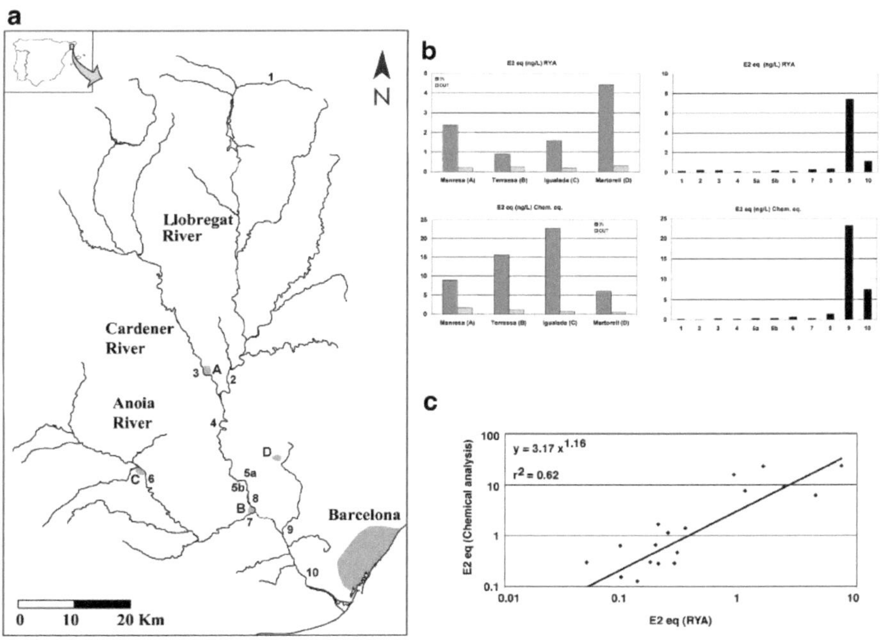

Fig. 4 Monitoring of putative endocrine disruptors in the Llobregat basin. (**a**) Map of sampling sites in the Llobregat basin. Numbers 1–10 indicate sampling points for river water, letters A–D correspond to STP locations. Grey areas indicate approximate locations of cities near to these STPs: A-Manresa; B-Terrassa, C-Igualada; D-Martorell. The position of the city of Barcelona is given to facilitate localisation of the area in the map. (**b**) Correlation between total estrogenicity values [(in E2 eq)] obtained from RYA (*x*-axis) and predicted from the chemical composition of the samples (*y*-axis). (**c**) Correlation between total estrogenicity values (in E2 eq) obtained from RYA (x-axis) and predicted from the chemical composition of the samples (y-axis). Note logarithmic scales and the power of regression. *Squares* represent data points for river samples, *solid triangles* correspond to SPT influent waters and *grey circles* show the corresponding values for effluent waters. Data from [60]

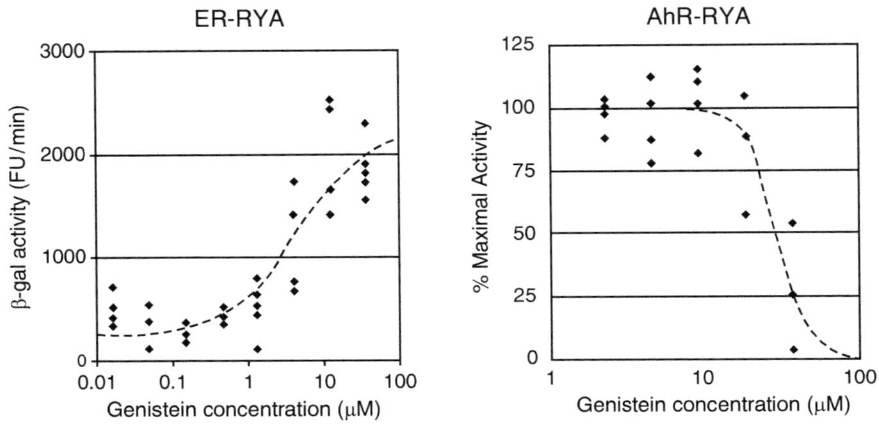

Fig. 5 Measurement of agonist and antagonist effects of genistein in RYA systems. (**a**) Dose–response curve for genistein in the ER-RYA system, in which it acts as a mild estrogen. (**b**) Inhibitory effect of genistein in the YCM-RYA (AhR), in which increasing concentrations of genistein resulted in a decrease of transcription of the reporter gene in spite of the presence of a saturating concentration of β-naphthoflavone (0.5 μM). Data from [60]

concentrations of the analysed compounds were enough to explain the total estrogenicity of water and STP samples. However, RYA systems are able to detect the presence of active substances not specifically targeted by chemical analysis.

The combination of different RYA systems can also help with the characterisation of specific receptor antagonists. In nature, ligand binding to nuclear receptors elicits a delicate imbalance of activatory and inhibitory effects central to the physiological homeostasis of the living being. Antagonist effects could be easily mimetized or masked by cytotoxic effects, which prevent transcriptional activation by inhibiting vital functions. As opposed to cytotoxic substances, true receptor antagonists for a given RYA system would display receptor specificity, that is they should be neutral or activators in a different RYA system. This is the case for the natural phytoestrogen genistein, which acts as an agonist in ER-RYA (Fig. 4), with a calculated EC_{50} value of 2.9 ± 0.7 μM (Table 3). However, when tested in an AhR-RYA system, genistein acted as an antagonist suppressing activation by β-naphtoflavone (Fig. 4).

2 Gene Expression Biomarkers in the Study of Endocrine Disruption

Biomarkers constitute an integral part of the ecotoxicological evaluation of risk in environmental management. Defined as changes in a biological (molecular) response that can be related to exposure to environmental pollutants, they are sensitive indicators of toxic effect of pollutants in exposed biota [78, 79]. A reliable biomarker should distinguish between natural variability and contaminant-induced stress and should have some toxicological significance, that is, its response has to be related to the impact to the organism. Biomarkers evaluate the effects of the observed pollution on the exposed biota and provide an early warning in the cases in which anthropogenic impact is suspected but not confirmed. They also provide an excellent instrument to monitor the recovery of former impacted sites in remediation schemes.

The main functional role of nuclear receptors is to modify transcription of the genes constituting their primary response (Figs 1 and 2). This response is thereafter transmitted to behavioural, immunological, metabolic and reproductive levels, with ultimate effects on the population and on the whole ecosystem. Therefore, one can monitor the very early response to a ligand (or a disrupter) by measuring the changes in the transcription rates of the primary hormone response genes. This approach can be utilised in at least two different aspects of the study of endocrine disruption: incorporated into a bioassay scheme, in which laboratory animals are treated with different substances and extracts to analyse their potential as disrupters; or use them as biomarkers in exposed populations to determine the possible impact of endocrine disrupters.

There are many designed biomarkers to detect endocrine disruption. Feminisating effects in fish and amphibians can be tested by measuring the levels of female-specific, egg-forming proteins in blood and/or liver from exposed males [80–83], including the egg-yolk precursor protein vitellogenin (Vtg) and several chorion-forming proteins, like the zona radiata proteins (Zrp). In mammals, the standard method is the uterotrophic test, which measures uterus growth in uterectomised rats or rabbits after injection of a given substance [35, 84]. There are fewer valid tests for androgenic disruption, one of the most reliable is monitoring the expression of the glue protein, spiggin, in kidneys of female sticklebacks (*Gasterosteus aculeatus*) [85, 86]. The presence of AhR ligands is usually revealed by the activation of Type I and II enzymes in liver or other organs or the increase of their associated enzymatic activities, like EROD [13, 87, 88]. Thyroid disruption can be monitored by the levels of thyroid hormones in the blood or by histology of the thyroid gland [15, 17, 89].

Detection of endocrine disruption by biochemical or histological biomarkers usually requires relatively large amounts of sample, obtained either by exanguination (blood) or by dissection (liver, kidney). In addition, in some cases they require specific antibodies that have to be obtained for each animal group and, sometimes, for each species to analyse. An alternative to these methods is the detection of the corresponding mRNA in the target tissue [42, 90–92] (Fig. 6). These methods require very little sample and benefit from the versatility and sensitivity of polymerase chain reaction (PCR)-based procedures. We would refer to these biomarkers based on measurement of the RNA as gene expression biomarkers.

Specific RNA quantification has become one of the most sensitive tools in Molecular Biology, especially since the introduction of the polymerase chain reaction, which allows detection of truly minimal amounts of RNA molecules by amplification of specific sequences. With appropriate extraction and analytical methodologies, as few as 10–100 RNA molecules can be detected and, with some limitations, quantified. The application of devices able to monitor the process of

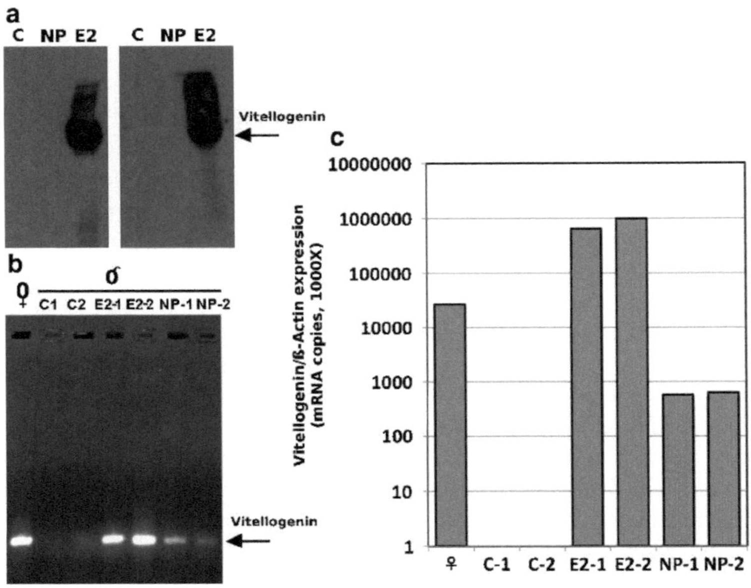

Fig. 6 Comparison of methods to monitor vitellogenin induction in *Fundulus heteroclitus* males upon injection of estradiol. (**a**) Immunological detection of vitellogenin in blood. (**b**) Semi-quantitative PCR analysis of vitellogenin expression in liver. (**c**) Relative quantitation of vitellogenin mRNA levels in liver qRT-PCR. Data from [42]

PCR in real time (quantitative Real-Time PCR, or qRT-PCR), makes more precise and reproducible the quantitation of amplified sequences, and allows processing of large numbers of samples [93, 94]. These characteristics are very useful when dealing with living populations. A high sensitivity allows both the application to small individuals (small animals, for example) and the use of dispensable parts of the body (scales, blood, blubber), avoiding the killing of larger animals [95–99]. On the other hand, the possibility of analysing many individuals allows the study of ecological impacts at the levels of populations rather than at the individual level.

The only true limitation for the application of gene expression biomarkers in biological monitoring is the knowledge of appropriate DNA sequences: any gene can be analysed in any species provided its sequence is known. RNA extraction protocols produce stable RNA solutions that can be stored at −80°C indefinitely and that can be re-tested and re-analysed should a new biomarker appear perhaps years after the original sampling. It is worthwhile considering that 1 mg of a soft tissue (e.g. liver) provides enough total RNA to perform 1,000 qRT-PCR tests; therefore, there is a considerable margin for re-evaluation of old, valuable samples. And lastly, total RNA samples contain information for all genes expressed in the sampled tissue, irrespectively of being enzymes, structural cell components or stress proteins. This allows simultaneous study of multiple effects and correlation with different sources of pollution with the same technology and sample preparation.

2.1 *Methodology of the Analysis of Gene Expression*

2.1.1 Sampling, Transport and Extraction

One of the critical points for the analysis of RNA from environmental samples is the choice of species and of tissue. Most (if not all, see below) RNA quantification methods rely on the knowledge of the exact sequence of at least a portion of the gene(s) to be monitored. Genomes of very different species are getting sequenced at amazing speed, but they are (and will forever be) restricted to a limited number of species, if we compare with the amazing biodiversity present in many environmentally relevant ecosystems. Even worse, there are entire Phyla and Classes that are severely under-represented in the list of sequencing projects, many of them of great environmental interest (for example, molluscs, annelids, and, among vertebrates, birds, reptiles and amphibians). This situation is bound to improve with time, but it would always be difficult to find one or two sequenced species in a particular environment. A plausible solution is to infer the unknown sequences from homologous genes from a close, better known species (see below); therefore, a suitable approach would be to choose species with well-known close relatives (rodents, salmonids, cyprinids, oysters, drosophilids, etc.) over those for which there are no "standard" relatives (gastropods, cartilaginous fish, annelids). In any case, most of the "orphan" taxa will be covered in the future, given the considerable effort being applied to cover wider portions of the extant biodiversity, especially for animals.

A second level of choice is what part of the organism is to be sampled. Some small organisms, like arthropods or very small fish, may be sampled whole, but in many cases the analysis has to be restricted to a defined organ or structure. This approach takes advantage of the differences in transcriptional regulation of many genes among tissues: a gene may respond to a particular hormone in liver, let's say, but not in brain or muscle —among other things, because the corresponding receptors may be absent. Therefore, limiting the sampling to the gene(s) and the tissues known to respond to a particular response increases the sensitivity and reliability of the results. As a disadvantage, only a fraction (some times minute) of the target tissue can be sampled for large to median-size individuals. Therefore, heterogeneous tissues, like brain, skin or even liver, if infiltrated with pancreas, require a special attention to ensure that the obtained sample is representative.

Nucleic acids in general, and RNA in particular, are highly unstable outside the living cells. This instability is not a consequence of their chemical structure, but of the ubiquitous presence of enzymes specifically designed to degrade them. These enzymes (DNases and RNases) come from a large variety of sources, the most important one being the very same tissue we pretend to extract nucleic acids from. There is only one method to really prevent RNA degradation, which is to get rid of RNases before they act. This is usually achieved by using phenol or/and mixes of chaotropic agents, such as guanidium salts, just after sampling. If this is not possible, liquid nitrogen is also used to quick-freeze samples and to transport them to the lab, where they can be safely processed. Recently, commercial mixtures of chaotropic agents have been introduced as RNA stabilisers at non-freezing temperatures; these mixtures allow collection and distribution of samples with a minimum of equipment and do not require deep-freezing equipment, like liquid nitrogen tanks, for example.

Once in the lab, there are many protocols to extract RNA from tissues. If an RNA preserver has been utilised, particular care is required to ensure that the extraction protocol is compatible with the preserver. Phenol/chloroform mixtures are probably the most reliable agents to purify nucleic acids; however, some less toxic, chaotropic-based methods are also available. Purity and integrity of samples can be checked by UV spectrophotometry and/or gel electrophoresis; given the small amounts of RNA usually obtained from field samples, miniaturised systems are most preferable (e.g. nanospectrophotometer and micro-fluidic electrophoresis).

2.1.2 Sequence Selection and Primer Design

Once RNA samples are obtained, the task of analysing them is comparatively simple, as far as one knows what genes are to be quantified. Up to now, most work on gene expression biomarkers uses counterparts of well-established biochemical biomar-kers, like cytochromes P450 for dioxin-like and other AhR ligands, or vitellogenin or zona radiata proteins for feminisating compounds. The rational underlying these assays is that the increase in expression and/or activity of the biochemical biomar-kers should be reflected by—or be a consequence of—changes in the expression of the corresponding genes. This is not necessarily true in all cases, so a previous validation of the intended genetic biomarkers is always required [92].

The requirement for exact sequences is also a limitation for this step. Many genes, including cytochromes and egg-forming proteins constitute multigenic families from which only some of the members may be hormone-responsive. Therefore, the analysis should be selective enough to distinguish very similar gene sequences. This is particularly complicated when the known sequence does not belong to the sampled species, but to a relative. In this case, clustering of homologous sequences from several species from close taxa (same family at least) may allow determining the most conserved regions, which are the ones that would presumably coincide with the sequence of the species of interest. Once the appropriate region of the gene is established, standard primer-design programs (available both on-line and as part of sequence-analysis program packs) will determine primer pairs suitable for amplification. In most cases, it is advisable to amplify (see below) and sequence a relatively large portion—let's say 1–2 Kb—of the actual gene in the tested species, and to redesign primers using the actual targeted sequence [83, 92].

The number of putative biomarkers is going to increase enormously with the advances in genomics and transcriptomics. The deciphering of the regulatory web controlling gene expression at the cellular level would provide a great deal of new hormone-responsive genes in different species.

2.1.3 RNA Quantification Methods

Once separated from the rest of the cellular components, RNA molecules are all very similar, at least from the point-of-view of their chemical and physical properties. This is especially true for the messenger RNA molecules (mRNA), which code for proteins and constitute the primary source for gene expression biomarkers. In fact, the only and crucial difference among them is their own sequence, and this is the character we should focus on to identify and quantify the selected mRNA species. In practice, this represents detecting a particular RNA sequence diluted in a mixture of other RNA molecules in a molar proportion of, at most, $1{:}10^5$ and, in many cases, several orders of magnitude lower; this can be likened to finding a needle in a haystack.

Two properties of nucleic acids and of nucleic acid-synthesising enzymes are central to the development of RNA (and DNA) quantifying methods. First, the notion of complementary strands: nucleic acid molecules with complementary sequences associate one to each other forming an extremely stable complex. This complex may be detected and quantified by labelling one of the complementary molecules and measuring the corresponding signal, either radioactive, fluorescent or enzymatic. This is the basis of many of the "classical" RNA quantification methods, like the Northern blot or the dot blot. In the first case, RNA molecules are separated by gel electrophoresis, blotted to a convenient membrane and exposed to a solution containing the labelled complementary molecule, usually obtained by cloning the corresponding DNA fragment and growing it in bacteria. The dot blot uses the reverse strategy: the complementary molecule is fixed to a membrane unlabelled, and the sample is the one that is marked with the appropriate labelling. In both cases, the amount of label remaining tightly associated to the membrane is assumed to reflect the relative concentration of the targeted sequence in the RNA sample.

A different strategy for RNA detection and quantification comes from the PCR technique. This method relies on the ability of DNA polymerases to make exact copies of any DNA molecule provided the necessary cofactors (a priming DNA molecule, deoxyribonucleotide triphosphates and magnesium) are present. The discovery of DNA polymerases able to survive temperatures higher than those required to separate the two strands of DNA (about 95°C) allowed the design of an iterative protocol in which a single DNA molecule can be duplicated 20, 30 or more consecutive times, until reaching concentrations detectable by basic DNA fluorescent dyes, like Ethidium bromide or similar ones. Semi-quantitative PCR methods estimate the amount of the original molecule by quantifying the total amount of the amplification product (amplicon) after a sub-saturating number of amplification cycles; in qRT-PCR, the amount of amplicon in each of the amplification cycles is continuously detected by a fluorescence-based monitoring system (Fig. 7). The sequence of the amplicon is determined by the sequence of the primers present in the reaction, which should be complementary to both ends of the sequence to be amplified. To apply this methodology to RNA, it should be first transcribed (technically, retro-transcribed) into a complementary DNA molecule (cDNA); there are different enzymes (retro-transcriptases) catalysing this reaction. The qRT-PCR reaction is summarised in Fig. 7.

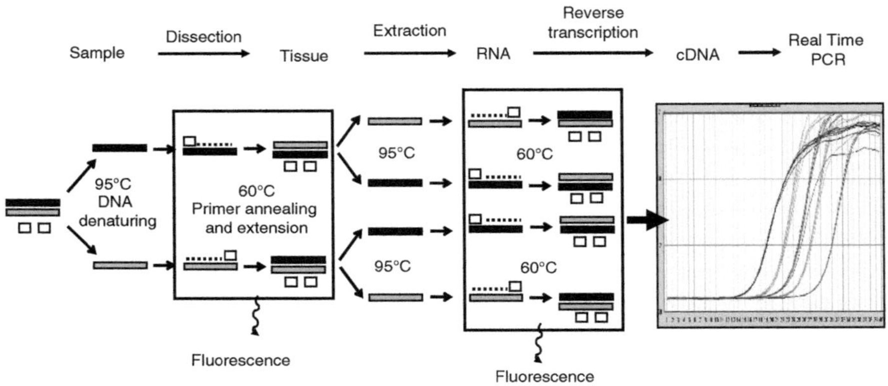

Fig. 7 Process of mRNA quantitation by qRT-PCR. Total RNA is extracted from various tissues (e.g. liver and scales) and copied to cDNA by the reverse-transcriptase reaction. Specific cDNA molecules are selectively amplified by a polymerase chain reaction; the process is monitored in real time by fluorescence methods. The number of cycles required for detection of the amplification product is inversely related to the initial number of molecules of mRNA codifying the amplified sequence

2.1.4 Real-Time PCR Quantitation

Quantification of the expression of a particular gene by real-time PCR is based on the exponential relationship between the number of molecules accumulated in the reaction well, N, and the initial amount of gene copies, N_0, according to (14):

$$N = N_0(1 + E)^c \tag{14}$$

In which E is the efficiency of the amplification reaction and c the cycle number. In optimal conditions, E equals 1.

During the amplification reaction, a fluorescent signal directly proportional to the amount of N is generated and monitored, giving rise to an amplification curve. The number of cycles required to reach the exponential increase in fluorescence is used to determine N_0 [see (14)]. Calculation of this cycle number depends on the real-time PCR platform, and there are essentially two methods: determination at a constant fluorescent level (Threshold Cycle or C_t, Applied Biosystems and Biorad) and calculating the second derivative maximum of the amplification curve (C_p, Roche Applied Science and Corbet Research) [93, 100, 101]. Both methods aim to determine how many cycles are required for each sample to reach the same number of copies of the PCR product (N); this information is used to calculate N_0 according to (14).

Quantification of copy number can be absolute, when relating the PCR signal to input copy number using a calibration curve [102–104], or relative, when measuring the relative changes of copy number between two or more samples. In this case, units used to express relative quantities are irrelevant, and the relative quantities can be compared across multiple real-time RT-PCR experiments. However, there are many factors that may influence the number of mRNA copies in a given sample and preclude adequate data comparison between samples. Changes in the integrity of the nucleic acid, in the amount of starting material and/or in the efficiency of the reverse transcription reaction may influence qRT-PCR results by even an order of magnitude. Therefore, and in order to compensate for these sources of variation, normalisation by an internal control is required. The most popular strategy is to refer data to a reference gene, whose abundance should be considered uniform across individuals and is not affected by the studied parameters [105].

Probably no gene fulfils completely these requisites in any living species, but a common approach is to select genes implicated in the formation and maintenance of the cell, the so-called "housekeeping" (HKG) genes. They fall mainly in three categories: structural proteins (actins, tubulins, etc.), enzymes involved in the glucolysis (GPD), and proteins involved in the formation and function of ribosomes (ribosomal proteins and translation factors) [102]. However, numerous studies have shown that mRNA levels of many HKG genes may vary under different experimental conditions [106, 107]. Then, the best strategy is to test some of these putative candidates in a relatively large subset of samples to assess which one better reflects the total amount of mRNA actually participating in the PCR reaction [105]. There are software tools capable of evaluating this kind of data and determine the most appropriate reference genes [108, 109]. Table 4 shows a list of reference genes developed for environmentally relevant species.

Amplification of reference genes can be carried out in separate reactions or co-amplified in the same tube (multiplex assay). Normalised copy number (R) of the target gene is calculated according to (15):

$$R = \frac{N_0^{\text{target}}}{N_0^{HKG}} = \frac{N_{\text{target}} / (1 + E_{\text{target}})^{C_t^{\text{target}}}}{N_{HKG} / (1 + E_{HKG})^{C_t^{HKG}}} \quad (15)$$

For simplicity, this and the following equations are formulated considering C_t values as the output from the qRT-PCR experiment, but they can also be applied to

Table 4 Housekeeping genes used in environmental biomonitoring

Housekeeping gene	Function	References
GAPDH	Important glycolytic pathway enzyme	[159, 160]
b-, g-actins	Essential for the structure and kinetics of the cytoskeleton	[161, 162]
a-, b-tubulins	Essential for the structure and kinetics of the cytoskeleton	
L8, S20	Ribosomal proteins	[160, 163]
16S, 18S rRNA	Ribosomal RNAs	[160, 164]
GAPDH, YWHAZ	Glyceraldehyde-3P-dehydrogenase, tyrosine 3-monooxygenase	[165]
EFIA (A and B isoforms)	Involved in translation by catalyzing GTP-dependent binding aminoacyl-tRNA	[160, 166]

C_p values without modification. Fold changes (RQ) in transcription levels between two samples, are simply calculated from the ratio between normalised copy number of the target gene in both samples, as described in (16).

$$RQ = \frac{R_{\text{sample}}}{R_{\text{control}}} = \frac{\left[\frac{(1+E_{HKG})^{C_t^{HKG}}}{(1+E_{\text{target}})^{C_t^{\text{target}}}}\right]_{\text{sample}}}{\left[\frac{(1+E_{HKG})^{C_t^{HKG}}}{(1+E_{\text{target}})^{C_t^{\text{target}}}}\right]_{\text{control}}} = \frac{(1+E_{\text{target}})^{\Delta C_t\,(\text{control - sample})}}{(1+E_{HKG})^{\Delta C_t\,(\text{control - sample})}}. \tag{16}$$

Two mathematical models are established to calculate R: the "delta C_t" or "delta-delta C_t" method without efficiency correction [110] and the efficiency-corrected methods [111]. "Delta C_t" methods assume 100% efficiencies for all genes in all samples, simplifying (16) to:

$$RQ = \frac{R_{\text{sample}}}{R_{\text{control}}} = 2^{-(\Delta C_{t,\text{sample}} - \Delta C_{t,\text{control}})}. \tag{17}$$

However, as can be observed from both (15) and (16), the PCR amplification efficiency (E) has an important influence on the accuracy of the quantification results, since small variation in the amplification kinetics will generate false differences in the expression ratio. Therefore, determination of E for all amplification reactions is essential and must be carried out before any relative quantification is done. This value can be estimated from the slope of the linear correlation existing between C_t and C_p values and the logarithm of the amount of template added in a serial dilution scheme, as described in (14). There are a variety of methods to calculate E for any given pair of primers (see http://efficiency.gene-quantification.info).

2.1.5 Statistical Analysis

Quantitation of gene expression by qRT-PCR poses several difficulties to traditional statistical analysis. Classical parametric tests (analysis of variance or T-test) assume a normal distribution of data, but this requisite cannot be expected in absolute or relative quantification analysis since quantities obtained are derived from ratios [see (4)]. To solve this problem, the usual approach is to perform these parametric tests on the logarithmic transformations of the same ratios, which follow a normal distribution in most cases—a procedure called normalisation. This is probably the best strategy when multiple sets of data are to be compared by using ANOVA (analysis of variance) or repeated measures ANOVA. Permutation or randomisation tests, like REST [112], overcome these problems by making no assumptions about the distribution of data. This test is based on the hypothesis that the observed effect in treated samples is as likely to have occurred in a control group. Then, C_t or C_p values for reference and target genes can be jointly reallocated to both groups and the effects on expression ratio are calculated, determining the probability of the hypothesis to be true. To assess the level of significance between control and treated samples REST 2005 software makes 50,000 allocations and calculates the mean and confidence interval of the expression ratio, and also the probability of the treatment effect to be at random (*p*-value). In addition, REST has the unique property of being capable of analysing multiple data points simultaneously. In this way, the analysis can integrate both variations between technical replicates (that is, independent qRT-PCR reaction from the same mRNA sample) and differences among biological replicates, from independent experiments or individuals. However, REST is not adequate when more than two sets of samples are to be compared. We strongly recommend the double approach (normalisation of data and randomisation tests) for analysing large sets of data from qRT-PCR experiments. There are available software packages to support statistical analysis of quantification results at http://www.gene-quantification.de/main-bioinf.shtml.

2.2 Practical Uses of Gene Expression Biomarkers: What are they Useful for?

Biomarkers are as diverse as the pathological effects of pollutants. No technique, however powerful, may conceivably cover all possible effects of the exposure of anthropogenic chemicals on biota. In this regard, gene expression biomarkers are restricted to a very specific subset of pollutants, the mode of action of which involves (or induces) changes in transcription of target genes. For this reason this technique is particularly suited for endocrine disruption, as hormone receptors are, almost by definition, regulatable transcription factors.

Within those limits, the application of gene expression biomarkers is extremely flexible. A single sample (a single RNA extraction) can be used to monitor any possible gene being expressed in the sampled tissue; by applying microarray techniques, all of them could theoretically be analysed at the same time. However, for most practical situations, analysis by qRT-PCR of a battery of target and reference genes is the option of choice. This method is relatively fast, inexpensive and precise; once the primers are optimised, the whole analysis could take a couple of hours or even less. The cost runs at under a US dollar for the qRT-PCR reaction, and 96 or 384 reactions (depending on the platform) can be performed simultaneously. It can be applied as part of a bioassay (e.g. exposing laboratory animals to

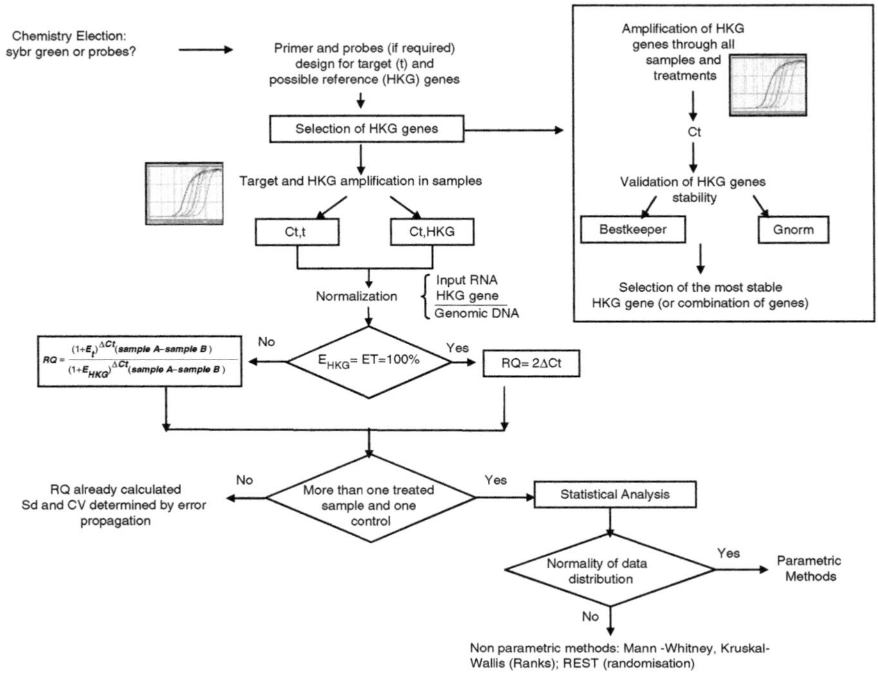

Fig. 8 Flow-chart of options and strategies required to develop a monitoring program based on quantitation of gene expression biomarkers by qRT-PCR

unknown substances or water samples) or as part of a field survey, taking samples of animals (or plants) in the wild, stabilising them with the appropriate solutions and analysing the samples in the lab. In this case, the variability present in nature imposes the requirement for a thorough analysis of the populations: 10–20 individuals per population are normally required for statistically relevant conclusions. It should be stressed here that small or non-vital parts of the specimen are frequently sufficient to perform the survey, allowing sampling of endangered or economically relevant populations. Figure 8 shows a flow-chart of options and strategies required to develop a monitoring program based on quantitation of gene expression biomarkers by qRT-PCR.

The list of available gene expression biomarkers and of species they can be applied to will surely increase enormously with genome sequencing programs and transcriptome analysis of model and non-model species under stress conditions. Table 4 shows a list of references using qRT-PCR techniques for evaluating endocrine disruption, either by analysing samples from the wild or by exposing controlled animals to known or unknown toxicants in the lab and analysing their responses (Table 5).

Table 5 Examples of the use of gene expression biomarkers for detection of endocrine disruption (laboratory experiments)

Species	Pollutant/Area or Condition studied[a)]	Genes analysed[b)]	Technique[c)]	Other methods[d)]	References
Carassius auratus	Cd2+, MeHg	ER, IGF-1, MT	RT-PCR	ALP activity, TRAP activity	[167]
Carassius auratus	BNF	CYP1A	qRT-PCR	EROD	[95]
Cyprinidon ariegatus	E2	Vtg	DD RT-PCR	not done	[90]
Cyprinus carpio	17b-estradiol.	Vtg	RT-cPCR	not done	[168]
Danio Rerio	B[a]P	Vtg,CYPIA	Q-RT-PCR	not done	[169]
Fundulus heteroclitus	PAH exposure	CYP1A, A12, XIR, DW	Q-PCR	Pyrene and PAH concentrations (HPLC)	[164]
Fundulus heteroclitus	E2	Vtg	qRT-PCR	None	[42]
Mugil soiuy	E2	Vtg	Q- RT-PCR	not done	[170]
Oncorhynchus mykiss	BNF	CYPIA	RT-PCR	EROD assay	[171]
Oncorhynchus mykiss	E2, CdCl2	Vtg, ER	RT-PCR Dot blot analysis	not done	[172]
Oncorhynchus mykiss	4-nonylphenol, EE2	Vtg	RT-cPCR	Western blot (Vtg/ZRP)	[91]
Oncorhynchus mykiss	Indole-3-carbinol	CYPIA, Vtg	Q-RT-PCR array	immunoblot analysis (CYP1A/ZR), Vtg by ELISA.	[173]
Oncorhynchus mykiss	a-ZEA, E2	ZR, Vtg	Q-RT-PCR	Vgt/ZR ELISA assay	[161]
Oncorhynchus tshawytscha	pulp mill effluent	CYP1A1	RT-cPCR	not done	[174]
Oreochromis mossambicus	Cd2+	MT	RT-PCR	Metal determination (AAS)	[175]
Oryzias latipes	NP, BPA, EE2	Vtg1,Vtg2, choriogenin, ER	RT-PCR	not done	[82]
Oryzias latipes	E2, NP, AdP, BPA, MBP, DHMS	VtgI, VtgII, ERa	RT-PCR	not done	[176]
Pimephales promelas	EE2, flutamide	Vtg, ERa, ERb, ERg, AR	Q- RT-PCR	Vtg by ELISA	[163]

(continued)

Table 5 (continued)

Species	Pollutant/Area or Condition studied[a)]	Genes analysed[b)]	Technique[c)]	Other methods[d)]	References
Platichthys flesus	Arochlor 1254	CYP1A	Q- RT-PCR	not done	[177]
Pleuronectes americanus	CdC12, NaAs3+	MT	RT-PCR	MT and Metal determination	[178]
Salmo salar	BNF	CYP1A	RT-PCR Slot-blot analysis Northern blot analysis	not done	[179]
Salvelinus namaycush	BNF	CYP1A1	Q-RT-PCR Northern blot analysis	Immunocytochemistry	[162]
Sparus aurata	TCDD	CYP1A	RT-PCR	EROD assay, Western blot analysis, CYP1A ELISA assay, Immunohistochemistry	[87]
Sparus aurata, Liza aurata	BNF	CYP1A1	RT-PCR	EROD assay, CYP1A ELISA, 2-AA	[180]
Cyprinus carpio	PCBs	CYP1A	Q- RT-PCR	Chemical analysis (GC-MS)	[181]

[a)]Abbreviations: BNF, ß-Naphthoflavone; BPA, Bisphenol A; E2, 17ß Estradiol; EE2, Ethynyl estradiol; MeHg, methyl mercury; NP nonylphenol; PAH, polycyclic aromatic hydrocarbons; PCB, polychloro biphenyl benzene

[b)]Abbreviations: CYP1A, cytochrome P4501A; ER, estrogen receptor; AR, androgen receptor; MT, metallothionein. Vtg, vitellogenine; EST, Expressed-sequence tag (not fully characterised); ZR, *Zona Radiata* protein.

[c)]Abbreviations: RT, reverse transcription; sqPCR, semiquantitative PCR; qPCR, quantitative Real-time PCR; cPCR competitive PCR.

Table 5 Examples of the use of gene expression biomarkers for detection of endocrine disruption (field surveys)

Species	Pollutant/Area or Condition studied[a)]	Biomarket[b)]	Technique[c)]	Other methods[d)]	References
Barbus graellsii	River pollution	CYP1A, MT-1, MT-2	qRT-PCR	EROD, heavy metal analysis, MT determination by differential pulse polarography	[182]
Fundulus heteroclitus	Spawning season	MT	RT-PCR QPCR (ECL)	not done	[183]
Gobio gobio	Belgium survey (Cd2+)	MT	qRT-PCR	MT determination (Cd saturation thiomolybdate) Metal determination (AES)	[184]
Gobiomorphus cotidianus	Cu(SO4), Zn(SO4)	MT	qRT-PCR	not done	[185]
Leuciscus cephalus	Metal mix (Ni^{2+}, Cd^{2+}, Zn^{2+}), lab exposure (Cu2+, Mn2+,Zn2+)	MT	RT-cPCR real time PCR (RotorGene)	MT determination (capillary electrophoresis)	[186]
Lithognathus mormyrus	no data	CYP1A	RT-cPCR	EROD assay CYPIA ELISA assay	[187]
Lithognathus mormyrus	estrogenic compounds	Vtg	qRT-PCR	not done	[188]
Mullus barbatus, Mugil cephalus, Gobius niger, Anguilla anguilla, Thunnus thynnus	estrogenic compounds	Vtg	RT-PCR	not done	[189]
Oncorhynchus mykiss	BaP; E2, Cd^{2+} PAHs, heavy metals and estrogenic compounds contamination	CYP1A, MT-B, Vtg	RT-PCR	Chemical analysis (HPLC)	[190]

(continued)

Table 5 (continued)

Species	Pollutant/Area or Condition studied[a)]	Biomarket[b)]	Technique[c)]	Other methods[d)]	References
Oncorhynchus mykiss, Oncorhynchus clarkii, Salmo salar	PAHs, heavy metals and estrogenic compounds contamination	CYP1A, MT, Vtg	qRT-PCR	not done	[191]
Platicthys flesus	PAHs, heavy metals and estrogenic compounds contamination	CYP1A, MT, Vtg	qRT-PCR	EROD assay, BILE FACS, VG ELISA assay, Cadmium-saturation assay (MT)	[192]
Salmo salar	PCB contamination BNF	CYPIA	qRT-PCR	not done	[193]
Salmo trutta	Airborne POPs	Cyp1A	qRT-PCR	Chemical analysis (GS-MS)	[194]
Salmo trutta	E2	Vtg	qRT-PCR	Vtg by ELISA	[195]
Trematomus bernacchii	PCB, PAHs	CYP1A1	RT-cPCR	Biliary FACs	[196]

[a)]Abbreviations: BNF, ß-Naphthoflavone; BPA, Bisphenol A; E2, 17ß Estradiol; EE2, Ethynyl estradiol; MeHg, methyl mercury; NP nonylphenol; PAH, polycyclic aromatic hydrocarbons; PCB, polychloro biphenyl benzene

[b)]Abbreviations: CYP1A, cytochrome P4501A; ER, estrogen receptor; AR, androgen receptor; MT, metallothionein, Vtg, vitellogenine: EST, Expressed-sequence tag (not fully characterised); ZR, *Zona Radiata* protein.

[c)]Abbreviations: RT, reverse transcription; sqPCR, semiquantitative PCR; qPCR, quantitative Real-time PCR; cPCR competitive PCR.

3 Conclusions

Recent advances in Molecular Biology have provided new approaches to develo-ping bioassays and biomarkers to evaluate the effects of the observed pollution on the exposed biota or to provide an early warning in cases in which anthropogenic impact is suspected but not confirmed. These instruments are particularly valuable in risk-assessment protocols (including QSAR), to monitor the recovery of formerly polluted sites in remediation schemes and to check possible environmental risk associated with any of the thousands of new compounds appearing each year in the market. The main—and perhaps the only—limitation to their development is our limited knowledge of how these compounds affect the physiology, genetics, embryology, and, in general, the biology of the impacted biota. Further research in biochemistry (including metabolism), genomics and transcriptomics would undoubtedly extend their applicability to monitoring of new environmental impacts arising from emerging contaminants and from the impending global change.

Acknowledgements This work has been supported by the Spanish Ministry for Education and Science (MEC, grants BIO2005-00840 and AGL2000-0133-P4-03).

References

1. Carlson R (1962) Silent spring. Boston, Houghton Mifflin
2. Colborn T, Dumanoski D, Myers J (1996) Our stolen future: are we threatening our fertility, intelligence, and survival?: a scientific detective story. New York, Dutton, p 306
3. European Commission (ed) (1996) European workshop on the impact of endocrine disrupters on human health and wildlife. European Commission, Weybridge, UK
4. Cooper RL, Kavlock RJ (1997) Endocrine disruptors and reproductive development: a weight-of-evidence overview. J Endocrinol 152(2):159–166
5. Colborn T vom Saal FS, Soto AM (1993) Developmental effects of endocrine-disrupting chemicals in wildlife and humans. Environ Health Perspect 101:378–384
6. Crisp TM, Clegg ED, Cooper RL, Wood WP, Anderson DG, Baetcke KP, Hoffmann JL, Morrow MS, Rodier DJ, Schaeffer JE, Touart LW, Zeeman MG, Patel YM (1998) Environmental endocrine disruption: an effects assessment and analysis. Environ Health Perspect 106(Suppl 1):11–56
7. Aranda A, Pascual A (2001) Nuclear hormone receptors and gene expression. Physiol Rev 81(3):1269–1304
8. Sampath, H, Ntambi JM (2005) Polyunsaturated fatty acid regulation of genes of lipid metabolism. Annu Rev Nutr 25:317–340
9. Zhou J, Suzuki T, Kovacic A, Saito R, Miki Y, Ishida T, Moriya T, Simpson ER, Sasano H, Clyne CD (2005) Interactions between prostaglandin E(2), liver receptor homologue-1, and aromatase in breast cancer. Cancer Res 65(2):657–663
10. Rivier M, Castiel I, Safonova I, Ailhaud G, Michel S (2000) Peroxisome proliferator-activated receptor-alpha enhances lipid metabolism in a skin equivalent model. J Invest Dermatol 114(4):681–687
11. Owen GI, Zelent A (2000) Origins and evolutionary diversification of the nuclear receptor superfamily. Cell Mol Life Sci 57(5):809–827
12. Wiens M, Batel R, Korzhev M, Muller WE (2003) Retinoid X receptor and retinoic acid response in the marine sponge *Suberites domuncula*. J Exp Biol 206(Pt 18):3261–3271

13. Waxman DJ (1999) P450 gene induction by structurally diverse xenochemicals: central role of nuclear receptors CAR, PXR, and PPAR. Arch Biochem Biophys 369(1):11–23
14. Davey JC, Nomikos AP, Wungjiranirun, M, Sherman JR, Ingram, L, Batki, C, Lariviere JP, Hamilton JW (2008) Arsenic as an endocrine disruptor: arsenic disrupts retinoic acid receptor- and thyroid hormone receptor-mediated gene regulation and thyroid hormone-mediated amphibian tail metamorphosis. Environ Health Perspect 116(2):165–172
15. DeVito M, Biegel L, Brouwer A, Brown S, Brucker-Davis F, Cheek AO, Christensen R, Colborn T, Cooke P, Crissman J, Crofton K, Doerge D, Gray E, Hauser P, Hurley P, Kohn M, Lazar J, McMaster S, McClain M, McConnell E, Meier C, Miller R, Tietge J, Tyl R (1999) Screening methods for thyroid hormone disruptors. Environ Health Perspect 107:407–415
16. Grun, F, Watanabe H, Zamanian Z, Maeda L, Arima K, Cubacha R, Gardiner DM, Kanno J, Iguchi T, Blumberg B (2006) Endocrine-disrupting organotin compounds are potent inducers of adipogenesis in vertebrates. Mol Endocrinol 20(9):2141–2155
17. Boas M, Feldt-Rasmussen U, Skakkebaek NE, Main KM (2006) Environmental chemicals and thyroid function. Eur J Endocrinol 154(5):599–611
18. Castro LF, Lima D, Machado A, Melo C, Hiromori Y, Nishikawa J, Nakanishi T, Reis-Henriques MA, Santos MM (2007) Imposex induction is mediated through the Retinoid X Receptor signalling pathway in the neogastropod *Nucella lapillus*. Aquat Toxicol 85(1):57–66
19. Janer G, Navarro JC, Porte C (2007) Exposure to TBT increases accumulation of lipids and alters fatty acid homeostasis in the ramshorn snail *Marisa cornuarietis*. Comp Biochem Physiol C Toxicol Pharmacol 146(3):368–374
20. Gonzalez FJ, Fernández-Salguero P (1998) The aryl hydrocarbon receptor: studies using the AHR-null mice. Drug Metab Dispos 26:1194–1198
21. Shimizu Y, Nakatsuru Y, Ichinose M, Takahashi Y, Kume H, Mimura J, Fujii-Kuriyama Y, Ishikawa T (2000) Benzo[a]pyrene carcinogenicity is lost in mice lacking the aryl hydrocarbon receptor. Proc Natl Acad Sci U S A 97:779–782
22. Hahn ME (2002) Aryl hydrocarbon receptors: diversity and evolution. Chem Biol Interact 141:131–160
23. Nebert DW, Puga A, Vasiliou V (1993) Role of the Ah receptor and the dioxin-inducible [Ah] gene battery in toxicity, cancer, and signal transduction. Ann N Y Acad Sci 685:624–640
24. Hankinson O (1995) The aryl hydrocarbon receptor complex. Annu Rev Pharmacol Toxicol 35:307–340
25. Poland A, Knutson JC (1982) 2,3,7,8-tetrachlorodibenzo-p-dioxin and related halogenated aromatic hydrocarbons: examination of the mechanism of toxicity. Annu Rev Pharmacol Toxicol 22:517–554
26. Abbott BD, Perdew GH, Buckalew AR, Birnbaum LS (1994) Interactive regulation of Ah and glucocorticoid receptors in the synergistic induction of cleft palate by 2,3,7,8-tetrachlorodibenzo-p-dioxin and hydrocortisone. Toxicol Appl Pharmacol 128(1):138–150
27. Safe S, Wang F, Porter W, Duan R, McDougal A (1998) Ah receptor agonists as endocrine disruptors: antiestrogenic activity and mechanisms. Toxicol Lett 102–103:343–347
28. Fang H, Tong W, Perkins R, Soto AM, Prechtl NV, Sheehan DM (2000) Quantitative comparisons of in vitro assays for estrogenic activities. Environmental Health Perspect 108(8):723–729
29. Garcia-Reyero N, Grau E, Castillo M, López de Alda MJ, Barceló D, Piña B (2001) Monitoring of endocrine disruptors in surface waters by the yeast recombinant assay. Environ Toxicol Chem 20(6):1152–1158
30. Mueller SO (2002) Overview of in vitro tools to assess the estrogenic and antiestrogenic activity of phytoestrogens. J Chromatogr B Analyt Technol Biomed Life Sci 777:155–165
31. Windal I, Denison MS, Birnbaum LS, Van Wouwe N, Baeyens W, Goeyens L (2005) Chemically activated luciferase gene expression (CALUX) cell bioassay analysis for the estimation of dioxin-like activity: critical parameters of the CALUX procedure that impact assay results. Environ Sci Technol 39(19):7357–7364
32. Legler J, Zeinstra LM, Schuitemaker F, Lanser PH, Bogerd J, Brouwer A, Vethaak AD, De Voogt P, Murk AJ, Van der Burg B (2002) Comparison of in vivo and in vitro reporter gene assays for short-term screening of estrogenic activity. Environ Sci Technol 36(20):4410–4415

33. Fox JE, Burow ME, McLachlan JA, Miller Iii CA (2008) Detecting ligands and dissecting nuclear receptor-signaling pathways using recombinant strains of the yeast *Saccharomyces cerevisiae*. Nat Prot 3(4):637–645
34. Mantovani A (2006) Risk assessment of endocrine disrupters: the role of toxicological studies. Ann N Y Acad Sci 1076:239–252
35. Coldham NG, Dave, M, Sivapathasundaram S, McDonnell DP, Connor C, Sauer MJ (1997) Evaluation of a recombinant yeast cell estrogen screening assay. Environ Health Perspect 105:734–742
36. Andersen H, Andersson A, Arnold S, Autrup H, Barfoed M, Beresford N, Bjerregaard P, Christiansen L, Gissel B, Hummel R, Jorgensen E, Korsgaard B, LeGuevel R, Leffers H, McLachlan J, Moller A, Nielsen J, Olea N, Oles-Karasko A, Pakdel F, Pedersen K, Perez P, Skakkeboek N, Sonnenschein C, Soto A, Sumpter J, Thorpe S, Grandjean P (1999) Comparison of short-term estrogenicity tests for identification of hormone-disrupting chemicals. Environ Health Perspect 107(Suppl. 1):89–108
37. Bovee TF, Helsdingen RJ, Rietjens IM, Keijer, J, Hoogenboom RL (2004) Rapid yeast estrogen bioassays stably expressing human estrogen receptors alpha and beta, and green fluorescent protein: a comparison of different compounds with both receptor types. J Steroid Biochem Mol Biol 91(3):99–109
38. Feldman D, Krishnan A (1995) Estrogens in unexpected places: possible implications for researchers and consumers. Environ Health Perspect 105(8):129–133
39. Thomas K, Hurst M, Matthiessen P, Waldock M (2001) Characterization of estrogenic compounds in water samples collected from United Kingdom estuaries. Environ Toxicol Chem 20:2165–2170
40. Garcia-Reyero N, Pina B, Grimalt JO, Fernandez P, Fonts R, Polvillo O, Martrat B (2005) Estrogenic activity in sediments from European mountain lakes. Environ Sci Technol 39(6):1427–1435
41. Gaido KW, Leonard LS, Lovell S, Gould JC, Babai D, Portier CJ, McDonnell DP (1997) Evaluation of chemicals with endocrine modulating activity in a yeast-based steroid hormone receptor gene transcription assay. Toxicol Appl Pharmacol 143(1):205–212
42. Garcia-Reyero N, Requena V, Petrovic M, Fischer B, Hansen PD, Díaz A, Ventura F, Barceló D, Piña B (2004) Estrogenic potential of halogenated derivatives of nonylphenol ethoxylates and carboxylates. Environ Toxicol Chem 23(3):705–711
43. Arulmozhiraja S, Shiraishi F, Okumura T, Iida M, Takigami H, Edmonds JS, Morita M (2005) Structural requirements for the interaction of 91 hydroxylated polychlorinated biphenyls with estrogen and thyroid hormone receptors. Toxicol Sci 84(1):49–62
44. Bovee TF, Lommerse JP, Peijnenburg AA, Fernandes EA, Nielen MW (2008) A new highly androgen specific yeast biosensor, enabling optimisation of (Q)SAR model approaches. J Steroid Biochem Mol Biol 108(1–2):121–131
45. Li Y, Zhang Z, Robinson GE, Palli SR (2007) Identification and characterization of a juvenile hormone response element and its binding proteins. J Biol Chem 282(52):37605–37617
46. De Loof A (2008) Ecdysteroids, juvenile hormone and insect neuropeptides: recent successes and remaining major challenges. Gen Comp Endocrinol 155(1):3–13
47. Bitter GA, Egan KM, Koski RA, Jones MO, Elliott SG, Giffin JC, Ray W, Lawrence G (1987) Expression and secretion vectors for yeast. In: Methods in Enzymology, vol 153. Academic, New York, pp 516–544
48. Rose AB, Broach JR, David VG (1990) Propagation and expression of cloned genes in yeast: 2-[mu]m circle-based vectors. In Methods in Enzymology, vol. 185. Academic, New York, pp 234–279
49. Schena M, Picard D, Yamamoto KR, Christine G, Gerald RF (1991) Vectors for constitutive and inducible gene expression in yeast. In: Methods in Enzymology, vol 194. Academic, New York, pp 389–398
50. Schneider JC, Guarente L (1991) Vectors for expression of cloned genes in yeast: regulation, overproduction, and underproduction. Methods Enzymol 194:373–388
51. Sherman F (1992) Getting started with yeast. Methods Enzymol 194(3):3–21

52. Sikorski RS, Hieter P (1989) A system of shuttle vectors and yeast host strains designed for efficient manipulation of DNA in *Saccharomyces cerevisiae*. Genetics 122(1):19–27
53. Olson M (1992) Genome structure and organization in *Saccharomyces cerevisiae*. In: Jones EW, Pringle JR, Broach JR (eds) The molecular and cellular biology of the yeast Saccharomyces. Gene expression, vol. 1. Cold Spring Harbor Laboratory Press, Cold Spring Harbor, pp 1–39
54. Broach JR, Volkert FC (1992) Circular DNA plasmids of Yeasts. In: Jones EW, Pringle JR, Broach JR (eds) The molecular and cellular biology of the yeast Saccharomyces. Gene Expression, vol 1. Cold Spring Harbor Laboratory Press, Cold Spring Harbor, pp 297–332
55. Taxis C, Knop M (2006) System of centromeric, episomal, and integrative vectors based on drug resistance markers for *Saccharomyces cerevisiae*. Biotechniques 40(1):73–78
56. Guarente L, Mason T (1983) Heme regulates transcription of the CYC1 gene of S. cerevisiae via an upstream activation site. Cell 32:1279–1286
57. Arnold SF, Collins BM, Robinson MK, Guillette LJ Jr, McLachlan JA (1996) Differential interaction of natural and synthetic estrogens with extracellular binding proteins in a yeast estrogen screen. Steroids 6:642–646
58. Boronat S, Piña B (2006) Development of RNR3- and RAD54-GUS reporters for testing genotoxicity in *Saccharomyces cerevisiae*. Anal Bioanal Chem 386:1625–1632
59. Routledge EJ, Sumpter JP (1996) Estrogenic activity of surfactants and some of their degradation products assessed using a recombinant yeast screen. Environ Toxicol Chem 15(3):241–248
60. Noguerol T, Boronat S, Jarque S, Barceló D, Piña B (2006) Detection of hormone receptor ligands in yeast by fluorogenic methods. Talanta 69:358–359
61. Bovee TF, Helsdingen RJ, Koks PD, Kuiper HA, Hoogenboom RL, Keijer J (2004) Development of a rapid yeast estrogen bioassay, based on the expression of green fluorescent protein. Gene 325:187–200
62. Collins BM, McLachlan JA, Arnold SF (1997) Transcriptional activation of the human estrogen receptor by DDT isomers and metabolites in yeast and MCF-7 cells. Biochem Pharmacol 62:1161–1172
63. Chen Q, Chen J, Sun T, Shen J, Shen X, Jiang H (2004) A yeast two-hybrid technology-based system for the discovery of PPARg agonist and antagonist. Anal Biochem 335(2):253–259
64. Shibata H, Spencer TE, Oñate SA, Jenster G, Tsai SY, Tsai MJ, O'Malley BW (1997) Role of co-activators and co-repressors in the mechanism of steroid/thyroid receptor action. Recent Prog Horm Res 14(1–2):141–164
65. Carver L, Jackiw V, Bradfield C (1994) The 90-kDa heat shock protein is essential for Ah receptor signaling in a yeast expression system. J Biol Chem 269:30109–30112
66. Boronat S, Casado S, Navas JM, Pina B (2007) Modulation of aryl hydrocarbon receptor transactivation by carbaryl, a nonconventional ligand. FEBS J 274:3327–3339
67. Jacobs MN, Dickins M, Lewis DF (2003) Homology modelling of the nuclear receptors: human oestrogen receptor beta (hERbeta), the human pregnane-X-receptor (PXR), the Ah receptor (AhR) and the constitutive androstane receptor (CAR) ligand binding domains from the human oestrogen receptor alpha (hERalpha) crystal structure, and the human peroxisome proliferator activated receptor alpha (PPARalpha) ligand binding domain from the human PPARgamma crystal structure. J Steroid Biochem Mol Biol 84(2–3):117–132
68. Giraldo J, Vivas NM, Vila E, Badia A (2002) Assessing the (a)symmetry of concentration-effect curves: empirical versus mechanistic models. Pharmacol Ther 95(1):21–45
69. Rehmann K, Schramm K-W, Krettrup AA (1999) Applicability of a yeast oestrogen screen for the detection of oestrogen-like activities in environmental samples. Chemosphere 38: 3303–3312
70. Cheng HC, Lai RW (2003) Use of the proportionality equations for analyses of dose-response curves. Pharmacol Res 47(2):163–173
71. Maloney EK, Waxman DJ (1999) Trans-activation of PPARalpha and PPARgamma by structurally diverse environmental chemicals. Toxicol Appl Pharmacol 161(2):209–218

72. Sugawara T, Nomura E (2006) Development of a recombinant yeast assay to detect Ah-receptor ligands. Toxicol Mech Meth 16(5):287–294
73. Noguerol TN, Boronat S, Casado M, Raldua D, Barcelo D, Pina B (2006) Evaluating the interactions of vertebrate receptors with persistent pollutants and antifouling pesticides using recombinant yeast assays. Anal Bioanal Chem 385:1012–1019
74. Céspedes R, Petrovic M, Raldúa D, Saura U, Piña B, Lacorte S, Viana P, Barceló D (2004) Integrated protocol for the determination of endocrine disrupting activity in the surface waters and sediments using biological techniques by recombinant yeast assay and chemical analysis by LC-ESI-MS. Anal Bioanal Chem 378:697–708
75. Wooge CH, Nilsson GM, Heierson A, McDonnell DP, Katzenellenbogen BS (1992) Structural requirements for high affinity ligand binding by estrogen receptors: a comparative analysis of truncated and full length estrogen receptors expressed in bacteria, yeast, and mammalian cells. Mol Endocrinol 62:861–869
76. Auriol M, Filali-Meknassi Y, Adams CD, Tyagi RD, Noguerol TN, Piña B (2008) Removal of estrogenic activity of natural and synthetic hormones from a municipal wastewater: efficiency of horseradish peroxidase and laccase from *Trametes versicolor*. Chemosphere 70(3):445–452
77. Céspedes R, Lacorte S, Raldúa D, Ginebreda A, Barcelo D, Piña B (2005) Distribution of endocrine disruptors in the Llobregat River basin (Catalonia, NE Spain). Chemosphere 61(11):1710–1719
78. Livingstone DR, Nasci C, Sole M, Da Ros L, O'Hara SCM, Peters LD, Fossato V, Wootton AN, Goldfarb PS (1997) Apparent induction of a cytochrome P450 with immunochemical similarities to CYP1A in digestive gland of the common mussel (*Mytilus galloprovincialis* L.) with exposure to 2,2′,3,4,4′,5′-hexachlorobiphenyl and Arochlor 1254. Aquat Toxicol 38:205–224
79. van der Oost R, Beyer J, Vermeulen NPE (2003) Fish bioaccumulation and biomarkers in environmental risk assessment: a review. Environ Toxicol Pharmacol 13:57–149
80. Jobling S, Sumpter J (1993) Detergent components in sewage effluent are weakly oestrogenic on fish: an in vitro study using rainbow trout (*Oncorhynchus mykiss*) hepatocytes. Aquat Toxicol 27:361–372
81. Lattier D, Gordon D, Burks D, Toth G (2001) Vitellogenin gene transcription: a relative quantitative exposure indicator of environmental estrogens. Environ Toxicol Chem 20:1979–1985
82. Lee C, Na JG, Lee KC, Park K (2002) Choriogenin mRNA induction in male medaka, *Oryzias latipes* as a biomarker of endocrine disruption. Aquat Toxicol 61:233–241
83. Rotchell JM, Ostrander GK (2003) Molecular markers of endocrine disruption in aquatic organisms. J Toxicol Environ Health B Crit Rev 6(5):453–496
84. Fukushima S, Freyberger A (2003) Simple, rapid assays for conventional definite testing of endocrine disruptor hazard: summary and recommendations. Pure Appl Chem 75(11–12): 2479–2482
85. Katsiadaki I, Morris S, Squires C, Hurst MR, James JD, Scott AP (2006) Use of the three-spined stickleback (Gasterosteus aculeatus) as a sensitive in vivo test for detection of environmental antiandrogens. Environ Health Perspect 114(Suppl 1):115–121
86. Allen YT, Katsiadaki I, Pottinger TG, Jolly C, Matthiessen P, Mayer I, Smith A, Scott AP, Eccles P, Sanders MB, Pulman KGT, Feist S (2008) Intercalibration exercise using a stickleback endocrine disrupter screening assay. Environ Toxicol Chem 27(2):404–412
87. Ortiz-Delgado J, Behrens A, Segner H, Sarasquete C (2008) Tissue-specific induction of EROD activity and CYP1A protein in Sparus aurata exposed to B(a)P and TCDD. Ecotoxicol Environ Saf 69:80–88
88. Taysse L, Chambras C, Marionnet D, Bosgiraud C, Deschaux P (1998) Basal level and induction of cytochrome P450, EROD, UDPGT, and GST activities in carp (Cyprinus carpio) immune organs (spleen and head kidney). Bull Environ Contam Toxicol 60:300–305
89. Brucker-Davis F, Thayer K, Colborn T (2001) Significant effects of mild endogenous hormonal changes in humans: considerations for low-dose testing. Environ Health Perspect 109(Suppl 1):21–26

90. Denslow ND, Garcia-Reyero N, Barber DS (2007) Fish ‘n’ chips: the use of microarrays for aquatic toxicology. Mol Biosyst 3(3):172–177
91. Ackermann GE, Schwaiger J, Negele RD, Fent K (2002) Effects of long-term nonylphenol exposure on gonadal development and biomarkers of estrogenicity in juvenile rainbow trout *Oncorhynchus mykiss*. Aquat Toxicol 60:203–221
92. Piña B, Casado M, Quirós L (2007) Analysis of gene expression as a new tool in ecotoxicology and environmental monitoring. Trends Anal Chem 26(11):1145–1154
93. Kubista M, Andrade JM, Bengtsson M, Forootan A, Jonak J, Lind K, Sindelka R, Sjoback R, Sjogreen B, Strombom L, Stahlberg A, Zoric N (2006) The real-time polymerase chain reaction. Mol Aspects Med 27(2–3):95–125
94. Valasek MA, Repa JJ (2005) The power of real-time PCR. Adv Physiol Educ 29:151–159
95. Quirós L, Raldúa D, Navarro A, Casado M, Barceló D, Piña B (2007) A noninvasive test of exposition to toxicants: quantitative analysis of cytochrome P4501A expression in fish scales. Environ Toxicol Chem 26(10):2179–2186
96. Stefanelli P, Ausili A, Di Muccio A, Fossi C, Di Muccio S, Rossi S, Colasanti A (2004) Organochlorine compounds in tissues of swordfish (*Xiphias gladius*) from Mediterranean Sea and Azores islands. Mar Pollut Bull 49(11–12):938–950
97. Arukwe A, Re K (2008) Molecular and cellular detection of expression of vitellogenin and zona radiata protein in liver and skin of juvenile salmon (*Salmo salar*) exposed to nonylphenol. Cell Tissue Res 331(3):701–712
98. Champoux L, Rodrigue J, Trudeau S, Boily MH, Spear PA, Hontela A (2006) Contamination and biomarkers in the great blue heron, an indicator of the state of the St. Lawrence River. Ecotoxicology 15(1):83–96
99. Kim IY, Hyun CK (2006) Comparative evaluation of the alkaline comet assay with the micronucleus test for genotoxicity monitoring using aquatic organisms. Ecotox Environ Saf 64(3):288–297
100. Nolan T, Hands RE, Bustin SA (2006) Quantification of mRNA using real-time RT-PCR. Nat Protoc 1(3):1559–1582
101. Pfaffl MW (2006) Relative quantification. In: Dorak T (ed) Real Time PCR (Bios Advanced Methods). Taylor and Francis, New York
102. Huggett J, Dheda K, Bustin S, Zumla A (2005) Real-time RT-PCR normalisation; strategies and considerations. Genes Immun 6(4):279–284
103. Bustin SA (2000) Absolute quantification of mRNA using real-time reverse transcription polymerase chain reaction assays. J Mol Endocrinol 25(2):169–193
104. Fronhoffs S, Totzke G, Stier S, Wernert N, Rothe M, Bruning T, Koch B, Sachinidis A, Vetter H, Ko Y (2002) A method for the rapid construction of cRNA standard curves in quantitative real-time reverse transcription polymerase chain reaction. Mol Cell Probes 16(2):99–110
105. Radonic A, Thulke S, Mackay IM, Landt O, Siegert W, Nitsche A (2004) Guideline to reference gene selection for quantitative real-time PCR. Biochem Biophys Res Commun 313(4):856–862
106. Arukwe A (2006) Toxicological housekeeping genes: do they really keep the house? Environ Sci Technol 40(24):7944–7999
107. Dheda K, Huggett JF, Bustin SA, Johnson MA, Rook G, Zumla A (2004) Validation of housekeeping genes for normalizing RNA expression in real-time PCR. Biotechniques 37(1):112–114, 116, 118–119
108. Vandesompele J, De Preter K, Pattyn F, Poppe B, Van Roy N, De Paepe A, Speleman F (2002) Accurate normalization of real-time quantitative RT-PCR data by geometric averaging of multiple internal control genes. Genome Biol 3(7):RESEARCH0034
109. Pfaffl MW, Tichopad A, Prgomet C, Neuvians TP (2004) Determination of stable housekeeping genes, differentially regulated target genes and sample integrity: BestKeeper–Excel-based tool using pair-wise correlations. Biotechnol Lett 26(6):509–515
110. Livak KJ, Schmittgen TD (2001) Analysis of relative gene expression data using real-time quantitative PCR and the 2(-Delta Delta C(T)) Method. Methods 25(4):402–408

111. Pfaffl M (2001) A new mathematical model for relative quantification in real-time RT-PCR. Nucl Acids Res 29:e45
112. Pfaffl MW, Horgan GW, Dempfle L (2002) Relative expression software tool (REST) for group-wise comparison and statistical analysis of relative expression results in real-time PCR. Nucleic Acids Res 30(9):e36
113. Mak P, McDonnell DP, Weigel NL, Schrader WT, O'Malley BW (1989) Expression of functional chicken oviduct progesterone receptors in yeast (*Saccharomyces cerevisiae*). J Biol Chem 264(36):21613–21618
114. Rudakoff B, Undisz K, Mayer G, Sobek L, Kaufmann G, Thiericke R, Grabley S, Munder T (1999) Dual reporter systems in yeast and mammalian cells for assessing progesterone receptor modulators. J Cell Biochem 73(1):126–136
115. Klotz DM, Ladlie BL, Vonier PM, McLachlan JA, Arnold SF (1997) o,p'-DDT and its metabolites inhibit progesterone-dependent responses in yeast and human cells. Mol Cell Endocrinol 129(1):63–71
116. Wang J, Xie P, Kettrup A, Schramm KW (2005) Inhibition of progesterone receptor activity in recombinant yeast by soot from fossil fuel combustion emissions and air particulate materials. Sci Total Environ 349(1–3):120–128
117. Chatterjee S, Kumar V, Majumder CB, Roy P (2008) Screening of some anti-progestin endocrine disruptors using a recombinant yeast based in vitro bioassay. Toxicol In Vitro 22(3):788–798
118. Poletti A, Weigel NL, McDonnell DP, Schrader WT, O'Malley BW, Conneely OM (1992) A novel, highly regulated, rapidly inducible system for the expression of chicken progesterone receptor, cPR(A), in *Saccharomyces cerevisiae*. Gene 114(1):51–58
119. Eldridge ML, Sanseverino J, Layton AC, Easter JP, Schultz TW, Sayler GS (2007) *Saccharomyces cerevisiae* BLYAS, a new bioluminescent bioreporter for detection of androgenic compounds. Appl Environ Microbiol 73(19):6012–6018
120. Rana S, Bisht D, Chakraborti PK (1998) Activation of rat androgen receptor by androgenic ligands is unaffected by antiandrogens in *Saccharomyces cerevisiae*. Gene 209(1–2):247–254
121. Michelini E, Leskinen P, Virta M, Karp M, Roda A (2005) A new recombinant cell-based bioluminescent assay for sensitive androgen-like compound detection. Biosens Bioelectron 20(11):2261–2267
122. Beck V, Reiter E, Jungbauer A (2008) Androgen receptor transactivation assay using green fluorescent protein as a reporter. Anal Biochem 373(2):263–271
123. Lee HJ, Lee YS, Kwon HB, Lee K (2003) Novel yeast bioassay system for detection of androgenic and antiandrogenic compounds. Toxicol In Vitro 17(2):237–244
124. Sohoni P, Sumpter JP (1998) Several environmental oestrogens are also anti-androgens. J Endocrinol 158(3):327–339
125. Kirk LA, Tyler CR, Lye CM, Sumpter JP (2002) Changes in estrogenic and androgenic activities at different stages of treatment in wastewater treatment works. Environ Toxicol Chem 21(5):972–979
126. Sohoni P, Lefevre PA, Ashby J, Sumpter JP (2001) Possible androgenic/anti-androgenic activity of the insecticide fenitrothion. J Appl Toxicol 21(3):173–178
127. Lyttle CR, Damian-Matsumura P, Juul H, Butt TR (1992) Human estrogen receptor regulation in a yeast model system and studies on receptor agonists and antagonists. J Steroid Biochem Mol Biol 6:677–685
128. Routledge E, Sumpter J (1996) Estrogenic activity of surfactants and some of their degradation products assessed using a recombinant yeast strain. Environ Toxicol Chem 15:241–248
129. Beck V, Pfitscher A, Jungbauer A (2005) GFP-reporter for a high throughput assay to monitor estrogenic compounds. J Biochem Biophys Methods 64(1):19–37
130. Liu JW, Jeannin E, Picard D (1999) The anti-estrogen hydroxytamoxifen is a potent antagonist in a novel yeast system. Biol Chem 380(11):1341–1345
131. Jungbauer A, Beck V (2002) Yeast reporter system for rapid determination of estrogenic activity. J Chromatogr B Anal Technol Biomed Life Sci 777(1–2):167–178

132. Graumann K, Jungbauer A (2000) Agonistic and synergistic activity of tamoxifen in a yeast model system. Biochem Pharmacol 59(2):177–185
133. Escher BI, Bramaz N, Maurer M, Richter M, Sutter D, Von Känel C, Zschokke M (2005) Screening test battery for pharmaceuticals in urine and wastewater. Environ Toxicol Chem 24(3):750–758
134. Beck V, Unterrieder E, Krenn L, Kubelka W, Jungbauer A (2003) Comparison of hormonal activity (estrogen, androgen and progestin) of standardized plant extracts for large scale use in hormone replacement therapy. J Steroid Biochem Mol Biol 84(2–3):259–268
135. Taneda S, Hayashi H, Sakata M, Yoshino S, Suzuki A, Sagai M, Mori Y (2000) Anti-estrogenic activity of diesel exhaust particles. Biol Pharm Bull 23(12):1477–1480
136. Promberger A, Dornstauder E, Fruhwirth C, Schmid ER, Jungbauer A (2001) Determination of estrogenic activity in beer by biological and chemical means. J Agric Food Chem 49(2):633–640
137. Klinge CM, Risinger KE, Watts MB, Beck V, Eder R, Jungbauer A (2003) Estrogenic activity in white and red wine extracts. J Agric Food Chem 51(7):1850–1857
138. Nakari T, Pessala P (2005) In vitro estrogenicity of polybrominated flame retardants. Aquat Toxicol 74(3):272–279
139. Hasenbrink G, Sievernich A, Wildt L, Ludwig J, Lichtenberg-Frate H (2006) Estrogenic effects of natural and synthetic compounds including tibolone assessed in *Saccharomyces cerevisiae* expressing the human estrogen alpha and beta receptors. FASEB J 20(9):1552–1554
140. Quirós L, Céspedes R, Lacorte S, Viana P, Raldúa D, Barceló D, Piña B (2005) Detection and evaluation of endocrine-disruption activity in water samples from Portuguese rivers. Environ Toxicol Chem 24(2):389–395
141. Garcia-Reyero N, Grimalt JO, Vives I, Fernandez P, Pina B (2007) Estrogenic activity associated with organochlorine compounds in fish extracts from European mountain lakes. Environ Pollut 145(3):745–752
142. Jisa E, Dornstauder E, Ogawa S, Inoue S, Muramatsu M, Jungbauer A (2001) Transcriptional activities of estrogen receptor alpha and beta in yeast properties of raloxifene. Biochem Pharmacol 62(7):953–961
143. Henry K, O'Brien ML, Clevenger W, Jow L, Noonan DJ (1995) Peroxisome proliferator-activated receptor response specificities as defined in yeast and mammalian cell transcription assays. Toxicol Appl Pharmacol 132(2):317–324
144. Nishikawa J, Mamiya S, Kanayama T, Nishikawa T, Shiraishi F, Horiguchi T (2004) Involvement of the retinoid X receptor in the development of imposex caused by organotins in gastropods. Environ Sci Technol 38(23):6271–6276
145. Leskinen P, Hilscherova K, Sidlova T, Kiviranta H, Pessala P, Salo S, Verta M, Virta M (2008) Detecting AhR ligands in sediments using bioluminescent reporter yeast. Biosens Bioelectron 23(12):1850–1855
146. Miller Iii CA (1999) A human aryl hydrocarbon receptor signaling pathway constructed in yeast displays additive responses to ligand mixtures. Toxicol Appl Pharmacol 160(3):297–303
147. Adachi J, Mori Y, Matsui S, Takigami H, Fujino J, Kitagawa H, Miller Iii CA, Kato T, Saeki K, Matsuda T (2001) Indirubin and Indigo are potent aryl hydrocarbon receptor ligands present in human urine. J Biol Chem 276(34):31475–31478
148. Alnafisi A, Hughes J, Wang G, Miller Iii CA (2007) Evaluating polycyclic aromatic hydrocarbons using a yeast bioassay. Environ Toxicol Chem 26(7):1333–1339
149. Chou PH, Matsui S, Misaki K, Matsuda T (2007) Isolation and identification of xenobiotic aryl hydrocarbon receptor ligands in dyeing wastewater. Environ Sci Technol 41(2): 652–657
150. Chou PH, Matsui S, Matsuda T (2006) Detection and identification of dyes showing AhR-binding affinity in treated sewage effluents. Water Sci Technol 53(11):35–42
151. Lee BC, Shimizu Y, Matsuda T, Matsui S (2005) Characterization of polycyclic aromatic hydrocarbons (PAHs) in different size fractions in deposited road particles (DRPs) from Lake Biwa area, Japan. Environ Sci Technol 39(19):7402–7409

152. Saeki KI, Matsuda T, Kato TA, Yamada K, Mizutani T, Matsui S, Fukuhara K, Miyata N (2003) Activation of the human Ah receptor by aza-polycyclic aromatic hydrocarbons and their halogenated derivatives. Biol Pharm Bull 26(4):448–452
153. Kato TA, Matsuda T, Matsui S, Mizutani T, Saeki KI (2002) Activation of the aryl hydrocarbon receptor by methyl yellow and related congeners: structure-activity relationships in halogenated derivatives. Biol Pharm Bull 25(4):466–471
154. Wright APH, Carlstedt-Duke J, Gustafsson JA (1990) Ligand-specific transactivation of gene expression by a derivative of the human glucocorticoid receptor expressed in yeast. J Biol Chem 265(25):14763–14769
155. Masuyama H, Hiramatsu Y, Kunitomi M, Kudo T, MacDonald PN (2000) Endocrine disrupting chemicals, phthalic acid and nonylphenol, activate Pregnane X receptor-mediated transcription. Mol Endocrinol 14(3):421–428
156. Kakizawa T, Miyamoto T, Kaneko A, Yajima H, Ichikawa K, Hashizume K (1997) Ligand-dependent heterodimerization of thyroid hormone receptor and retinoid X receptor. J Biol Chem 272(38):23799–23804
157. Berghöfer-Hochheimer Y, Zurek C, Langer G, Munder T (1997) Expression of the vitamin D and the retinoid X receptors in *Saccharomyces cerevisiae*: alternative in vivo models for ligand-induced transactivation. J Cell Biochem 66(2):184–196
158. Zhao XY, Eccleshall TR, Krishnan AV, Gross C, Feldman D (1997) Analysis of vitamin D analog-induced heterodimerization of vitamin D receptor with retinoid X receptor using the yeast two-hybrid system. Mol Endocrinol 11(3):366–378
159. Craft JA, Robertson FE, McPhail ME, Brown E, Stagg RM (2001) Measurement of cytochrome P4501A induction in dab (*Limanda limanda*) and other teleosts with species-specific cDNA probes: isolation and characterisation of dab cDNA and its use in expression studies with beta-naphthoflavone-treated fish. Comp Biochem Physiol C Toxicol Pharmacol 129:115–127
160. Olsvik PA, Lie KK, Jordal AE, Nilsen TO, Hordvik I (2005) Evaluation of potential reference genes in real-time RT-PCR studies of Atlantic salmon. BMC Mol Biol 6:21
161. Celius T, Matthews JB, Giesy JP, Zacharewski TR (2000) Quantification of rainbow trout (*Oncorhynchus mykiss*) zona radiata and vitellogenin mRNA levels using real-time PCR after in vivo treatment with estradiol-17 beta or alpha-zearalenol. J Steroid Biochem Mol Biol 75:109–119
162. Chung-Davidson YW, Rees CB, Wu H, Yun SS, Li W (2004) beta-naphthoflavone induction of CYP1A in brain of juvenile lake trout (*Salvelinus namaycush* Walbaum). J Exp Biol 207(Pt 9):1533–1542
163. Filby AL, Tyler CR (2007) Appropriate 'housekeeping' genes for use in expression profiling the effects of environmental estrogens in fish. BMC Mol Biol 8:10
164. Roling JA, Bain LJ, Baldwin WS (2004) Differential gene expression in mummichogs (*Fundulus heteroclitus*) following treatment with pyrene: comparison to a creosote contaminated site. Mar Environ Res 57(5):377–395
165. Spinsanti G, Panti C, Lazzeri E, Marsili L, Casini S, Frati F, Fossi CM (2006) Selection of reference genes for quantitative RT-PCR studies in striped dolphin (*Stenella coeruleoalba*) skin biopsies. BMC Mol Biol 7:32
166. Hansen BH, Romma S, Garmo OA, Pedersen SA, Olsvik PA, Andersen RA (2007) Induction and activity of oxidative stress-related proteins during waterborne Cd/Zn-exposure in brown trout (*Salmo trutta*). Chemosphere 67:2241–2249
167. Suzuki N, Yamamoto M, Watanabe K, Kambegawa A, Hattori A (2004) Both mercury and cadmium directly influence calcium homeostasis resulting from the suppression of scale bone cells: the scale is a good model for the evaluation of heavy metals in bone metabolism. J Bone Miner Metab 22(5):439–446
168. Lattier DL, Gordon DA, Burks DJ, Toth GP (2001) Vitellogenin gene transcription: a relative quantitative exposure indicator of environmental estrogens. Environ Toxicol Chem 20:1979–1985
169. Hoffmann JL, Oris JT (2006) Altered gene expression: a mechanism for reproductive toxicity in zebrafish exposed to benzo a pyrene. Aquat Toxicol 78:332–340

170. An L, Hu J, Zhang Z, Yang M (2006) Quantitative real-time RT-PCR for determination of vitellogenin mRNA in so-iuy mullet (*Mugil soiuy*). Anal Bioanal Chem 386:1995–2001
171. Navas JM, Segner H (2001) Estrogen-mediated suppression of cytochrome P4501A (CYP1A) expression in rainbow trout hepatocytes: role of estrogen receptor. Chem Biol Interact 138:285–298
172. Vetillard A, Bailhache T (2005) Cadmium: an endocrine disrupter that affects gene expression in the liver and brain of juvenile rainbow trout. Biol Reprod 72(1): 119–126
173. Tilton SC, Givan SA, Pereira CB, Bailey GS, Williams DE (2006) Toxicogenomic profiling of the hepatic tumor promoters indole-3-carbinol, 17beta-estradiol and beta-naphthoflavone in rainbow trout. Toxicol Sci 90:61–72
174. Fernandez MP, Campbell PM, Ikonomou MG, Devlin RH (2007) Assessment of environmental estrogens and the intersex/sex reversal capacity for chinook salmon (*Oncorhynchus tshawytscha*) in primary and final municipal wastewater effluents. Environ Int 33(3):391–396
175. Wong CK, Wong MH (2000) Morphological and biochemical changes in the gills of Tilapia (*Oreochromis mossambicus*) to ambient cadmium exposure. Aquat Toxicol 48(4):517–527
176. Yamaguchi A, Ishibashi H, Kohra S, Arizono K, Tominaga N (2005) Short-term effects of endocrine-disrupting chemicals on the expression of estrogen-responsive genes in male medaka (*Oryzias latipes*). Aquat Toxicol 72:239–249
177. Dixon TJ, Taggart JB, George SG (2002) Application of real time PCR determination to assess interanimal variabilities in CYP1A induction in the European flounder (*Platichthys flesus*). Mar Environ Res 54:267–270
178. Jessen-Eller K, Crivello JF (1998) Changes in metallothionein mRNA and protein after sublethal exposure to arsenite and cadmium chloride in juvenile winter flounder. Environ Toxicol Chem 17:891–896
179. Arukwe A (2002) Complementary DNA cloning, sequence analysis and differential organ expression of beta-naphthoflavone-inducible cytochrome P4501A in Atlantic salmon (*Salmo salar*). Comp Biochem Physiol C Toxicol Pharmacol 133:613–624
180. Cousinou M, Nilsen B, Lopez-Barea J, Dorado G (2000) New methods to use fish cytochrome P4501A to assess marine organic pollutants. Sci Total Environ 247:213–215
181. Fisher MA, Mehne C, Means JC, Ide CF (2006) Induction of CYP1A mRNA in Carp (*Cyprinus carpio*) from the Kalamazoo River polychlorinated biphenyl-contaminated superfund site and in a laboratory study. Arch Environ Contam Toxicol 50:14–22
182. Quirós L, Piña B, Solé M, Blasco J, López MA, Riva MC, Barceló D, Raldúa D (2007) Environmental monitoring by gene expression biomarkers in *Barbus graellsii*: laboratory and field studies. Chemosphere 67:1144–1154
183. Van Cleef-Toedt KA, Kaplan LA, Crivello JF (2001) Killifish metallothionein messenger RNA expression following temperature perturbation and cadmium exposure. Cell Stress Chaperones 6(4):351–359
184. Knapen D, Reynders H, Bervoets L, Verheyen E, Blust R (2007) Metallothionein gene and protein expression as a biomarker for metal pollution in natural gudgeon populations. Aquat Toxicol 82:163–172
185. Laurie AD (2004) Quantitation of metallothionein mRNA from the New Zealand common bully (*Gobiomorphus cotidianus*) and its implications for biomonitoring. N Z J Mar Freshwater Res 38:869–877
186. Hayes RA, Regondi S, Winter MJ, Butler PJ, Agradi E, Taylor EW, Kevin Chipman J (2004) Cloning of a chub metallothionein cDNA and development of competitive RT-PCR of chub metallothionein mRNA as a potential biomarker of heavy metal exposure. Mar Environ Res 58:665–669
187. Tom M, Shmul M, Shefer E, Chen N, Slor H, Rinkevich B, Herut B (2003) Quantitative evaluation of hepatic cytochrome P4501A transcript, protein, and catalytic activity in the striped sea bream (*Lithognathus mormyrus*). Environ Toxicol Chem 22:2088–2092
188. Funkenstein B, Dyman A, Levavi-Sivan B, Tom M (2004) Application of real-time PCR for quantitative determination of hepatic vitellogenin transcript levels in the striped sea bream, *Lithognathus mormyrus*. Mar Environ 58:659–663

189. Barucca M, Canapa A, Olmo E, Regoli F (2006) Analysis of vitellogenin gene induction as a valuable biomarker of estrogenic exposure in various Mediterranean fish species. Environ Res 101:68–73
190. McClain JS, Oris JT, Burton GA Jr, Lattier D (2003) Laboratory and field validation of multiple molecular biomarkers of contaminant exposure in rainbow trout (*Oncorhynchus mykiss*). Environ Toxicol Chem 22:361–370
191. Roberts AP, Oris JT, Burton GA Jr, Clements WH (2005) Gene expression in caged fish as a first-tier indicator of contaminant exposure in streams. Environ Toxicol Chem 24:3092–3098
192. George S, Gubbins M, MacIntosh A, Reynolds W, Sabine V, Scott A, Thain J (2004) A comparison of pollutant biomarker responses with transcriptional responses in European flounders (*Platicthys flesus*) subjected to estuarine pollution. Mar Environ Res 58:571–575
193. Rees CB, McCormick SD, Li W (2005) A non-lethal method to estimate CYP1A expression in laboratory and wild Atlantic salmon (*Salmo salar*). Comp Biochem Physiol C Toxicol Pharmacol 141:217–224
194. Quirós L, Jarque S, Lackner R, Fernández P, Grimalt JO, Piña B (2007) Physiological response to persistent organic pollutants in fish from mountain lakes: analysis of Cyp1A gene expression in natural populations of *Salmo trutta*. Environ Sci Technol 41:5154–5160
195. Burki R, Vermeirssen EL, Korner O, Joris C, Burkhardt-Holm P, Segner H (2006) Assessment of estrogenic exposure in brown trout (*Salmo trutta*) in a Swiss midland river: integrated analysis of passive samplers, wild and caged fish, and vitellogenin mRNA and protein. Environ Toxicol Chem 25:2077–2086
196. Miller HC, Mills GN, Bembo DG, Macdonald JA, Evans CW (1999) Induction of cytochrome P4501A (CYP1A) in *Trematomus bernacchii* as an indicator of environmental pollution in Antarctica: assessment by quantitative RT-PCR. Aquat Toxicol 44:183–193

Hdb Env Chem (2009) 5J: 115–160
DOI:10.1007/698_2008_13

Biosensors for Aquatic Toxicology Evaluation

Marinella Farré and Damià Barceló

Abstract There is an increasing need for effective tools to estimate the risks derived from the large number of pollutants released to the environment by human activities. These studies need the combination of chemical analysis and ecotoxicological evaluation. Environmental toxicology is the qualitative and quantitative study of the adverse effects of anthropogenic and naturally occurring chemical stressors.

Initial aquatic ecotoxicology studies were based on acute toxicity measurements of vertebrates. However, these methods suffer some standardization problems, are expensive, time-consuming, and moreover, are associated with ethical problems. For this reason new technologies for ecotoxicological studies, such as embryo tests, micro invertebrates, algae, micro organisms, cell lines and tissues, biochemical reactions and nucleic acids, were launched.

Biological tools like bio-analytical systems, bioassays, biomarkers and biosensors provide us with detection systems for signalling a potential damage in the environment. Early recognition will prevent eventual damage to environmental matrices. Ideally, early warning signals in ecosystems using sensing systems (biosensors) and biochemical responses (biomarkers) as well as the classical effect-related bioassays would not only tell us the initial levels of damage, but these signals will also provide us with answers for the development of control strategies and precautionary measures.

Herein new trends in ecotoxicity assessment will be reviewed with special emphasis on principles, advantages and limitations of biosensors for environmental toxicology assessment in the aquatic environment.

Keywords Bioassay, Biosensor, Ecotoxicology, Environmental monitoring, Sensor, Toxicity

---

M. Farré
Department of Environmental Chemistry, IDAEA-CSIC
c/Jordi Girona 18-26, 08034, Barcelóna, Spain

D. Barceló (✉)
Department of Environmental Chemistry, IDAEA-CSIC
c/Jordi Girona 18-26, 08034, Barcelóna, Spain
Catalan Institute of Water Research, ICRA C/Pic de Peguera, 15, 17003, Girona, Spain
e-mail: dbcqam@iiqab.csic.es

Contents

1 Introduction

Environmental toxicology is the qualitative and quantitative study of the adverse effects of anthropogenic and naturally occurring chemical stressors.

The publication in 1962 of Rachel Carson's Silent Spring [1] catalyzed the separation of environmental toxicology – and, subsequently, ecotoxicology – from classical toxicology.

Initial aquatic ecotoxicology studies were based on acute toxicity measurements of vertebrates such as fish [2–4]. However, these methods suffer some standardization problems, are expensive, time-consuming, and moreover, are associated with ethical problems [2].

The development of in vitro technologies occurred for scientific, ethical and economical reasons. Ecotoxicological tests involving higher organisms have been progressively replaced by different technologies, such as embryo tests, micro invertebrates, algae, micro organisms, cell lines and tissues, biochemical reactions and nucleic acids [5, 6].

During the last four decades a wide variety of biological methods have been developed to investigate different end points, sub-lethal effects, and biochemical responses, evaluating: acute toxicity, cytotoxicity, genotoxicity, neurotoxicity or endocrine disrupting effects [5].

About 80,000 chemicals (including 600 pesticides) are currently used, and more than 1,000 enter the market every year [7]. The increasing need for effective tools to estimate the risks from the large number of pollutants released into the environment by human activities has led to the development of sensitive techniques for toxicological and chemical assessment. In recent years, a growing number of initiatives and legislative actions for environmental pollution control, with particular emphasis on water quality control, have been adopted in parallel with increasing scientific and social concern in this area. These new policies such as the Water Framework Directive (WFD) (Directive 2000/60/EC of the European Parliament and of the Council) of 23rd October 2000, establishing a framework for Community action in the field of water policy [8], forces us to increase toxicological and chemical monitoring studies based on the use of new methodologies able to obtain responses in short times, or even on-line.

Biological tools like bio-analytical systems, bioassays, biomarkers and biosensors provide us with detection systems for signalling a potential damage in the environment. Early recognition will prevent eventual damage to the environmental matrices. Ideally, early warning signals in ecosystems using sensing systems (biosensors) and biochemical responses (biomarkers) as well as the classical effect-related bioassays would not only tell us the initial levels of damage, but these signals will also provide us with answers for the development of control strategies and precautionary measures.

Figure 1 presents a representation of the relationship between environmental distress signal detectability and ecological relevance.

The implementation of safety programs calls for environmental analysis comprising two parts [9, 10]:

Aquatic ecotoxicological studies based on screening methods capable of predicting the possible dangerous biological effects of the cocktail of pollutants;

Analysis of the positive samples with chromatographic methods, such as LC/MS-MS, LC-QqTOF/MS, GC/MS or ICP-MS for the identification of the compounds causing the effects.

Bio-response-linked instrumental analysis combines two processes, biological recognition initiating a biological effect, and chemical analysis. The biomolecular components are used as targets for active substances. Although it is possible to

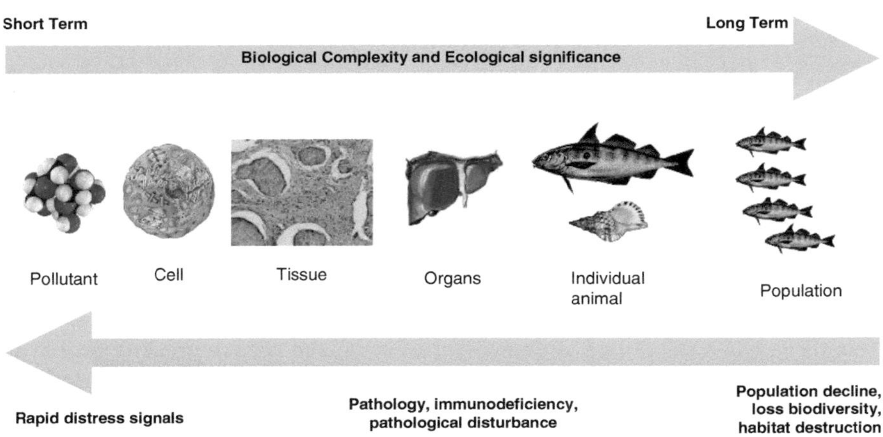

Fig. 1 Scheme of relationship between environmental distress signal detectability and ecological relevance

apply, for example, binding assays that provide effects-linked equivalents, information on the responsible contaminants is only accessible by chemical analysis. Therefore, a subsequent step is provided by the chemical analysis of the substances that are bound by the biorecognition components and which are therefore bio-effective.

In order to allow toxicant identification both effect-directed analysis (EDA) and toxicity identification evaluation (TIE), are shown to be appropriate [10–13]. EDA and TIE are analytical procedures combining biological and chemical analysis with fractionation techniques. While TIE originates from effluent control in a regulatory context in the U.S., EDA is a more scientific approach developed by analytical chemists to identify unknown hazardous compounds in the environment.

In addition to bioassays based on microorganisms, cell lines, recombinant organisms, or biochemical responses, biosensors have been demonstrated to have great potential and thus arise as proposed tools for effective ecotoxicology monitoring programs. According to the IUPAC definition a biosensor is an integrated device that is capable of providing specific quantitative or semi-quantitative analytical information using a biological recognition element (biochemical receptor), immobilized onto a transduction element. This is the main difference with bioassays.

The main biosensor advantages are that they allow rapid responses, portable, miniaturized, on-line, on-site, and remote control configurations, and cost-effective measurements. These technologies are excellent alarm tools [14].

Herein new trends in ecotoxicity assessment will be reviewed with special emphasis on principles, advantages and limitations of biosensors for environmental toxicology assessment in the aquatic environment.

2 Bioassays for Aquatic Ecotoxicity Assessment

Among aquatic toxicity bioassays should be emphasized the tests based on invertebrates such us *Daphnia magna* [15], micro algae, such as *Skeletonema costatum* [16] and *S. capricornotum* [17], the marine bacteria *Vibrio fischeri* and *Photobacterium phosphoreum* [18], because these are well-established tests using standardized organisms, and are available from different commercial firms. The use of macro invertebrates, such as *D. magna* or *Cerodaphnia*, presents many advantages, such as high sensitivity and short reproductive cycle.

Chronic tests with members of the genus *Daphnia* were initiated in the early 1970s19 by Biesinger and Christensen [] and are currently broadly accepted. Despite the diversity of test species available, in many regulatory schemes, the only, or preferred, invertebrate species recommended for acute or chronic toxicity testing is the *cladoceran*, *D. magna.* Presumably, the focus on *D. magna* results from its sensitivity to environmental contaminants relative to other invertebrate species [20]. This focus has resulted in a large number of acute and chronic toxicity tests with this species. Nevertheless, the advantages of *D. magna* over other species such as *Ceriodaphnia* has been a widely debated question, and some authors support that the acute and chronic toxicity of a broad range of materials, including metals, organic compounds and effluents, are approximately similar for members of the genera *Ceriodaphnia* and *Daphnia* [21]. These toxicity tests can be carried out for lethality, growth, reproduction, mobility and population growth rate by exposing the organisms to the tested substances under controlled conditions, and measuring the acute toxicity or the chronic toxicity.

Experiments using *D. magna* are usually carried out using genetically identical populations and using 3rd–5th brood offspring. While the standardized acute lethality toxicity test based on *D. magna* is conducted over 21 days [22], other alternative tests at screening level are carried out with short exposure times, such as the 24–48-h screening test with *Daphnia* [23].

The effects produced by pollutants on *D. magna* at suborganismal and organismal levels, as well as, the implications for population dynamics are of great interest. A high number of studies have established the responses using *Daphnids* to different pollutants using different conditions and end points. Some examples are the studies of pesticides [24, 25], polyaromatic compounds [26] and anionic surfactants [27, 28] in *Daphnia.*

The use of algae as bio-indicators of toxicity is also well established, because they represent an important component of any aquatic community, besides being the lowest link in the food chain. For this reason, a substance toxic for the algae of an ecosystem can produce a strong impact in the ecosystem. Microalgae, such as *Chlorella fusca, Selenastrum capricornotum, Dunaliellasalina* or *Dunaliella tertiolecta* are used as toxicity indicators because of their sensitivity to pesticides [29, 30] and metals [31].

Hernando et al. [32] reported the utility and validity of toxicity tests based on *V. fischeri, S. capricornotum* and *D. magna* for monitoring of wastewater treatments.

Nevertheless, the more widely used bioassays in routine laboratories for evaluating water toxicity are based on inhibition of the bioluminescence of marine bacteria. The better-known species of luminescent marine bacteria are *V. fischeri* and *Photobacterium phosphoreum*, which naturally emit light due to an enzyme, the bacterial luciferase. Any substance that affects the bacterial metabolism produces a proportional inhibition of the luminescence.

The use of luminescence organisms to assess toxicity has been known for more than 40 years [33]. In 1979 a toxicity bioassay using luminescent bacteria was developed by Bulich et al. [34] to assess toxicity of wastewater effluents and industrial discharges. This technique allows the easy screening of large numbers of aqueous samples in a quick, reliable and inexpensive way. This toxicity assay was commercialized for the first time by Microtox and described in Beckman's Operating Manual.

The bioluminescence inhibition of *V. fischeri* has been exploited to investigate organic [35–38] and inorganic pollutants [39, 40], but one of the more relevant applications is for the rapid evaluation of real samples [41–45].

Over the years, many researchers have studied the physiological, biochemical and genetic control of bacterial bioluminescence. These discoveries have revolutionized the area of Environmental Microbiology using luminescent genes as biosensors for environmental studies. The ability to introduce the *lux* phenotype, coupled to specific promoters allowing its expression only in the presence of the proper analyte, into different bacterial species, provides a convenient method to multiply the possibilities for rapid, simple and sensitive screening of the environmental conditions. The *lux* operons, employed in these genetic manipulations, include those encoding for the bacterial luminescent system, the *lux* genes of firefly luciferase as well as those of *gfp*, the green fluorescent protein (GFP) from *A. victoria.*

A number of recombinant plasmids, carrying the *lux* operon expressed constitutively in many Gram– and Gram+ bacteria, have been constructed and used to transform a wide variety of bacteria. A promoter is usually required on the plasmid to obtain a high expression. The possibility to transform in luminescent organisms the bacteria previously isolated from the media under analysis allows one to produce biosensors with the maximum of suitability, since the vitality of the transformed organisms in their environment will be optimal, as well as its response to the presence of extraneous, noxious compounds [46].

A large number of procedures, testing the wastewater toxicity, are based on the GMB Shk1. A quick estimation of the toxicities of organic chemicals with different functional groups has been proposed, developing quantitative structure-activity relationship (QSAR) models, based on the logarithm of the octanol-water partition coefficient, as well as an overall QSAR model without discriminating the functional groups [47]. Model predictions were compared to experimental data and the model accuracy was found to be one order of magnitude from the experimental values.

The performance of the construct obtained by fusing the reporter genes of green fluorescent proteins (*A. victoria gfp*) and bioluminescence (*V. fischeri luxCDABE*) genes, to either SOS (*recA*) or heat shock (*grpE*) promoters, was compared [48]. In both cases, bacterial bioluminescence allowed faster and more sensitive detection. The fluorescent (FL) reporter proteins were much more stable and in long-term tests allowed detection at levels similar to that of the bioluminescence.

To evaluate the nitrification inhibition activity in wastewaters, a recombinant, tryptophan-dependent *Nitrosomonas europaea* luciferase gene, has been employed. In the case where the ammonia monooxygenase activity of *Nitrosomonas* was inhibited, a rapid drop in NADH or ATP concentration occurs. This reduction in the overall energy level of the cell was revealed by the reduction of the light emission intensity.

Various *Escherichia coli* strains, containing plasmid-borne fusions of *V. fischeri lux* to the *recA* promoter-operator region, have been created: (1) modifying the host cell's toxicant efflux capacity via a *tolC* mutation; (2) incorporating the *lux* fusion onto the bacterial chromosome, rather then on a plasmid; (3) changing the reporter element to a different *lux* system (*Photorhabdus luminescens*), with a broader temperature range; (4) using *Salmonella typhimurium* instead of an *E. coli* host [49].

Bacterial bioluminescence can be combined with other techniques to enhance its response in specific conditions. A bioluminescent *Campylobacter jejuni* organism, generated from the fusion of *lux*CDABE genes from *Xenorhabdus luminescens* to the *flaA* promoter of *C. jejuni*, was used in conjunction with two-dimensional gradient gels to map the responses of this organism to pH, NaCl concentration, temperature, and to various concentrations of l-fucose, d-fucose and sodium desoxycholate [50]. To monitor the bio-active fraction of Hg, instead of the total Hg, bioluminescent sensor systems were developed by fusion of a Hg-resistant operon (*mer* operon) from *Pseudomonas* sp. K-6y4 with the *lux* operon from *V. fischeri*. The resulting recombinant plasmids were cloned in *E. coli* cells for environmental applications [51].

A number of genotoxicity assays have been also developed based on the use of recombinant microorganisms. Genotoxicity is associated with different structures, such as phenols, chlorophenols, polychlorinated biphenyls (PCBs), or polyaromatic hydrocarbons (PAHs), and constitutes an early screening for possible cancer-inducing activity of pollution. The most widespread is the Ames test that was established as a routine method of analysis. It is based on the retro mutation of *Salmonella typhimurium* TA98 (histidine dependent). The umu test is also based on *S. typhimurium*, and the genotoxicity is detected measuring the activation of the bacterial SOS repair response of genetic damage in the bacterium, through measuring β-galactosidase activity [52]. This is a standardized method that is validated for water control. Other genotoxicity bioassays that should be mentioned are those based on *Saccharomyces cerevisiae* [53–56], *Bacillus subtilis* [57–61] or *Escherichia coli* [15, 62–64].

In vitro systems such as primary cells and cell lines are of growing importance in ecotoxicology. Cells from different tissues and species of fish are used for the assessment of toxic action of chemicals and evaluation of environmental samples. For different organic pollutants such as organotins and substituted phenols, it has been found that the in vitro cytotoxicity is positively correlated with the acute toxicity in vivo, and therefore cytotoxicity assays may serve as an alternative for acute fish toxicity testing. A high number of works have been published, and correlation between toxicity, cytotoxicity and estrogenicity with in vivo tests has been presented. The cytotoxicity of eight aromatic products of toluene aerobic biotransformation was investigated by Shen et al. [65] in bluegill sunfish BF-2 cells. The concentrations of these products causing 50% inhibition in cell replication, protein content, uptake of natural red, and colony formation were evaluated and compared. The results of the study indicate a direct relationship between the exposure concentration of these products and observed cytotoxic effects. In descending order of cytotoxicity, the compounds were 3-methylcatechol, 4-methylcatechol, catechol, *o*-cresol, *p*- cresol, *m*-cresol, benzaldehyde and methyl benzoate.

Okamura et al. [66] developed a simple, rapid toxicity test using the suspension-cultured fish cell line CHSE-sp derived from chinook salmon *Oncorhynchus tshawytscha* embryos in order to assess the toxicity of new marine antifouling compounds. Good correlation was found with in vivo tests.

Kannan et al. [67] assessed the genotoxicity and cytotoxicity using an in vitro system with a permanent cell line *Epithelioma papulosum cyprini* (EPC) derived from a skin tumour of carp (Cyprinus carpio L.). EPC cells were exposed to different concentrations of organic sediment extracts from the North Sea for 24 h. After incubation the cells were analyzed for viability and DNA strand breaks with the comet assay or single-cell gel electrophoresis (SCGE). The results confirm the sensitivity of this assay.

Vevers et al. [68] studied the intrinsic genotoxic and cytotoxic potential of titanium dioxide (TiO_2) engineered nanoparticles (ENPs). The biological effects were evaluated in a metabolically competent assay based on a fish cell line derived from rainbow trout gonadal tissue (i.e. RTG-2 cells).

In recent years, many natural and synthesized substances have been associated with endocrine disruptions in wildlife, especially aquatic fauna [69]. Several bioassays have been developed to assess the activity of alleged estrogenic substances. In vivo assays are widely used. They are based on a wide variety of end points, including cell differentiation and enzyme activities. Nevertheless, they are unsuitable for large-scale screening; moreover, in vivo assays are based on complex responses that may be modulated through mechanisms that do not directly involve the estrogenic receptor. In order to carry out environmental monitoring of endocrine disrupting chemicals (EDCs), a great number of in vitro assays have been developed. The assays, based on the strength of binding of a substance to the estrogen receptor (ER), are well established [70]. Their main drawback is the fact that the "binding activity" is not sufficient to determine the estrogenicity of a substance, because that also depends on the ability of the ligand to elicit an estrogenic-receptor response.

Yeast reporter gene assays and MCF-7 cell-based proliferation assays (E-screen) are particularly popular. The E-screen, in which proliferation of human breast cancer cells (MCF-7) is measured as a response to estrogen, has also been used to determine the estrogenicity of sewage effluent and surface water [71]. A correlation of an enzyme-linked receptor assay (ELRA) with a yeast reporter gene assay is shown. Several assays have been reported in the literature, such as the ultra-sensitive luminescent ELRA developed by Seifert [72] with a detection limit of 20 ng L^{-1} for 17β-estradiol.

Although these assays are simple, many factors affect the potency of estrogenic substances, such as differences in culture conditions, cell density or cell-line clones, and that makes the standardization of these methods difficult.

The induction of several proteins or enzyme activities has also been used to study estrogenicity (e.g. the increasing levels of alkaline phosphatase, cathepsin D, prolactin and vitellogenin as a consequence of progestogens). However, expression of these proteins or enzyme activities is restricted to specific cell lines and cannot be extrapolated to other tissues or species.

Over the past decade, a great number of recombinant receptor-reporter gene assays have been developed. Basically, these assays are based on simple cell models that express a gene under the control of defined promoters responding to specific substances and that produce an easily quantified signal. These assays are of great interest for rapid screening of the estrogenicity of artificial or natural compounds. In integrated systems, these cell models are very useful to study the synergy or antagonism of different substances and, in the field of environmental research, they are excellent tools to identify EDCs. Some of them are based on the bioluminescent gene reporter assay. Among these, the YES assay using yeast cells [73] should be emphasized.

Those papers mentioned above and additional papers, concerning new ecotoxicity bioassays, are listed in Table 1.

Beyond the practical advantages of new bioassays for toxicity assessment, they should be comparable to conventional analytical systems in terms of reliability and robustness. There is a lack of validation of new bioassays. To date just a few bioassays are standardized; some examples are *D. magna*, *V. fischeri*, *S. capricornutum*, and the umu-test. The implementation of new bioassays therefore needs to be verified and validated before being accepted. The validation procedures can vary depending on the technology that is evaluated and include organization and participation in

Table 1 Organisms widely used for toxicity assessment

Species	Parameter measured	References
Plants and algae		
Selenastrum capricornutum	Growth inhibition	[28–30, 74–77]
Dunaliella tertiolecta	Growth inhibition	[29, 75, 77, 78]
Skeletonema costatum	Growth inhibition	[79, 80]
Daphnids		
Daphnia magna	Immobility	[24–26, 32, 81–83]
Daphnia magna	Mortality	[84–88]
Cerodaphnia duvia	Immobility	[86, 89]
Microorganism		
Vibrio fischeri	Bioluminescence	[36, 43, 44, 90, 91]
Escherechia coli	Metabolic status	[92–94]
Pseudomonas putida	Metabolic status	[37, 98–100]
Pseudomonas fluorescens	Bioluminescence	[97, 101]
Bacillus subtilis	Growth	[102–105]

interlaboratory trials, use and development of certified reference materials (CRMs) or comparison with conventional chemical analysis. Validation may involve some difficulties since, in addition to general quality parameters, it may be necessary to evaluate some other aspects such as stability of the biological reagents. In addition, there is a lack of guidelines established for validation of different groups of bioassays.

On the other hand, although there are several validation procedures accepted for quantitative methods, in the case of biological methods there are no established guidelines [106, 107].

3 Biosensor Classification

Biosensors are usually classified either according to:

- the transduction element (for example, electrochemical, optical, piezoelectrical or thermal) and bioreceptor (for example, enzymatic, immunoaffinity recognition, whole-cell sensors, or DNA)

3.1 Optical Transducers

Optical transducers are based on various technologies of optical phenomena, which are the result of an interaction of an analyte with the receptor part. The various types of optical transducers exploit properties such as simple light absorption, fluorescence/ phosphorescence, bio/chemiluminescence, reflectance, Raman scattering and refractive index. Apart from speed, sensitivity and robustness, other attractive features of optical sensors include their suitability for component miniaturization, remote sensing and their multi-analyte sensing capabilities. In addition, the increasing market of telecommunications has supported new optical materials research.

This group may be further subdivided according to the type of optical properties which have been applied in sensing:

Absorbance, measured in a transparent medium.
Reflectance is measured in non-transparent media, usually using an immobilized indicator.
Luminescence, based on the measurement of the intensity of light emitted by a chemical reaction in the receptor system.
Fluorescence, measured as the positive emission effect caused by irradiation. Also, selective quenching of fluorescence may be the basis of such devices.
Refractive index, measured as the result of a change in solution composition. This may include also a surface plasmon resonance effect (SPR).
Optothermal effect, based on a measurement of the thermal effect caused by light absorption.
Light scattering, based on effects caused by particles of definite size present in the sample.

The application of many of these phenomena in sensors became possible because of the use of optical fibres in various configurations. Such devices have also been called optodes.

A high number of optical transduction techniques can be used for biosensor development. These may employ linear optical phenomenon, including adsorption, fluorescence, phosphorescence, polarization, rotation, interference, etc., or non-linear phenomena, such as second harmonic generation.

Total internal reflection fluorescence (TIRF) has been used with planar and fibre optic waveguides as signal transducers in a number of reported biosensors. In these transducers, light is propagated down a waveguide which generates an electromagnetic wave (evanescent wave) at the surface of the optically denser medium of the waveguide and the adjacent less optically dense medium. The amplitude of the standing wave decreases exponentially with distance into the lower refractive index material. The fluorescence of a fluorophore excited within the evanescent field can be collected.

Surface plasmon resonance biosensors exploit special electromagnetic waves—surface plasmon polaritons—to probe interactions between an analyte in solution and a biomolecular recognition element immobilized on the SPR sensor surface. A surface plasmon wave can be described as a light-induced collective oscillation in electron density at the interface between a metal and a dielectric. At SPR, most incident photons are either absorbed or scattered at the metal/dielectric interface and, consequently, reflected light is greatly attenuated. The resonance wavelength and angle of incidence depend upon the permittivity of the metal and dielectric (Fig. 2).

The general advantages of optical techniques involve the speed and reproducibility of the measurement, and the main drawback is the high cost of the apparatus.

Most of the works based on SPR have been focused on the rapid identification and quantification of different groups of toxicant pollutants.

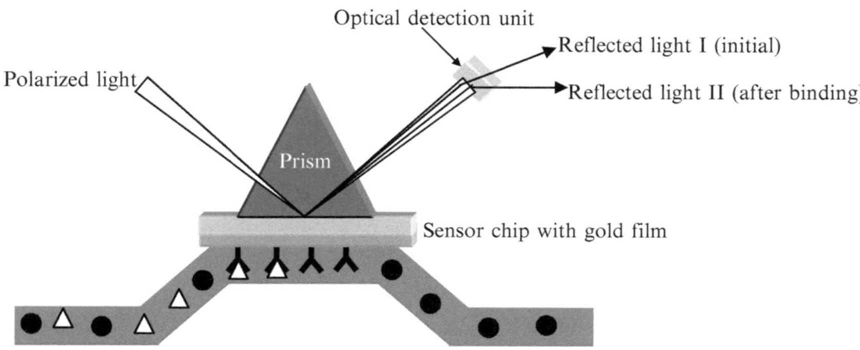

Fig. 2 General scheme of a surface plasmon resonance sensor

3.2 Electrochemical Transduction

Electrochemical transduction transforms the effect of the electrochemical interaction between an analyte and the electrode into a primary signal. Such effects may be stimulated electrically or may result in a spontaneous interaction at the zero-current condition. Most of the biosensors described in the literature are electrochemical biosensors. Electrochemistry offers high sensitivity, compatibility with modern microfabrication technologies, portability, low cost and minimal power requirements

The following subgroups may be distinguished:

Voltammetric sensors, including amperometric devices, in which current is measured in the d/c or a/c mode. This subgroup may include sensors based on chemically inert electrodes, chemically active electrodes and modified electrodes. In this group are included sensors with and without (galvanic sensors) an external current source.

Potentiometric sensors, in which the potential of the indicator electrode (ion-selective electrode, redox electrode, metaVmetal oxide electrode) is measured against a reference electrode.

Chemically sensitized field effect transistor (CHEMFET) in which the effect of the interaction between the analyte and the active coating is transformed into a change of the source-drain current.

Potentiometric solid electrolyte gas sensors, differing from potentiometric sensors because they work in high temperature solid electrolytes and are usually applied for gas sensing measurements.

Electrochemical biosensors are based on monitoring electroactive species that are produced or consumed by the action of the biological elements (e.g. enzymes), and can be performed basically under potentiometric and amperometric measurements.

The principle operation of amperometric biosensors is defined by a constant potential applied between a working and a reference electrode. The imposed potential promotes a redox reaction, which produces a current. The magnitude of this current is proportional to the concentration of electroactive species present in solution.

Potentiometric biosensors are based on the monitoring of the potential at a working electrode, with respect to a reference electrode.

3.3 Electrical Sensors

Electric sensors are based on measurements, where no electrochemical processes take place, but the signal arises from the change of electrical properties caused by the interaction of the analyte. Examples are shown below:

Metal oxide semiconductor (MOS) sensors are used principally as gas phase detectors, based on reversible redox processes of analyte gas components.
Organic semiconductor sensors, based on the formation of charge transfer complexes, which modify the charge carrier density.
Electrolytic conductivity sensors.
Electric permittivity sensors.

3.4 Mass-Sensitive Sensors

Mass-change sensors transform the mass change at a specially modified surface into a change of a property of the support material. The mass change is caused by accumulation of the analyte.

Piezoelectric devices used mainly in the gaseous phase, but also in solutions, are based on the measurement of the frequency change of the quartz oscillator plate caused by adsorption of a mass of the analyte at the oscillator.

Surface acoustic wave devices depend on the modification of the propagation velocity of a generated acoustical wave affected by the deposition of a definite mass of the analyte.

The vibration of piezoelectric crystals produces an oscillating electric field in which the resonant frequency of the crystal depends on its chemical nature, size, shape and mass. By placing the crystal in an oscillating circuit, the frequency can be measured as a function of the mass. When the change in mass (m) is very small compared to the total mass of the crystal, the change in frequency (f) relates to m as follows:

$$\Delta f = Cf^2 \, \Delta m / A$$

where f is the vibration frequency of the crystal in the circuit, A is the area of the electrode and C is a constant determined in part by the crystal material and thickness. Piezoelectric crystals, sometimes referred to as quartz crystal microbalances (QCM), are typically made of quartz and operate at frequencies between 1 and 10 MHz. These devices can operate in liquids with a frequency determination limit of 0.1 Hz, the detection limit of mass bound to the electrode surface is about 10^{-10} to 10^{-11} g.

Microcantilevers: Micromechanical devices comprise emerging sensor platforms with straightforward sensing mechanisms. Molecular adsorption onto the sensing element, typically a cantilever, shifts its resonance frequency and changes its surface forces (surface stress). Adsorption onto the sensing element composed of two chemically different surfaces produces a differential stress between the two surfaces and induces bending. The analyte that induces the mechanical response may be physi- or chemisorbed onto the cantilever in a reversible or irreversible process. Devices that respond to chemical stimuli in this manner are more commonly referred to as microcantilever (MC) sensors. A compelling feature of microcantilever sensors is that they can be operated in air, vacuum or liquid. The rapid growth in microcantilever-based sensor technology parallels advancements in micromachining methodologies and is in response to the need for more sensitive and selective detection of airborne and waterborne toxic and pathogenic substances.

Vertical, lateral or torsional movement of a cantilever changes its position. This movement ranges from several angstroms to a micrometer or more, depending upon the dimensionality of the cantilever and the magnitude of surface stress.

This group of sensors can be based on optical, interferometric, the electrical conductivity of a piezoresistive material when stress is applied to it, or under capacitive modes when the cantilever acts as one of the parallel plates of a capacitor.

3.5 *Thermometric Sensors*

Thermometric sensors exploit the absorption or evolution of heat in biological reactions [108]. This is reflected as a change in the temperature within the reaction medium and is transduced by a change in the resistance of a thermistor, which acts as a temperature transducer.

As reviewed by Ramanathan and Danielsson [108], thermistor-based calorimetric biosensors have mainly been applied to clinical and industrial process monitoring [108].

In the environmental field, only a few applications have been described up to now: several pesticides [109], for example, have been measured by thermometric biosensors.

4 Ecotoxicity Biosensors

The main classes of biosensors for ecotoxicity assessment are summarized in Fig. 3

4.1 *Enzyme Biosensors*

Enzyme biosensors have been described using a range of transduction elements (amperometry, potentiometry, optical and photo-thermal).

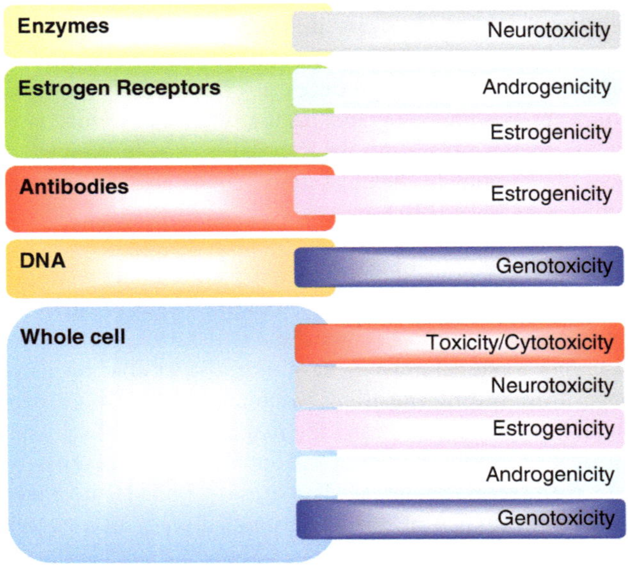

Fig. 3 Summary of main classes of biosensors for ecotoxicity assessment

The first biosensor was described in the literature by Clarck and Lyons (1962) based on the use of glucose oxidase with electrochemical detection [110]. This principle has been widely exploited for the detection of some groups of organic compounds, such as phenols [111–115] and pesticides [116–122].

Enzymes typically used in amperometric biosensors are oxidases. The main advantage of this class of transducer is the low cost; a high degree of reproducibility and disposable electrodes are often available. The instrumentation is also very easy to obtain and can be inexpensive and compact; this allows for the possibility of on-site measurements. Limitations for amperometric measurements include potential interferences to the response if electroactive compounds are present in the sample, because they can generate false current values. An amperometric biosensor for hydrogen peroxide can also be used to measure organophosphate (OP) pesticides at concentrations as low as 10^{-9} M.

One of the main contributions of enzyme biosensors to ecotoxicology studies is related to neurotoxins. Neurotoxin detection has been accomplished using acetylcholinesterase [123], acetylcholine receptors, and butylcholine estearase (BchE) [124].

The combined inhibition effects in mixtures of organophosphates and carbamates (such as paraoxon/carbaryl, diisopropyl fluorophosphate (DFP)/carbaryl, paraoxon/DFP/carbaryl) were studied by Simonian et al. [125]. Mutual interactions of various neurotoxins did lead to competition for acetylcholinesterase (AChE) binding sites, and the overall inhibition effects were not additive but dependent on the types of chemicals present.

Neuropathy target esterase is the target protein for neuropathic organophosphorus compounds that produce organophosphorus compound-induced delayed neurotoxicity (OPIDN). Inhibition/aging of brain neuropathy target esterase within hours of exposure predicts the potential for development of OPIDN in susceptible animal models. Lymphocyte NTE has also found limited use as a biomarker of human exposure to neuropathic organophosphorus compounds. Makhaeva et al. [126] developed a highly sensitive biosensor for neuropathy target esterase activity using a tyrosinase carbon-paste electrode for amperometric detection of phenol produced by hydrolysis of the substrate, phenyl valerate.

During the last years new nanotechnological materials have led to the development of improved characteristic biosensors. Some examples are those based on nanoparticles, as a novel electrochemical method for the investigation of pesticides using AChE. The sol–gel-derived silicate network assembling gold nanoparticles (AuNPs–SiSG) provided a biocompatible microenvironment around the enzyme molecule to stabilize its biological activity and prevented them from leaking out of the interface. AuNPs promoted a conductive pathway for electron transfer and improved electrochemical reactions at a lower potential. In this work monocrotophos, methyl parathion and carbaryl were selected for the sensitivity tests [127]. The same group in another work developed a simple immobilization method of AChE on multi-wall carbon nanotube (MWNT)–chitosan composite which was proposed to increase sensitivity and stability of amperometric sensors for quantitative determination of organophosphorous insecticides [128]. In another work by Shulga et al. [129] the electrodeposition of colloidal gold nanoparticles onto planar gold electrodes was used to create a more favourable surface for the attachment of the enzyme AChE.

Several examples of enzyme biosensors have been reported using optical transduction [130, 131]. During the last few years, optodes, such as fibre optic biosensors, have been demonstrated to be of great interest because they provide some advantages, such as no direct electric connection, ease of miniaturization, possibility of remote sensing, and in situ monitoring, and different examples have been reported such as a sol–gel acetyl cholinesterase fibre optic biosensor for organophosphorus neurotoxins [130].

In order to improve the storage stability on enzyme-based biosensors, different immobilizations and electrodes have been assayed, including carbon-paste electrodes (CPEs), solid graphite electrodes [132] and surface-modified electrodes [112].

4.2 Estrogen Receptor

Biosensor estrogenic potency of real samples polluted with endocrine disruptor compounds has been assessed by means of different biosensor approaches based on the immobilization of estrogen receptors.

Steroid hormones induce different effects in mammalian cells after binding to specific intercellular receptors, which are ligand-dependent transcription factors.

Many endocrine disruptors can bind to estrogen receptors either as agonists or antagonists. Thus, the binding ability of the chemicals toward the ER is used to test their potential environmental toxicity. The advantage of receptor assays is that they are quite simple to perform and allow the identification of all endocrine disrupters that act through the estrogen receptor [133]. The natural sensing element most commonly used is the human ER [134]. Some of these ER-based biosensors are reported in Table 2.

By using surface plasmon resonance, Usami et al. [143] developed a simple competitive assay for the evaluation of different chemicals using human recombinant ER. The system measured the binding between estradiol immobilized on the sensor chip and an injected human recombinant ER.

In another example using an optical SPR sensor, and the human estrogen α receptor, Hock et al. [144] performed binding studies with estradiol and xenoestrogens. A rapid enzyme-linked receptor assay using the BIAcore system was developed and validated by Seifert et al. [145] for detection of estrogens and xenoestrogens. ELRA can be used to estimate the estrogenic potential of chemicals, for drinking water control and environmental monitoring.

Sesay and Cullen described the detection of hormone mimics [146] by a portable SPR. Recently, Butala et al. [141] described a fractal analysis used to model the binding and dissociation kinetics between analytes in solution and the estrogen receptor immobilized on a SPR sensor chip.

A novel evanescent-based biosensor [147] was developed with laser-based fibre optics using fluorescent dye-labelled recombinant human estrogen receptor-alpha

Table 2 ER-based biosensors

Transduction	Analytes	Effect measured	References
Nanomechanical	Diethylstilbestrol, 17-β-estradiol, 17-α-estradiol, 2-OH-estrone, bisphenol A, p,p′-dichlorodiphenyldichloroethylene	Estrogenicity	[135]
Electrophoretic mobility shift assay	17-β-estradiol	Estrogenicity	[136]
Fluorescence	17-β-testosterone	Androgenicity	[137]
Fluorescence	17-β-estradiol	Estrogenicity	[138]
SPR	Estrogens	Estrogenicity	[139]
Fluorescence	Ligands of nuclear hormone receptors	Estrogenicity and androgenicity	[140]
SPR	Estrogens	Estrogenicity	[141]
Cyclic voltammetry	17β-Estradiol	Estrogenicity	[142]
SPR	Estrogens, progestogens, bisphenol A, 4-nonylphenol and tamoxifen.	Estrogenicity	[143]
SPR	17β-Estradiol, synthetic estrogens and xenostrogens	Estrogenicity	[144]
SPR	Estrogens and xenoestrogens	Estrogenicity	[145]

(rhERalpha) and hERbeta as probes. A three-tiered approach evaluating various steps in the formation of the estrogen-receptor complex and its subsequent activity was developed for instrument calibration to detect estrogen mimics in biological samples, water and soil. Using this approach, binding affinities and activities of certain known estrogen mimics were determined for their use as calibrator molecules. Results indicated rhERalpha and rhERbeta may be employed as probes to distinguish estrogen mimics with a broad range of affinities.

In another example, Dutta et al. [135] developed a nanomechanical transducer to detect endocrine disrupting chemicals combining fluidic sample injection and delivery with bioreceptor protein functionalized microcantilevers. The combination of protein receptors, which include estrogen receptor alpha (ER-α) and estrogen receptor beta (ER-β), as well as monoclonal antibodies (Ab), with MC systems employing modified nanostructured surfaces provides for excellent nanomechanical response sensitivity and the inherent selectivity of biospecific receptor-EDC interactions.

Murata et al. [142] proposed a bioaffinity sensor based on the specific binding of estrogens to the receptor immobilized on a gold disk with cyclic voltammetry detection. The biosensor was applied to the detection of estradiol.

Finally, Zhihong et al. [148] developed a new sandwich-type assay for estrogens using a piezoelectric biosensor. The principle of the detection is that the estrogen receptor captures the estrogen and then the complex is bound with an estrogen response element (ERE) immobilized on the sensor.

4.3 *Immunosensors*

The recognition elements are immunochemical interactions antibody-antigen. This type of device combines the principles of solid-phase immunoassay with physicochemical transduction elements (electrochemical, optical, piezoelectric, evanescent wave and surface plasmon resonance).

Most of the applications of immunosensors are to determine target analytes or groups of structurally related compounds, in contrast with applications reporting whole biological effects. However, several applications have been developed for estrogenicity. Biomarkers, such as vitellogenin, can be used to show effects of endocrine substances on whole organisms [133]. Vitellogenin (Vtg) is a serum phospholipo-glycoprotein precursor to egg yolk, which is produced in high amounts by fish exposed to estrogenic compounds. Thus, the presence of Vtg in male fish is a useful biomarker to identify estrogenic activity of natural or anthropogenic substances. By using a biosensor to detect vitellogenin in fish, the estrogenicity of surface waters was determined by Kroger et al. [134].

Darain et al. [149] demonstrated a disposable amperometric immunosensor for the rapid detection of carp (*Carassius auratus*) vitellogenin. The sensor was fabricated based on screen-printed carbon arrays (SPCAs) containing eight carbon working electrodes and an integrated carbon counter electrode. The sensor arrays exhibit a linear range of Vtg concentration from 0.25 to 7.8 ng ml^{-1} and the detection limit was determined to be 0.09 ng ml^{-1}. Furthermore, the performance of the immunosensor for the determination of Vtg was evaluated by a standard addition method performed in fish serum samples.

4.4 DNA Biosensors

DNA's structure is very sensitive to the influence of environmental pollutants, such as heavy metals [150, 151], polychlorinated biphenyls [152] and polyaromatic compounds [153]. These substances are characterized by a great affinity to DNA, causing mutagenesis and carcinogenesis. So, it is very attractive to use DNA-based biosensors [154, 155] to perform genotoxicity assays or for rapid testing of pollutants for mutagenic and carcinogenic activity. Genotoxicity sensors should thus be considered alternatives to the classical and more involved genotoxicity tests and will simplify the analysis of large numbers of samples.

Most nucleic acid biosensors are based on the highly specific hybridization of complementary strands of DNA and also RNA molecules. In addition, toxicity tests based on the interaction of toxic compounds with DNA molecules have been developed especially for environmental monitoring and toxicity screening assays. In an earlier review by Mascini et al. [156] DNA biosensors were described including biosensors for toxicity screening, PCB and aflatoxin detection with specific discussion of application to food and environmental samples. Improvement of the immobilization of DNA probes onto transducer surfaces has been reported by Tombelli et al. [157].

Electrochemical biosensors based on immobilized DNA and its parts integrate detection sensitivity with a high specificity of biomolecules. This reduces the consumption of DNA and gives rise to the development of modern methods of analysis of DNA effectors, including toxic ones, in environmental and biological objects [158, 159].

Disposable electrochemical DNA-based biosensors have been used for the determination of low-molecular weight compounds with affinity for nucleic acids and for the detection of the hybridization reaction. The rapid screening of genotoxic compounds using the molecular interaction between surface-linked DNA and the target pollutants or drugs has been applied in different configurations.

The determination of genotoxic compounds was measured by their effect on the oxidation signal of the guanine peak of calf thymus DNA immobilized on the electrode surface and investigated by chrono-potentiometric analysis by Mascini et al. [156]. This type of DNA biosensor is able to detect known intercalating compounds, such as daunomycin, polychlorinated biphenyls, aflatoxin B1 and aromatic amines. Applicability to river and wastewater samples is also demonstrated [160]. Disposable electrochemical sensors for the detection of a specific sequence of DNA were realized by immobilising synthetic single-stranded oligonucleotides onto a graphite screen-printed electrode. The probes became hybridized with different concentrations of complementary sequences present in the sample. The hybrids formed on the electrode surface are evaluated by chrono-potentiometric analysis using daunomycin as indicator of the hybridization reaction.

An interesting article by Lu et al. investigates the use of catalytic RNA and DNA for the detection of metals and their potential application in environmental and pharmaceutical fields [161]. Lu et al. analyzed 27 different metals ranging from Ag to Zn with a DNA-based sensor in an effort to develop broad range sensing elements for environmental analysis [161].

In another example the electrochemical detection of cadmium using surface-immobilized DNA was reported by Wong et al. [162].

Recently, a dsDNA-electrochemical in situ biosensor, employing differential pulse voltammetry, was used for the in situ evaluation of Pb^{2+}, Cd^{2+} and Ni^{2+} interaction with dsDNA. The results confirm that Pb^{2+}, Cd^{2+} and Ni^{2+} bind to dsDNA, and that this interaction leads to different modifications in the dsDNA structure. These modifications were electrochemically recognized as changes in the oxidation peaks of guanosine and adenosine bases [163].

Chiti et al. described a chrono potentiometric DNA biosensor for amine detection. The sensor was able to quantify concentrations of those molecules with two or more aromatic rings in the submicromolar range after only two minutes of accumulation. Wastewater samples were tested and biosensor signals compared favourably with classical genotoxicity tests like the Toxalert 100 based on bioluminescence signals [164, 165].

In a similar approach, Oliveira-Brett et al. utilized a DNA biosensor for the detection of pesticides using ten derivatives of the 1,3,5-triazine herbicides as model analytes. A time-dependent interaction of all herbicides with the immobilized DNA was observed with the electrochemical sensor [166].

A variety of optical DNA and RNA biosensors using surface plasmon resonance have been reported. DNA or RNA aptamers are single-stranded oligonucleotide sequences, which can be produced to bind to various molecular targets such as small molecules, proteins, nucleic acids, and even cells, tissues and organisms [123]. Moreover, the synthesis of aptamers is straightforward and reproducible. However, few examples of application reported the use of SPR biosensors for toxicological measurements such as genotoxicicty.

So et al. [167] recently presented a simultaneous optical and voltammetric detection of bioactivated genotoxicity for the first time employing ultrathin films of DNA, model metabolic enzymes, and electrochemiluminescence (ECL) generating metallopolymer $[Ru(bpy)_2PVP_{10}]^{2+}$ on pyrolytic graphite (PG) electrodes. Cytochrome $P450_{cam}$ and myoglobin were used as model monoxygenase enzymes to mimic in vivo processes. Sensor film growth and component amounts were monitored using a quartz crystal microbalance. Subsequent to the enzyme reaction, DNA damage in the sensor films was measured simultaneously using a simple apparatus combining a standard voltammetry cell coupled with an optical fibre and photomultiplier tube.

Recent improvements in DNA biosensors arise from the use of nanomaterials for the biosensors construction, and several examples have been published during recent years. The highly sensitive and sequence-specific detection of single-stranded oligonucleotides using non-oxidized silicon nanowires (SiNWs) has been demonstrated by Zhang et al. [168]. To maximize device sensitivity, the surface of the SiNWs was functionalized with a densely packed organic monolayer via hydrosilylation, subsequently immobilized with peptide nucleic acid (PNA) capable of recognizing the label-free complementary target DNA. Because of the selective functionalization of the SiNWs, binding competition between the nanowire and the underlying oxide is avoided.

Gold nanoparticle/polyaniline nanotube membranes on the glassy carbon electrode (Au/nanoPAN/GCE) were used by Feng et al. [169] to construct electrochemical sensors. The properties of the Au/nanoPAN/GCE, the characteristics of the immobilization and hybridization of DNA were studied by cyclic voltammetry, differential pulse voltammetry and electrochemical impedance spectroscopy. According to the results of this work the synergistic effect of the two kinds of nanomaterials, nanogold and nanoPAN, could enhance dramatically the sensitivity for the DNA hybridization recognition.

4.5 Whole-Cell Biosensors

Main classes of whole-cell biosensors applied to ecotoxicological studies are:

Bacterial biosensors
Algal biosensors
Yeast biosensors
Fish-cell biosensors

Because of their use as genotoxicity and toxicity sensors, biosensors based on whole cells as the biorecognition element form the largest group among the publications targeting ecotoxicity.

Good general review articles providing an overview about the technology and publications can be found [170].

Genetic engineering of bacteria or yeast cells to bear the *lux* or *luc* gene operon, expressing luminescence proteins such as the green fluorescence protein (GFP), and similar techniques have been investigated for years. Good review articles giving an overview of this technology in recent years can be found, which focus on hydrocarbon pollutants [171], stress [172, 173], or include descriptions of other signal reporter systems such as bacterial luciferase, insect luciferase, β-galactosidase, green fluorescent protein, alkaline phosphatase etc. Depending on the genetic engineering approach and the cell type chosen, biosensors specific for one analyte can be constructed, or those that exhibit general stress response such as toxicity and genotoxicity sensors. A few examples of the different types of biosensors and the different analytes targeted are given in the following discussion.

Fabrication of a whole-cell biosensor requires immobilization of microorganisms on transducers. Since whole-cell biosensor response, operational stability and long-term use are, to some extent, a function of the immobilization strategy used, immobilization technology plays a very important role and the choice of immobilization technique is critical. Microorganisms can be immobilized on transducer or support matrices by chemical or physical methods [174–176].

Chemical methods: Chemical methods of bacteria immobilization include covalent binding and cross-linking [175–177]. Covalent binding methods are based on a covalent bond between functional groups of the microorganism's cell wall components such as amine, carboxylic or sulphydryl and the transducer such as amine, carboxylic, epoxy or tosyl. To achieve this goal, whole cells are exposed to harmful chemicals and harsh reaction conditions, which may damage the cell membrane and decrease the biological activity. For that reason, this method has not been successful for immobilization of viable microbial cells [174].

Cross-linking involves bridging between functional groups on the outer membrane of the cells by multifunctional reagents such as glutaraldehyde and cyanuric chloride, to form a network. This method has found a wide acceptance for immobilization of microorganisms. The cells may be cross-linked directly onto the transducer surface or on a removable support membrane, which can then be placed on the transducer. The ability to replace the membrane with the immobilized cells is an advantage of the latter approach.

While cross-linking has advantages over covalent binding, the cell viability and/or the cell membrane biomolecules can be affected by the cross-linking agents. Thus, cross-linking is suitable for constructing microbial biosensors where cell viability is not important and only the intracellular enzymes are involved in the detection [178].

Physical methods: Adsorption and entrapment are the two widely used physical methods for microbial immobilization, because they produce a relatively small perturbation of microorganism native structure and function [174, 179].

Physical adsorption is the simplest one. Typically, a microbial suspension is incubated with the electrode or an immobilization matrix, such as alumina and glass bead [175, 177, 178, 180, 181], followed by rinsing with buffer to remove unadsorbed cells. The microbes are immobilized due to adsorptive interactions such as ionic, polar or hydrogen bonding and hydrophobic interaction. However, immobilization using adsorption alone generally leads to poor long-term stability because of desorption of microbes.

The immobilization of microorganisms by entrapment can be achieved by either retention of the cells in close proximity of the transducer surface using dialysis or filter membrane or in chemical/biological polymers/gels such as (alginate, carrageenan, agarose, chitosan, collagen, polyacrylamide, polyvinylachohol, poly(ethylene glycol), polyurethane, etc. [174, 182]. A major disadvantage of entrapment immobilization is the additional diffusion resistance offered by the entrapment material, which will result in lower sensitivity and detection limit.

4.5.1 Bacterial Biosensors

Most reported transduction elements in bacterial biosensors are electrochemical and optical.

Amperometric whole-cell biosensors: The most relevant application is the determination of the biochemical oxygen demand (BOD) [183–185]. However, bacterial biosensors using amperometry as the transduction method have also been relevant in other fields such as environmental toxicity assessment.

Neurotoxic organophosphate compounds have found wide application as pesticides and insecticides in agriculture and as chemical warfare agents in military practice [186].

Amperometric biosensors based on genetically engineered *Moraxella* sp. and *Pseudomonas putida* with surface-expressed organophosphorus hydrolase (OPH) have been developed by Lei at al. [186, 187] for the detection of organophosphorus pesticides. The biosensor consisted of recombinant PNP-degrading/oxidizing bacteria *P. putida* JS444 anchoring and displaying organophosphorus hydrolase on its cell surface as the biological sensing element and a dissolved oxygen electrode as the transducer. Surface-expressed OPH catalyzed the hydrolysis of fenitrothion and EPN to release 3-methyl-4-nitrophenol and p-nitrophenol, respectively, which were oxidized by the enzymatic machinery of *P. putida* JS444 to carbon dioxide while consuming oxygen, which was measured and correlated to the concentration of organophosphates. Under the optimum operating conditions, the biosensor was able to measure as low as 277 ppb of fenitrothion.

Recently, a novel single-walled carbon nanotube (SWNT) based biosensor for real-time detection of organophosphate has been developed by Liu et al. [188]. Horizontally aligned SWNTs were assembled to desirable electrodes using AC dielectrophoresis technique. Organophosphorus hydrolase immobilized on the SWNTs by non-specific binding triggers enzymatic hydrolysis of organophosphates, such as paraoxon, consequently causing a detectable change in the conductance of the SWNTs. The conductance change is found to be correlated to the concentration of organophosphate.

Amperometric bacterial biosensors based on monitoring of cell respiration have been used for surfactants analysis. Some examples are the use of surfactant-degrading bacteria [189, 190], or *Acetobacter peroxydans* [191].

The characterization of wastewater toxicity by means of a whole-cell bacterial biosensor using screen-printed electrodes with *E. coli* was reported by Farré et al. [93]. In another study, the same group determined the acute toxicity of wastewaters using *P. putida* electrodes, in conjunction with chemical analysis [192].

Potentiometric bacterial biosensors: Conventional potentiometric microbial biosensors consist of an ion-selective electrode (pH, ammonium, chloride and so on) or a gas-sensing electrode (*p*CO2 and *p*NH3) coated with an immobilized microbe layer. Microbe consuming analyte generates a change in potential resulting from ion accumulation or depletion. Potentiometric transducers measure the difference between a working electrode and a reference electrode, and the signal is correlated to the concentration of analyte. Because of a logarithmic relationship between the potential generated and analyte concentration, a wide detection range is possible.

However, this method requires a very stable reference electrode, which may be a limitation of these transducers. A few examples of biosensors based on potentiometric transducers have been developed for toxicological purposes in environmental research. This type of biosensor has been applied for neurotoxicity evaluation. Several microbial biosensors based on modification of a glass pH electrode with genetically engineered *E. coli* expressing organophosphorus hydrolase intracellularly and on the outer surface of cells and wild-type organophosphorus pesticides degrading bacteria *Flavobacteium* sp. have been reported [193–195].

Conductimetric biosensors: Many microbe-catalyzed reactions involve a change in ionic species. Associated with this change is a net change in the conductivity of the reaction solution. Even though the detection of solution conductance is non-specific, conductance measurements are extremely sensitive. Nevertheless, few applications have been developed. An example is the single-use conductivity and microbial sensor developed by Bhatia et al. [196] to investigate the metabolic activity of *E. coli*. This sensing system combines physico-chemical and biological sensing and greatly increases the ease with which comparative data could be assimilated.

The modulation in optical properties such as UV-VIS absorption, bio- and chemi-luminescence, reflectance and fluorescence brought by the interaction of the biocatalyst with the target analyte is the basis for optical microbial biosensors. Optical-based biosensors offer advantages of compactness, flexibility, resistance to electrical noise, and a small probe size.

Bioluminescent bacteria systems: Most of the systems developed for ecotoxicity assessment using whole-cell biosensors and optical transduction have been based on the use of luminescence (natural or transgenic microorganisms), because bacterial luciferase immobilized systems can offer a general tool, and when the luminescent reagents and enzymes, or the whole cells, are immobilized on solid supports their peculiarities change, usually in a positive way.

Several solid supports and procedures are available to immobilize macromolecules or whole cells. The chemical methods give better yields, in terms of active luciferase, than the physical ones. Nevertheless, chemical procedures involve covalent coupling, and usually lead to some protein inactivation. Agarose, collagen, epoxy methacrylate and nylon have proved to be the most effective among the different solid supports that have been investigated. The gel entrapment technique has the advantage of better protein stability and enables the co-immobilization of luciferase and other enzymes with their substrates through an easy process. An important, general, feature of these immobilized enzymes is the possibility to incorporate them into flow cells, used for multiple assays, to be recycled and reused in automated devices. The continuous flow format offers greater possibilities than a single-batch system, and it leads to rapid and sensitive assays. Flow assays are characterized by extremely accelerated kinetics: a very high surface-area-to-volume ratio is obtained and the reactions do not have to rely on passive diffusion to bring reagents together. Many analytes can be detected at pmol levels, with good precision and a wide range of linearity [197]. The immobilization of biological components on a solid support is quite regularly the preliminary step to the creation of a biosensor.

In fact, biosensors are generally described as probe-type devices made up of a selective biological layer, with very sharp molecular recognition capacity, and of a physicochemical transducer, often an optical system. Among these transducers, the (bio- or chemi-) luminescent ones have the advantages that they do not require light sources and monochromators [198].

Any kind of luminescent bacteria can undergo immobilization, including the genetically modified ones. A systematic review of the methods to immobilize the different luminescent enzymes and microorganisms (natural and genetically engineered) for use in biosensors is that of Kratasyuk and Esimbekova [197].

An immobilized film of *Photobacterium phosphoreum* has been used to prepare a sensor based on the determination of acute toxicity effects of molecules that are difficult or even impossible to be measured by traditional analytical chemistry methods [199].

Recently, a biodetector was designed based on a bioluminescence test with the bacterium *V. fischeri* performed in a liquid continuous flow-through system. In this system a new flow-cell holder and a new case including a "top cover" to connect the flow cell with the waste and the incubation capillary in a light-proof manner are described [200].

Genotoxic agents, such as mitomycin C, have also been studied by a stable dark variant of *P. phosphoreum* (A2), fixed in agar-gel membranes immobilized onto the exposed end of optic fibres, linked with a bioluminometer [201]. A bio-MEMS based cell-chip can be fabricated by immobilizing bioluminescent bacteria in a microfluidic chip. The pattern of the recombinant *E. coli* strain GC2 was successfully generated by photolithography, utilizing the PVA-SbQ (polyvinyl alkyl-styrylpyridinium) polymer, as the immobilization material [202]. A hydrogen peroxide detection system was obtained simply immobilizing bioluminescent bacteria, the DK1 strain, which increased bioluminescence in the presence of oxidative damage in the cells [203]. *Pseudomonas fluorescens* HK44, which emits light when in contact with naphthalene and its metabolites, has been immobilized into a silica matrix, on glass slides, by the sol–gel technique [204]. The test slides could be used for multiple determinations, since the bacteria responded to the inductor at least 8 months after immobilization, and to more than 50 induction cycles.

A series of two-stage bioreactor systems, connected by a fibre optic probe to a luminometer, have been assembled to set up a multi-channel system for continuous monitoring and classification of toxicants. Each channel was used for cultivating different recombinant bacterial strains: TV1061 (*grpE::luxCDABE*), DPD2794 (*recA::luxCDABE*) and DPD2540 (*fabA::luxCDABE*), which are induced by protein-, DNA- and cell-membrane damaging agents, respectively. GC2(*lac::luxCDABE*) is a bacterium expressing bioluminescence constitutively [205]. Each channel showed a specific bioluminescent response according to the chemicals contained in wastewater samples, while GC2 showed a general response to cellular toxicity [205]. A portable format of the previously described biosensor consists of three parts, a freeze-dried biosensing strain within a vial, a small lightproof test chamber, and an optic-fibre connecting the sample chamber to a luminometer. It can be used for field sample analyses and monitoring on-site of various water systems. A different arrangement of genetically engineered luminescent organisms has been used in a bacterial cell chip, creating the possibility to emit light of different colours from each well, by using quantum dots (QDs). Quantum dots have several advantages such as broad absorption spectrum, narrow emission spectra and stability [206].

Each well gives off the specific signal (light) according to the recombinant strain exciting the quantum dots. A multi-strain bacterial cell array chip has been developed also by Lee et al., as a revealing test for the presence of reactive oxygen species (superoxide radical and hydrogen peroxide), generated by different chemicals in the sample [207]. The previously described multi-channel system has been successfully implemented in the form of computer-based data acquisition. The bioluminescent signatures are delivered from four channels by switching one at once, while the data are automatically logged to a personal computer [208]. Improvements of the system have been the manipulation of the dilution rate and the use of *thermo-lux* fusion strains. The system is now being implemented in a drinking water reservoir and river for remote sensing as an early warning system [208].

The use of quantum dots opens a new window in biosensor development. Another example is an electrogenerated precursor that has been developed for green synthesis of highly luminescent aqueous CdTe quantum dots with unique quantum yield and strong electrogenerated luminescence, which can access cellular targets via specific binding and have potential application as biolabels in highly sensitive biosensing and cell imaging [209].

The papers reported above and additional papers concerning biosensors based on luminescent and fluorescent bacteria are listed in Table 3.

Table 3 Bioluminescence and fluorescence bacterial biosensors

Target	Bacteria	Natural o induced luminescence	Transducer	References
Acute toxicity	*Photobacterium phosphoreum*	Natural	Bioluminescence	[202]
Acute toxicity	*Vibrio fischeri*	Natural	Bioluminescence	[210]
Genotoxicity	*Photobacterium phosphoreum*	Dark variant	Bioluminescence	[204]
Genotoxicity	*Escherechia coli*	GC2	Luminescence	[205]
Acute toxicity	*Pseudomonas fluorescens*	10568	Fluorescence	[211]
Acute toxicity	*Vibrio fischeri*	Natural	Bioluminescence	[211]
Acute toxicity	*Escherichia coli*	HB101	Luminescence	[211]
Acute toxicity	*Pseudomonas putida*	TVA8	Luminescence	[211]
Acute toxicity	*Escherichia coli*	DH5α	Luminescence	[211]
Toxicity and genotoxicity	*Escherichia coli*	DPD2511, DPD2540, DPD2794 and TV1061	Luminescence	[212]
Genotoxicity	*Pseudomonas fluorescens*	HK44	Luminescence	[207]
Acute toxicity	*Photobacterium phosphoreum*	Natural	Bioluminescence	[213]
Genotoxicity	*Escherichia coli*		Luminescence	[214]
Genotoxicity	*Escherichia coli*	recA′::lux V. fischeri	Luminescence	[215]
Genotoxicity	*Escherichia coli*	DPD1718 recA′::lux	Luminescence	[216]
Genotoxicity	*Salmonella typhmurium TA1535*	TL210 and TL210ctl	Luminescence	[217]

Different examples of the application of biosensors based on bioluminescent bacteria for wastewater monitoring have been reported. The construction of whole cell, genetically modified, bioluminescent biosensors and their immobilization on thin films of poly(vinyl alcohol) cryogels has been carried out by Philp et al. [218]. The biosensor was designed for use in monitoring the toxicity of industrial wastewaters containing phenolic materials. It has been proved that they operate predictably with pure toxicants, within the wastewater treatment plant [218]. Several recombinant bioluminescent bacteria have been employed to set up a multi-channel, continuous, water toxicity monitoring system. That system developed by the research group of Kim was based on channels, each one hosting a different recombinant bacterial strain, and was composed of two mini-bioreactors, to enable a continuous operation, that is without system interruption due to highly toxic samples [219]. The luminescent strains were: DPD2540 (*fabA::luxCDABE*), DPD2794 (*recA::luxCDABE*) and TV1061 (*grpE::luxCDABE*), induced by cell membrane-, DNA- and protein-damaging agents, respectively; GC2 (lac::luxCDABE) was a constitutive strain. Field samples were waters discharged from a nuclear power plant and a thermoelectric power plant. Each channel showed specific luminescent response profiles and by comparing the luminescent bacteria signals of the standard chemicals with those of discharged water samples, the equivalent toxicity of the field water could be estimated [219].

A procedure developed for rapid toxicity finger printing of polluted waters involves a single constitutive *lux* bacterial biosensor, thanks to the fact that the different toxicants elicit highly characteristic light response curves in the biosensor. The technique can be applied to any pollutant or pollutant mixture that elicits a toxic response in the biosensor.

Two bioluminescent bioassays, one based on lyophilized marine luminous bacteria, the *Microbiosensor*-B17-677F, and the other on a genetically modified luminous strain of *E. coli*, the *Microbiosensor-ECK*, have been employed to reveal areas of impaired water quality in the river and sewage waters of different regions of Siberia, showing the same dependence on the concentration of the toxicants. Nevertheless, the sensitivity to phenol compounds of the *Microbiosensor-ECK* was higher, and corresponded to that determined on intact cells of *P. phosphoreum* and of various hydrobionts.

Aqueous extracts of soils have been tested by the *lux*-marked bacterial biosensor *E. coli* HB101, to reveal the presence and toxicity of four commonly used herbicides (atrazine, diuron, mecoprop and paraquat) [220]. Toxic responses, for all four herbicides, were stronger in the extracts than in the corresponding spiked water samples, suggesting that intrinsic soil factors may be altering the bio available fraction of herbicides, making them more toxic than equivalent concentrations in water.

Recently, two toluene bacterial biosensors have been reported. The toluene bacterial biosensors comprised of two reporter genes, gfp and luxCDABE, characterized by green fluorescence and luminescence, respectively, and their abilities were compared to detect bioavailable toluene and related compounds. The bacterial luminescence biosensor allowed faster and more sensitive detection of toluene; the fluorescence biosensor strain was much more stable and thus more applicable for long-term exposure. Both biosensors were field-tested to measure the relative bioavailability of BTEX in contaminated groundwater and soil samples [221].

4.5.2 Algal Toxicity Biosensors

Fluorescence transduction has been widely applied in ecotoxicity biosensors using immobilized microalgae. Bozeman et al. [222], in a pioneering work, compared the toxicity of seven pollutants of different origin (Cd, Cu, Glyphosate, Hydrothol, Paraquat, pentachlorophenol and sodium dodecyl sulphate) to free and immobilized cells of the green microalga *Selenastrum capricornutum*, suggesting the possibility of the use of immobilized systems in in situ toxicity experiments. Following those authors, differences in toxicity for free and immobilized algae varied from no significant differences for copper and pentachlorophenol to nearly four times more sensitive for free cells in the case of Glyphosate or Paraquat. Admiraal et al. [223] performed an experiment on sand and natural glass-attached microbentic assemblages of algae and bacteria in a metal polluted stream in the river Dommel system (Belgium). The authors compared the sensitivity of those assemblages to zinc, finding different sensitivities as a function of the origin of the assemblages (the most polluted origin, the lower sensitivity).

Protection against toxicity in immobilized cells is reported in different works [224]. Awasthi and Rai [225] demonstrated lower inhibition of nitrate uptake in *Scenedesmus quadricauda* immobilized (with respect to free cells) when exposed to Ni, Zn or Cd. In this work, no metal measurements were performed in the media, and the easiest explanation is the removal of part of the metals by the entrapping matrix, being thus less available for cells. But removal of toxicants by immobilizing matrixes would not explain all cases of less toxicity of immobilized cells. Surfactants are not so selectively adsorbed by Ca-alginates. Moreno-Garrido et al. [226] found less toxicity for immobilized cells of *Phaeodactylumtricornutum* exposed to sediments spiked with surfactant lineal alkylbenzene sulphonate (LAS) than for free cells.

Immobilization by gel entrapment is also a very interesting topic for in situ microalgal experiment design because it provides protection to microalgae fromgrazers [227]. Microalgal grazers cannot be easily eliminated from waters or sediments due to the small size that those organisms can achieve: nematodes and, over all, amoebas or ciliates can predate on microalgal cells slightly smaller than them.

Twist et al. [227] developed a method of in situ biomonitoring using Ca-alginate immobilized *Scenedesmus subspicatus* for the assessment of eutrophication in surface waters.

A clear advantage of this technique is that local flora can be isolated and incorporated into the biomonitor. Of course, when Ca-alginate is used in a natural environment, a limitation of the technique is degradation of the beads. In freshwater streams the timescale for bead degradation seems to be a couple of weeks. In marine environments the timescale is quite a lot shorter (a few days) [228]. Improvements in bead fabrication are being made in an effort to avoid these problems [229].

Frense et al. [230] used *S. subspicatus* immobilized on filter paper and covered with alginate in an optical biosensor based on the chlorophyll a fluorescence as the biomarker for pollutants in water and soil extracts, as a pre-selection protocol before sending samples to high-cost standard analysis. *C. vulgaris* has also been used in optic biosensors in order to determine the toxicity of herbicides [231]. Algal immobilization was performed on GF/C Whatman filters. Filter paper disks immobilizing algae have also been used by Sanders et al. [232] in biosensors designed to detect chemical warfare agents.

Electrochemical transduction has also been applied in some examples. An amperometric algal-based sensor has been designed by Shitanda et al. [233], taking advantage of variations in the photosynthetically produced oxygen.

Another example is the development of novel conductimetric biosensors based on immobilization of the microalgae *C. vulgaris* by Chouteau et al. [234]. In this study, the detection of the local conductivity variations caused by algae enzymatic reactions could be achieved. The inhibition of *C. vulgaris* alkaline phosphatase activities in the presence of cadmium ions was measured. These results were compared with measurements in bioassays. It finally appeared that conductometric biosensors using algae seemed more sensitive than bioassays to detect low levels of cadmium ions (the detection limit for the first experiment was 1 ppb of Cd^{2+}).

Recent efforts have been made in the field of co-immobilization [233]. De-Bashan et al. [235] co-immobilized *Chlorella* with a microalgae growth-promoting bacterium (*Azospirillum brasiliense*) in Ca-alginate beads. This bacterium is not able to remove nutrients from wastewaters, but enhances growth of immobilized algae.

Limitations of toxicity testing using free or immobilizing microalgae are restricted to those toxicants that affect structures present in the algal cells. But these types of biosensor are highly sensitive to toxicants affecting photosynthesis (such as copper ions or herbicides).

4.5.3 Yeast-Based Biosensors

Most whole-cell sensors have used bacterial cells as the sensing element. Fungal cells, however, can provide all of the advantages bacterial cells offer but in addition they can provide information that is more relevant to other eukaryote organisms. These cells are easy to cultivate and manipulate for sensor configurations and are amenable to a wide range of transducer methodologies. An overview of the use of yeast and filamentous fungi as the sensing element of some biosensors is presented in this section.

There are several reports on the use of wild-type fungal cells and yeast to detect toxicity. As explained above the use of eukaryote cells as the detection element is preferable because their response to the toxin more accurately predicts the response in plant and animal cells than does the prokaryote response. The naturally bioluminescent fungi *Armillaria mellea* and *Mycena citricolor* were used as the sensing element in toxicity bioassays [236], and the basis for the development of toxicity biosensors. The bioluminescence inhibition during 60 min exposure to the toxin was monitored to provide EC_{50} values.

Immobilized *Sacharomices cerevisiae* was used to assess the toxicity of four free, and three conjugated cholanic acids and a toxicity scale was constructed from the results [237]. The authors concluded that this sensor could therefore be considered a valid instrument for the preliminary evaluation of the toxicity of organic compounds or drugs. Campanella et al. also used *S. cerevisiae* as the sensing element of four general toxicity biosensors. Toxicity was determined by detecting the retardation of metabolic activity.

Many of the current yeast sensors are for genotoxicity and are based on genetically modified *S. cerevisiae*. Other genetically modified yeast sensors have been constructed to detect specific molecules or groups of molecules. Some modified cell sensors function as model eukaryote cells to detect toxicity and in particular genotoxicity, either of single specific molecules or of mixtures of molecules such as in environmental samples. While their use in the laboratory is in general not problematic, the use of genetically modified sensors outside the laboratory, for example for on-site environmental monitoring, may be restricted by biosafety requirements in some jurisdictions.

The green fluorescent protein reporter gene (GFP) has been incorporated into a plasmid, yEGFP, to optimize its expression in *S. cerevisiae*. This reporter gene was fused to the promoter region of the yeast gene, RAD54, to create a yeast genotoxicity biosensor. The reporter responds to the genetic regulation of DNA repair by RAD54 but does not respond to chemicals that delay mitosis and thus is specific for DNA damaging molecules [238].

Further work incorporated a human cytochrome P450 gene into the modified *S. cerevisiae* cells. Exposure of these cells to genotoxic molecules causes in vivo metabolic activation of promutagens, replicating the human response to these agents more accurately than the Salmonella-based Ames test [239].

The construction and characterization of dual stress-responsive bacterial biosensors was reported by Mitchel et al. [240] Using the genes for the green fluorescence protein and *Xenorhabdus luminescens* luciferase operon and the promoters for the recA and katG genes, two stress-responsive *E. coli* biosensor strains have been constructed that can individually or concurrently respond to oxidative and genotoxic conditions.

A yeast-based promoter-reporter biosensor was constructed by Benton et al. [241] for the detection of genotoxic compounds within a cell's local environment. In this work a fusion containing the HUG1 promoter and GFP was performed and it was incorporated into the yeast genome creating a stable, sensitive genotoxicity indicator.

Firefly luciferase has been cloned into *S. cerevisiae* to create a bioluminescent yeast strain. The presence of any toxic chemical that interferes with the cell's metabolism results in a quantitative decrease in bioluminescence. Hollis et al. demonstrated that the luminescent yeast strain senses chemicals known to be toxic to eukaryotes in samples assessed as non-toxic by prokaryotic biosensors [242]. The authors noted that this biosensor complements the GFP biosensor designed by

Walmsley et al. in that it responds to a wide range of toxins, not just genotoxic agents.

Terziyska et al. [243] used a *S. cerevisiae* mutant to increase the sensitivity of this organism to mutagens and carcinogens present in environmental samples. The cells carry a mutation which increases the permeability of the outer mannoprotein layer of the cell wall. The responses of these cells to environmental samples were compared to the Ames test (*Salmonella typhimurium*) and the D7 test of Zimmerman (*S. cerevisiae*). The authors conclude that their test (D7ts1) detects mutagenic/carcinogenic activity undetectable by the D7 test. Furthermore, all samples positive in the Ames test were positive in the D7ts1 test and some samples negative in the Ames test were D7ts1 positive.

Other genetically modified yeast have been constructed to detect a specific molecule or group of molecules. A strain of *S. cerevisiae* that lacked the gene for dihydrofolate reductase (DHFR) was complemented with mouse DHFR containing a ligand-binding domain inserted in a flexible loop. This construct was used to identify mutations in the ligand-binding region. Additionally, the cells could discriminate between analogues of the ligand, in this case oestrogen, and was used as a screening protocol to identify the various analogues [244].

Sakaki et al. [245] constructed a modified strain of *S. cerevisiae* by cloning rat CYP1A1 and CYP1A2 genes to create recombinant yeast cells capable of metabolizing polychlorinated dibenzo-*p*-dioxins (PCDD's).

Lehmann et al. [246] have constructed sensors for copper by transforming *S. cerevisiae* with a plasmid containing the Cu2+-inducible promoter of the CUP1 gene of *S. cerevisiae* fused to the lacZ gene of *E. coli*. The constructed lac+ve yeast responded quantitatively to the presence of copper. The response to copper was detected by monitoring the catabolism of lactose.

Webb et al. [247] transformed *A. pullulans* with a vector containing the GFP reporter gene. The level of fluorescence was directly related to the number of live cells. Exposure of the organism to biocides such as sodium hypochlorite and 2-*n*-octylisothiozolin-3-one (OIT) caused a rapid loss of fluorescence that was highly correlated with a decrease in the number of viable cells.

A yeast-based cytotoxicity and genotoxicity assay for environmental monitoring using novel portable instrumentation was presented by Knight et al. [248]. This assay uses eukaryotic (yeast) cells, genetically modified to express a green fluorescent protein whenever DNA damage, as a result of exposure to genotoxic agents, is repaired. A measure of the reduction in cell proliferation is used to characterize general toxicity producing lowest effect concentration (LOEC) and EC_{50} (50% effective concentration) data. The results of sensitivity to a wide range of substances and effluents suggested that the assay can be useful for environmental toxicity monitoring.

Cell suspension, membrane entrapment and cell immobilization are all used in the construction of yeast biosensors. The choice of cell location is dependant on the biosensor format. For example, where the cell response to the analyte is slow, incubation of the cells with the substrate could be separated from the transducer whereas if the cell response is rapid, cells may be situated on the transducer surface. Cells situated on the surface of the transducer may be immobilized on a membrane that is attached to the surface, immobilized in a double membrane sandwich that is attached to the transducer surface or immobilized in a polymer directly on the surface. Membranes used to retain cells at the transducer surface include dialysis membranes, membrane filters of various materials such as cellulose acetate, nitrocellulose, polycarbonate and Teflon. The pore size can be relatively large given the size of yeast cells and fungal filaments.

Various polymers have been used to entrap yeast either directly on the transducer surface, on the surface of a membrane or to form structurally robust beads, discs, etc.

4.5.4 Fish-Cell Biosensors

Cell cultures, in particular those derived from fish, have been successfully employed as a biological alternative to the use of whole animals in ecotoxicity studies. The use of fish cell lines in conventional bioassays such as neutral red retention assays is however labour intensive, lengthy and costly. To date several transduction elements have been explored, however, most biosensor schemes are based on optical transduction. While vertebrate cell lines in biosensor configurations can be monitored electrochemically [249] stability is poor, compared with microbial cells. This may reflect stressor damage to the cells during immobilization, or adverse sensitivity to the mediator regimen, or both. Therefore, an alternative approach to monitor cellular activity using luminescence reported genes is being developed. Different works report biosensors using this strategy. Fibroblastic cells of blue gill sunfish (BF-2) have been transfected with a plasmid containing the luciferase gene Luc, which allows the luminescence of the transgenic cells following environmental challenge to be monitored via an optical transducer [250, 251].

Fish chromatophores have been shown to be promising biosensors for the detection of hostile agents in the environment [252, 253].

5 Current Trends and Future Perspectives

Biosensors and especially bio-effect related sensors like whole-cell sensors and receptor sensors constitute a very promising field of research that has been transformed over the past two decades through advancements such as new material technologies, novel transduction approaches, and powerful computer software to control devices. However, in order to exploit the principal advantage of biosensors as early warning systems for ecotoxicological purposes, standardization and harmonization are needed.

Current research activities to improve and to obtain more reliable devices are focused on continuous monitoring systems and remote controlled devices, to obtain more stable bio-receptors, more robust transduction elements, and miniaturized devices.

Some systems have been developed for use as continuous monitoring systems that can provide easy, rapid and on-site measurements. They may also be useful for mapping of contamination when it is important to obtain rapid results in the field, such as after accidental spills or pollution events. Gu et al. [254] have recently applied a novel early-warning protocol for monitoring the toxicity of the effluents of a water treatment plant.

The emergence of large-scale widely deployed chemo/biosensing networks for environmental monitoring will grow from the top-down, with initial activity focusing on relatively sophisticated platforms deployed at relatively large distances (km). Clusters of these devices may be located at particular pollution hotspots, but the realization of a densely deployed heterogeneous sensor network requires significant effort to understand the level of information content that can be acquired from relatively dumb sensors used in a collaborative manner. On the other hand, developments in wearable sensors will occur much more quickly, as issues like power maintenance, length of deployment, communication distance and availability of a reliable platform infrastructure are more solvable [255, 256].

Nanotechnology is also playing a relevant role in the development of biosensors. Different studies reported the improvement of sensitivities and robustness in final biosensors developed using these new materials [257–259, 260].

Self-assembly of biomaterials, such as proteins, lipids or nucleic acids, has inspired the development of new biosensors, and molecular self assembly has been proposed for the synthesis of nanostructures capable of performing unique functions.

Several research groups have begun to explore alternative strategies for the development of optical SPR biosensors based on the extraordinary optical properties of metal nanoparticles.

Endo et al. [261] reported a novel label-free cell-based assay that is based on a localized surface plasmon resonance (LSPR) biosensor, which is excited using a core-shell structured nanoparticle layer substrate. The detection of the cell metabolites was achieved in this work using the isolated cells from mouse thymus. For detection of the cellular metabolites, the refractive index change by the specific interaction between the antigen and antibody was detected on the antibody immobilized LSPR-based biosensor.

Cao et al. [262] showed how oligonucleotides and presumably other biomolecules can be used to modify the surfaces of such particles, thereby imparting useful biorecognition properties to them. In another example, Banerjee et al. [263] presented a novel and simple cell-based detection system with a collagen-encapsulated B-lymphocyte cell line as a biosensor for rapid detection of pathogens and toxins. This approach is based on a cell-based sensing system using collagen-encapsulated mammalian cells for rapid detection of pathogenic bacteria or toxins, and demonstrates a potential for on-site use as a portable detection system.

Souza et al. [264] recently reported a novel approach for fabrication of spontaneous, biologically active molecular networks consisting of bacteriophage (phage) directly assembled with gold (Au) nanoparticles. The spontaneous organization of these targeted networks can be manipulated further by incorporation of imidazole, which induces changes in fractal structure and near-infrared optical properties. The networks can be used as labels for enhanced fluorescence and dark-field microscopy, surface-enhanced Raman scattering detection, and near-infrared photon-to-heat conversion. Together, the physical and biological features within these targeted networks offer convenient multifunctional integration within a single entity with potential for nanotechnology-based biomedical and environmental applications.

Newly developed methods for spatially resolved immobilization of biomolecular recognition elements in transducer surfaces are playing an important role in the development of multi-analyte devices.

In addition, the advances in microelectronics and microfluidics allowed the miniaturization of analytical systems, reducing sample volumes and reagent consumption. Biosensor miniaturization has a particular significance in in vivo monitoring [265–268].

The current generation of biosensors is taking advantage of novel biosensing materials based on gene engineering. Gene engineering focuses on two main fields in the area of biosensors: genetically transformed cells and genetically engineered receptor molecules.

Modified cell biosensors are being constructed by fusing a reporter gene to a promoter element that is induced by the presence of a target compound. In the case of bioluminescence sensors, the recombinant bacteria employed contain bacterial luminescence genes and gene fusions between the regulatory regions of the mer operon (merR) that respond to contaminants. Many biosensors based on this principle have been already discussed.

6 Conclusions

Biosensors have great potential as smart, cost-effective devices for on-time and on-site determinations, especially at the screening level.

Biological approaches applied in conjunction with chemical analysis give more complete information about the possible biological impact of complex, polluted effluents.

Nevertheless, validation of new biosensors is an important issue for their progressive integration in pollution-control programs, and biological detector systems, such as biosensors and related techniques, must overcome a number of obstacles, such as the limited life times for some biorecognition components, and the lack of validation and correlation studies.

In addition, new environmental issues call for on-time analysis, miniaturized equipment and more robust methods. Many technical obstacles have been overcome thanks to the progress in optics, microelectronics, and nanotechnology that should permit new improvements in detection systems, such as the miniaturization of diode lasers and a decrease in component costs.

However, stabilization of the properties of biorecognition elements continues to be a crucial aspect in future development of biological techniques.

Furthermore, the development of genetic engineering offers the possibility of obtaining genetically engineered receptor molecules and genetically transformed cells.

References

1. Carson R (2008) Silent spring. (First edition 1962)
2. Farre M, Barcelo D (2003) Toxicity testing of wastewater and sewage sludge by biosensors, bioassays and chemical analysis. Trends Anal Chem 22:299–310
3. Munkittrick KR, McMaster ME, McCarthy LH, Servos MR, Van Der Kraak GJ (1998) An overview of recent studies on the potential of pulp-mill effluents to alter reproductive parameters in fish. J Toxicol Environ Health B Crit Rev 1:347–371
4. Solomon KR, Baker DB, Richards RP, Dixon KR, Klaine SJ, La Point TW, Kendall RJ, Weisskopf CP, Giddings JM, Giesy JP, Hall J, Williams WM (1996) Ecological risk assessment of atrazine in North American surface waters. Environ Toxicol Chem 15:31–76
5. Hansen PD (2008) Biosensors and ecotoxicology. Eng Life Sci 8:26–31
6. Tothill IE, Turner APF (1996) Developments in bioassay methods for toxicity testing in water treatment. Trends Anal Chem 15:178–188
7. Landrigan PJ (2001) Pesticides and polychlorinated biphenyls (PCBs): an analysis of the evidence that they impair children's neurobehavioral development. Mol Genet Metab 73:11–17
8. Directive 2000/60/EC of the European Parliament and of the Council of 23 October 2000 establishing a framework for Community action in the field of water policy (2000) Offic J Eur Commun L327:73
9. Brack W (2003) Effect-directed analysis: a promising tool for the identification of organic toxicants in complex mixtures? Anal Bioanal Chem 377:397–407
10. Brack W, Schmitt-Jansen M, MacHala M, Brix R, Barcelo D, Schymanski E, Streck G, Schulze T (2008) How to confirm identified toxicants in effect-directed analysis. Anal Bioanal Chem 390:1959–1973
11. Burgess RM (2000) Characterizing and identifying toxicants in marine waters: a review of marine toxicity identification evaluations (TIEs). Int J Environ Pollut 13:2–33
12. Hewitt LM, Marvin CH (2005) Analytical methods in environmental effects-directed investigations of effluents. Mutat Res Rev Mutat Res 589:208–232
13. Schuetzle D, Lewtas J (1986) Anal Chem 58
14. Rodriguez-Mozaz S, Lopez De Alda MJ, Barcelo D (2006) Biosensors as useful tools for environmental analysis and monitoring. Anal Bioanal Chem 386:1025–1041
15. ISO (2008) Water quality: determination of the inhibition of the mobility of Daphnia magna Straus (Cladocera, Crustacea), ISO 6341. International Standardization Organization, Geneva, Switzerland, 1996
16. ISO (2008) Water quality: marine algal growth inhibition tests with Skeletonema costatum and phaeodactylum tricornutum, ISO 10253. International Standardization Organization, Geneva, 1995
17. EPA (2008) The Selenastrum capricornutum Prinzt algal assay bottle test Experimental design, application and data interpretation protocol, EPA-600/9-78-018, Environmental Protection Agency, Corvallis, Oregon, USA, 1978. Second US/USSR symposium: biological aspects of pollutants effects on marine organisms, EPA-600/3-82-034. Environmental Protection Agency, Corvallis, Oregon, USA, 1982, pp 112–122
18. ISO 2008 Water quality: determination of the inhibitory effect of water samples on the light emission of Vibrio fischeri (Luminescent bacteria test), ISO 11348-1, 2 and 3, International Standardization Organization, Geneva, Switzerland, 1998
19. Biesinger KE, Christensen GM (1972) J Fish Res Board Can 29:1691–1700
20. Kimerle RA (1986) Managing uncertainty and margins of safety for aquatic organisms. Environmental Science and Technology 5:113–115
21. Versteeg DJ, Stalmans M, Dyer SD, Janssen C (1997) Ceriodaphnia and daphnia: A comparison of their sensitivity to xenobiotics and utility as a test species. Chemosphere 34:869–892
22. OECD Environmental Health and Safety Publication. Series on Testing and Assessment No. 6 Report of the Final Ring Test of the *Daphnia magna* Reproduction Test. Paris 1997
23. Toussaint MW, Shedd TR, Van der Schalie WH, Leather GR (1995) A comparison of standard acute toxicity tests with rapid-screening toxicity tests. Environ Toxicol Chem 14:907–915

24. Barata C, Baird DJ, Nogueira AJA, Soares AMVM, Riva MC (2006) Toxicity of binary mixtures of metals and pyrethroid insecticides to *Daphnia magna* Straus. Implications for multi-substance risks assessment. Aquat Toxicol 78:1–14
25. Duquesne S (2006) Effects of an organophosphate on *Daphnia magna* at suborganismal and organismal levels: Implications for population dynamics. Ecotoxicol Environ Saf 65:145–150
26. Ikenaka Y, Eun H, Ishizaka M, Miyabara Y (2006) Metabolism of pyrene by aquatic crustacean, *Daphnia magna*. Aquat Toxicol 80:158–165
27. Hodges G, Roberts DW, Marshall SJ, Dearden JC (2006a) Defining the toxic mode of action of ester sulphonates using the joint toxicity of mixtures. Chemosphere 64:17–25
28. Hodges G, Roberts DW, Marshall SJ, Dearden JC (2006b) The aquatic toxicity of anionic surfactants to *Daphnia magna*-A comparative QSAR study of linear alkylbenzene sulphonates and ester sulphonates. Chemosphere 63:1443–1450
29. DeLorenzo ME, Serrano L (2006) Mixture toxicity of the antifouling compound irgarol to the marine phytoplankton species *Dunaliella tertiolecta*. J Environ Sci Health B Pesticides Food Contamin Agric Wastes 41:1349–1360
30. Gatidou G, Thomaidis NS (2007) Evaluation of single and joint toxic effects of two antifouling biocides, their main metabolites and copper using phytoplankton bioassays. Aquat Toxicol 85:184–191
31. Lim CY, Yoo YH, Sidharthan M, Ma CW, Bang IC, Kim JM, Lee KS, Park NS, Shin HW (2006) Effects of copper (I) oxide on growth and biochemical compositions of two marine microalgae. J Environ Biol 27:461–466
32. Hernando MD, Fernandez-Alba AR, Tauler R, and Barcelo D (2005) Toxicity assays applied to wastewater treatment. Talanta 65:358–366
33. W.F.Serat, FEBPKM (1965) J Bacteriol 832–833
34. BulichAA (2008a) Use of luminescent bacteria for determining toxicity in aquatic environments. In Markings LL, Kimerle RA (eds) Aquatic toxicology, ASTM 667. American Society for Testing and Materials, Philadelphia, PA, pp 98–106
35. Farre M, Asperger D, Kantiani L, Gonzalez S, Petrovic M, Barcelo D (2008) Assessment of the acute toxicity of triclosan and methyl triclosan in wastewater based on the bioluminescence inhibition of *Vibrio fischeri*. Anal Bioanal Chem 390:1999–2007
36. Farre M, Fernandez J, Paez M, Granada L, Barba L, Gutierrez HM, Pulgarin C, Barcelo D (2002) Analysis and toxicity of methomyl and ametryn after biodegradation. Anal Bioanal Chem 373:704–709
37. Farre M, Gonzalves C, Lacorte S, Barcelo D, Alpendurada MF (2002b) Pesticide toxicity assessment using an electrochemical biosensor with *Pseudomonas putida* and a bioluminescence inhibition assay with *Vibrio fischeri*. Anal Bioanal Chem 373:696–703
38. Park S, Choi K (2008) Hazard assessment of commonly used agricultural antibiotics on aquatic ecosystems. Ecotoxicology 17:526–538
39. Mortimer M, Kasemets K, Heinlaan M, Kurvet I, Kahru A High throughput kinetic *Vibrio fischeri* bioluminescence inhibition assay for study of toxic effects of nanoparticles. Toxicol In vitro 22:1412–1417
40. Heinlaan M, Ivask A, Blinova I, Dubourguier HC, Kahru A (2008) Toxicity of nanosized and bulk ZnO, CuO and TiO2 to bacteria *Vibrio fischeri* and crustaceans *Daphnia magna* and *Thamnocephalus platyurus*. Chemosphere 71:1308–1316
41. Escher BI, Bramaz N, Quayle P, Rutishauser S, Vermeirssen ELM (2008) Monitoring of the ecotoxicological hazard potential by polar organic micropollutants in sewage treatment plants and surface waters using a mode-of-action based test battery. J Environ Monit 10:622–631
42. Farre M, Garcia M-J, Castillo M, Riu J, Barcelo D (2001) Identification of surfactant degradation products as toxic organic compounds present in sewage sludge. J Environ Monit 3:232–237
43. Farre M, Kloter G, Petrovic M, Alonso MC, Lopez de Alda MJL, Barcelo D (2002) Identification of toxic compounds in wastewater treatment plants during a field experiment. Anal Chim Acta 456:19–30
44. Hernando MD, De Vettori S, Martinez Bueno MJ, Fernandez-Alba AR (2007) Toxicity evaluation with *Vibrio fischeri* test of organic chemicals used in aquaculture. Chemosphere 68: 724–730

45. Park S, Choi K (2008) Hazard assessment of commonly used agricultural antibiotics on aquatic ecosystems.
46. da Silva Nunes-Halldorson V, Duran NL (2003) Braz J Microbiol 34:91
47. Ren S, Frymier PD (2002) Estimating the toxicities of organic chemicals to bioluminescent bacteria and activated sludge. Water Res 36:4406–4414
48. Sagi E, Hever N, Rosen R, Bartolone AJ, Premkumar JR, Ulber R, Lev O, Scheper T, and Belkin S (2003) Fluorescence and bioluminescence reporter functions in genetically modified bacterial sensor strains. Sens Actuators B Chem 90:2–8
49. Davidov Y, Rozen R, Smulski DR, Van Dyk TK, Vollmer AC, Elsemore DA, LaRossa RA, Belkin S (2000) Improved bacterial SOS promoter::lux fusions for genotoxicity detection. Mutat Res Genet Toxicol Environ Mutagen 466:97–107
50. Kelana LC, Griffiths MW (2003) Growth of autobioluminescent *Campylobacter jejuni* in response to various environmental conditions. J Food Prot 66:1190–1197
51. Yamagata T, Ishii M, Narita M, Huang GC, Endo G (2002) Bio-affecting mercury detection using mercury resistance gene module fused with bioluminescence reporter genes. Water Sci Technol 46:253–256
52. Ames BN (2008) FDLWEDPNASU717
53. Anderson D, Bishop JB, Garner RC, Ostrosky-Wegman P, Selby PB (1995) Cyclophosphamide: Review of its mutagenicity for an assessment of potential germ cell risks. Mutat Res Fundam Mol Mech Mutagen 330:115–181
54. Ischenko AA, Saparbaev MK (2002) Alternative nucleotide incision repair pathway for oxidative DNA damage. Nature 415:183–187
55. Longhese MP, Clerici M, Lucchini G (2003) The S-phase checkpoint and its regulation in *Saccharomyces cerevisiae*. Mutat Res Fundam Mol Mech Mutagen 532:41–58
56. Magdaleno A, Mendelson A, de Iorio AF, Rendina A, Moretton J Genotoxicity of leachates from highly polluted lowland river sediments destined for disposal in landfill. Waste Manage 28:2134–2139
57. Lee JW, Helmann JD (2006) The PerR transcription factor senses H2O2 by metal-catalysed histidine oxidation. Nature 440:363–367
58. Liwarska-Bizukojc E, Miksch K, Malachowska-Jutsz A, Kalka J (2005) Acute toxicity and genotoxicity of five selected anionic and nonionic surfactants. Chemosphere 58:1249–1253
59. Matsui M, Matsui K, Kawasaki Y, Oda Y, Noguchi T, Kitagawa Y, Sawada M, Hayashi M, Nohmi T, Yoshihira K, Ishidate J, Sofuni T (1996) Evaluation of the genotoxicity of stevioside and steviol using six in vitro and one in vivo mutagenicity assays. Mutagenesis 11:573–579
60. Muzio HE, Magdaleno A, Moretton J (2005) Genotoxicity of radiographic photofilm wastewater: influence of the treatment with a metal exchange unit. Bull Environ Contamin Toxicol 74:86–93
61. Schafer B, Neffgen A, Klinner U (2008) A novel yeast-based tool to detect mutagenic and recombinogenic effects simultaneously. Mutat Res Genet Toxicol Environ Mutagen 652:20–29
62. Flegrova Z, Bartos T, Cupr P, Skarek M (2008) Sensitivity analysis of RecA genotoxicity test. Ann Microbiol 58:1–5
63. Moreau P, ABRD (1976) Proc Natl Acad Sci U S A 3700
64. Kitano M, Mizuhashi F, Kubo H, Kishida H, Fujii K, Kitahara M, Hosoe K (2007) Evaluation of the mutagenic and genotoxic potential of ubiquinol. Int J Toxicol 26:533–544
65. Shen Y, West C, Hutchins SR (2000) In vitro cytotoxicity of aromatic aerobic biotransformation products in bluegill sunfish BF-2 cells. Ecotoxicol Environ Saf 45:27–32
66. Okamura H, Watanabe T, Aoyama I, Hasobe M (2002) Toxicity evaluation of new antifouling compounds using suspension-cultured fish cells. Chemosphere 46:945–951
67. Kammann U, Bunke M, Steinhart H, Theobald N (2001) A permanent fish cell line (EPC) for genotoxicity testing of marine sediments with the comet assay. Mutat Res Genet Toxicol Environ Mutagen 498:67–77
68. Vevers WF, Jha AN (2008) Genotoxic and cytotoxic potential of titanium dioxide (TiO2) nanoparticles on fish cells in vitro. Ecotoxicology 17:410–420
69. Colborn T, Vom Saal FS, Soto AM (1993) Developmental effects of endocrine-disrupting chemicals in wildlife and humans. Environ Health Perspect 101:378–384

70. White R, Jobling S, Hoare SA, Sumpter JP, Parker MG (1994) Environmentally persistent alkylphenolic compounds are estrogenic. Endocrinology 135:175–182
71. Spengler P, Kórner W, Metzger JW (2001) Substances with estrogenic activity in effluents of sewage treatment plants in southwestern Germany. 1. Chemical analysis. Environ Toxicol Chem 20:2133–2141
72. Seifert M (2004) Luminescent enzyme-linked receptor assay for estrogenic compounds. Anal Bioanal Chem 378:684–687
73. García-Reyero N, Grau E, Castillo M, De Alda MJL, Barcelo D, Piña B (2001) Monitoring of endocrine disruptors in surface waters by the yeast recombinant assay. Environ Toxicol Chem 20:1152–1158
74. Baun A, Sorensen SN, Rasmussen RF, Hartmann NB, Koch CB (2008) Toxicity and bioaccumulation of xenobiotic organic compounds in the presence of aqueous suspensions of aggregates of nano-C60. Aquat Toxicol 86:379–387
75. Colomb A, Yassaa N, Williams J, Peeken I, Lochte K (2008) Screening volatile organic compounds (VOCs) emissions from five marine phytoplankton species by head space gas chromatography/mass spectrometry (HS-GC/MS). J Environ Monitor 10:325–330
76. Paixao SM, Silva L, Fernandes A, O'Rourke K, Mendonca E, Picado A (2008) Performance of a miniaturized algal bioassay in phytotoxicity screening. Ecotoxicology 17:165–171
77. DeLorenzo ME, Keller JM, Arthur CD, Finnegan MC, Harper HE, Winder VL, Zdankiewicz DL (2008) Toxicity of the antimicrobial compound triclosan and formation of the metabolite methyl-triclosan in estuarine systems. Environ Toxicol 23:224–232
78. Mecozzi M, Onorati F, Oteri F, Sarni A (2008) Characterisation of a bioassay using the marine alga *Dunaliella tertiolecta* associated with spectroscopic (visible and infrared) detection. Int J Environ Pollut 32:104–120
79. Macedo RS, Lombardi AT, Omachi CY, rig LR (2008) Effects of the herbicide bentazon on growth and photosystem II maximum quantum yield of the marine diatom *Skeletonema costatum*. Toxicol In vitro 22:716–722
80. Zhang AQ, Leung KMY, Kwok KWH, Bao VWW, Lam MHW (2008) Toxicities of antifouling biocide Irgarol 1051 and its major degraded product to marine primary producers. Mar Pollut Bull 57:575–586
81. Barata C, Damasio J, pez MA, Kuster M, De Alda ML, Barceló D, Riva MC, Raldúa D (2007) Combined use of biomarkers and in situ bioassays in *Daphnia magna* to monitor environmental hazards of pesticides in the field. Environ Toxicol Chem 26:370–379
82. Kim BC, Park KS, Kim SD, Gu MB (2003) Evaluation of a high throughput toxicity biosensor and comparison with a *Daphnia magna* bioassay. Biosens Bioelectron 18:821–826
83. Rosa E, Barata C, sio J, Bosch MP, Guerrero A (2006) Aquatic ecotoxicity of a pheromonal antagonist in *Daphnia magna* and *Desmodesmus subspicatus*. Aquat Toxicol 79:296–303
84. Barbosa IR Nogueira AJA Soares AMVM (2008) Acute and chronic effects of testosterone and 4-hydroxyandrostenedione to the crustacean *Daphnia magna*.Ecotoxicol Environ Saf (E pub ahead of print)
85. Bianchini A, Wood CM (2008) Does sulfide or water hardness protect against chronic silver toxicity in *Daphnia magna*? A critical assessment of the acute-to-chronic toxicity ratio for silver. Ecotoxicol Environ Saf 71:32–40
86. Cropek DM, Esarey JC, Conner CL, Goran JM, Smith T, Soucek DJ (2008) Toxicological effects of military fog oil obscurant on *Daphnia magna* and *Ceriodaphnia dubia* in field and laboratory exposures. Ecotoxicology 17:517–525
87. Hermens J, Canton H, Janssen P, De Jong R (1984) Quantitative structure-activity relationships and toxicity studies of mixtures of chemicals with anaesthetic potency: Acute lethal and sublethal toxicity to *Daphnia magna*. Aquat Toxicol 5:143–154
88. LeBlanc GA (1980) Acute toxicity of priority pollutants to water flea (*Daphnia magna*). Bull Environ Contam Toxicol 24:684–691
89. Ingle T, Alexander R, Hannigan R, Buchanan R, Bouldin J (2007)Detection, measurement and toxicology of semiconductor nanocrystals in *Ceriodapnia dubia*. Proceedings of NSTI Nanotechnology, Santa Clara, CA, pp 612615

90. Farre M, Asperger D, Kantiani L, Gonzalez S, Petrovic M, Barcelo D (2007) Assessment of the acute toxicity of triclosan and methyl triclosan in wastewater based on the bioluminescence inhibition of Vibrio fischeri. Anal Bioanal Chem 390:1999–2007
91. Farre M, Martinez E, Hernando MD, Fernandez-Alba A, Fritz J, Unruh E, Mihail O, Sakkas V, Morbey A, Albanis T, Brito F, Hansen PD, Barcelo D (2006) European ring exercise on water toxicity using different bioluminescence inhibition tests based on *Vibrio fischeri*, in support to the implementation of the water framework directive. Talanta 69: 323–333
92. Chang CT, Chen BY (2008) Toxicity assessment of volatile organic compounds and polycyclic aromatic hydrocarbons in motorcycle exhaust. J Hazard Mater 153:1262–1269
93. Farre M, Pasini O, Carmen Alonso M, Castillo M, Barcelo D (2001) Toxicity assessment of organic pollution in wastewaters using a bacterial biosensor. Anal Chim Acta 426:155–165
94. Gomes De Moraes S, Sanches Freire R, Duran N (2000) Degradation and toxicity reduction of textile effluent by combined photocatalytic and ozonation processes. Chemosphere 40:369–373
95. Hurst JK, Lymar SV (1997) Toxicity of peroxynitrite and related reactive nitrogen species toward Escherichia coli. Chem Res Toxicol 10:802–810
96. Juvonen R, Martikainen E, Schultz E, Joutti A, Ahtiainen J, Lehtokari M (2000) A battery of toxicity tests as indicators of decontamination in composting oily waste. Ecotoxicol Environ Saf 47:156–166
97. Mekki A, Dhouib A, Feki F, Sayadi S (2008) Assessment of toxicity of the untreated and treated olive mill wastewaters and soil irrigated by using microbiotests. Ecotoxicol Environ Saf 69:488–495
98. Erlenkaemper B, Bugiel C, Brecher C, Werth S, Eisentraeger A (2008) Environmental assessment of ester-based lubricants after application. Environ Sci Pollut Res Int 15:68–74
99. Hahn D, Cozzolino A, Piccolo A, Armenante PM (2007) Reduction of 2,4-dichlorophenol toxicity to *Pseudomonas putida* after oxidative incubation with humic substances and a biomimetic catalyst. Ecotoxicol Environ Saf 66:335–342
100. Sutterlin H, Alexy R, Summerer K (2008) The toxicity of the quaternary ammonium compound benzalkonium chloride alone and in mixtures with other anionic compounds to bacteria in test systems with *Vibrio fischeri* and *Pseudomonas putida*. Ecotoxicol Environ Saf 71:498–505
101. Paton GI, Viventsova E, Kumpene J, Wilson MJ, Weitz HJ, Dawson JJC (2006) An ecotoxicity assessment of contaminated forest soils from the Kola Peninsula. Sci Total Environ 355:106–117
102. Lyon DY, Adams LK, Falkner JC, Alvarez PJJ (2006) Antibacterial activity of fullerene water suspensions: Effects of preparation method and particle size. Environ Sci Technol 40:4360–4366
103. Ogawa T, Shibata M, Yatome C, Idaka E (1988) Growth inhibition of Bacillus subtilis by basic dyes. Bull Environ Contamin Toxicol 40:545–552
104. Rasmussen JP, Cursaro M, Froscio SM, Saint CP (2008) An examination of the antibiotic effects of cylindrospermopsin on common gram-positive and gram-negative bacteria and the protozoan *Naegleria lovaniensis*. Environ Toxicol 23:36–43
105. Takigami H, Taniguchi N, Shimizu Y, Matsui S (1998) Toxicity assays and their evaluation on organic polymer flocculants used for municipal sludge dewatering. Water Sci Technol 38:207–215
106. Cardenas S, Valcarcel M (2005) Analytical features in qualitative analysis. Trends Anal Chem 24:477–487
107. Farre M, Martinez E, Barcelo D (2007) Validation of interlaboratory studies on toxicity in water samples. Trends Anal Chem 26:283–292
108. Ramanathan K, Danielsson B (2001) Biosens Bioelectron 16:417–423
109. Borrebaeck C, Mattiasson B (1979) Recent developments in heterogeneous enzyme immunoassay. J Solid Phase Biochem 4:57–67
110. Clarck LC (2008) CLANYAS112

111. Gianfreda L, Xu F, Bollag JM (1999) Laccases: a useful group of oxidoreductive enzymes. Bioremed J 3:1–25
112. Nistor C, Emneus J, Gorton L, Ciucu A (1999) Improved stability and altered selectivity of tyrosinase based graphite electrodes for detection of phenolic compounds. Anal Chim Acta 387:309–326
113. Teke M, rk MK, kaya E, Telefoncu A (2008) Two biosensors for phenolic compounds based on mushroom (*Agaricus bisporus*) homogenate: Comparison in terms of some important parameters of the biosensors. Prep Biochem Biotechnol 38:51–60
114. Yuan J, Guo W, Wang E (2008) Utilizing a CdTe quantum dots-enzyme hybrid system for the determination of both phenolic compounds and hydrogen peroxide. Anal Chem 80:1141–1145
115. Zejli H, Hidalgo-Hidalgo de Cisneros JL, Naranjo-Rodriguez I, Liu B, Temsamani KR, Marty JL (2008) Phenol biosensor based on Sonogel-Carbon transducer with tyrosinase alumina sol-gel immobilization. Anal Chim Acta 612:198–203
116. Cortina M, Del Valle M, Marty JL (2008) Electronic tongue using an enzyme inhibition biosensor array for the resolution of pesticide mixtures. Electroanalysis 20:54–60
117. Kuswandi B, Fikriyah CI, Gani AA (2008) An optical fiber biosensor for chlorpyrifos using a single sol-gel film containing acetylcholinesterase and bromothymol blue. Talanta 74:613–618
118. Mionetto N, Marty JL, Karube I (1994) Acetylcholinesterase in organic solvents for the detection of pesticides: biosensor application. Biosens Bioelectron 9:463–470
119. Mulchandani A, Chen W, Mulchandani P, Wang J, Rogers KR (2001) Biosensors for direct determination of organophosphate pesticides. Biosens Bioelectron 16:225–230
120. Palchetti I, Cagnini A, Del Carlo M, Coppi C, Mascini M, Turner APF (1997) Determination of anticholinesterase pesticides in real samples using a disposable biosensor. Anal Chim Acta 337:315–321
121. Skladal P, Mascini M (1992) Sensitive detection of pesticides using amperometric sensors based on cobalt phthalocyanine-modified composite electrodes and immobilized cholinesterases. Biosens Bioelectron 7:335–343
122. Valdés-Ramírez, Fournier D, Ramírez-Silva MT, Marty JL (2008) Sensitive amperometric biosensor for dichlorovos quantification: Application to detection of residues on apple skin. Talanta 74:741–746
123. Wang Z, Wilkop T, Xu D, Dong Y, Ma G, Cheng Q (2007) Surface plasmon resonance imaging for affinity analysis of aptamer-protein interactions with PDMS microfluidic chips. Anal Bioanal Chem 389:819–825
124. Ng JH, Ilag LL (2003) Biochips beyond DNA: technologies and applications. Biotechnol Annu Rev 9:1–149
125. Belkin S (2003) Microbial whole-cell sensing systems of environmental pollutants. Curr Opin Microbiol 6:206–212
126. Makhaeva GF, Sigolaeva LV, Zhuravleva LV, Eremenko AV, Kurochkin IN, Malygin VV, Richardson RJ (2003) Biosensor detection of neuropathy target esterase in whole blood as a biomarker of exposure to neuropathic organophosphorus compounds. J Toxicol Environ Health A 66:599–610
127. Du D, Chen S, Cai J, Zhang A (2008) Electrochemical pesticide sensitivity test using acetylcholinesterase biosensor based on colloidal gold nanoparticle modified sol-gel interface. Talanta 74:766–772
128. Du D, Huang X, Cai J, Zhang A (2007) Amperometric detection of triazophos pesticide using acetylcholinesterase biosensor based on multiwall carbon nanotube-chitosan matrix. Sens Actuators B Chem 127:531–535
129. Shulga O, Kirchhoff JR (2007) An acetylcholinesterase enzyme electrode stabilized by an electrodeposited gold nanoparticle layer. Electrochem Commun 9:935–940
130. Choi JW, Kim YK, Lee IH, Min J, Lee WH (2001) Optical organophosphorus biosensor consisting of acetylcholinesterase/viologen hetero Langmuir-Blodgett film. Biosens Bioelectron 16:937–943

131. Doong RA, Tsai HC (2001) Immobilization and characterization of sol-gel-encapsulated acetylcholinesterase fiber-optic biosensor. Anal Chim Acta 434:239–246
132. Choi CK, Margraves C, English AE, Kihm KD (2007) Opto-electric biosensor to examine in vitro toxicity stimuli to endothelial cell motility and morphology. Proceedings of the ASME Summer Bioengineering Conference, SBC, pp. 61–62
133. Oosterkamp AJ, Hock B, Seifert M, Irth H (1997) Novel monitoring strategies for xenoestrogens. Trends Anal Chem 16:544–553
134. Kroeger S, Piletsky S, Turner APF (2002) Biosensors for marine pollution research, monitoring and control. Mar Pollut Bull 45:24–34
135. Dutta P, Hill K, Datskos PG, Sepaniak MJ (2007) Development of a nanomechanical biosensor for analysis of endocrine disrupting chemicals. Lab on a Chip Miniatur Chem Biol 7:1184–1191
136. Andres SA, Kerr DA 2nd, Bumpus SB, Kruer TL, Thieman JW, Smolenkova IA, Wittliff JL (2008) A three-tiered approach for calibration of a biosensor to detect estrogen mimics. Adv Exp Med Biol 614:305–313
137. Bovee TFH, Lommerse JPM, Peijnenburg AACM, Fernandes EA, Nielen MWF (2008) A new highly androgen specific yeast biosensor, enabling optimisation of (Q)SAR model approaches. J Steroid Biochem Mol Biol 108:121–131
138. Wozei E, Hermanowicz SW, Holman HYN (2006) Developing a biosensor for estrogens in water samples: Study of the real-time response of live cells of the estrogen-sensitive yeast strain RMY/ER-ERE using fluorescence microscopy. Biosens Bioelectron 21:1654–1658
139. Ramakrishnan A, Tan Y, Sadana A (2005) A kinetic study of analyte-receptor binding and dissociation for surface plasmon resonance biosensors applications. IEEE Sens J 5:356–364
140. Muddana SS, Peterson BR (2003) Fluorescent cellular sensors of steroid receptor ligands. ChemBioChem 4:848–855
141. Butala HD, Sadana A (2003) A fractal analysis of analyte-estrogen receptor binding and dissociation kinetics using biosensors: environmental effects. J Colloid Interf Sci 263:420–431
142. Murata M, Nakayama M, Irie H, Yakabe K, Fukuma K, Katayama Y, Maeda M (2001) Novel biosensor for the rapid measurement of estrogen based on a ligand-receptor interaction. Anal Sci 17:387–390
143. Usami M, Mitsunaga K, Ohno Y (2002) Estrogen receptor binding assay of chemicals with a surface plasmon resonance biosensor. J Steroid Biochem Mol Biol 81:47–55
144. Hock B, Seifert M, Kramer K (2002) Biosens Bioelectron 17:239–249
145. Seifert M, Haindl S, Hock B (1999) Development of an enzyme linked receptor assay (ELRA) for estrogens and xenoestrogens. Anal Chim Acta 386:191–199
146. Sesay AM, Cullen DC (2001) Detection of hormone mimics in water using a miniturised SPR sensor. Environ Monitor Assess 70:83–92
147. Wittliff JL, Andres SA, Kruer TL, Kerr DA 2nd, Smolenkova IA, Erb JL (2008) Biosensors for detecting estrogen-like molecules and protein biomarkers. Adv Exp Med Biol 614:315–322
148. Zhihong M, Xiaohui L, Weiling F (1999) Anal Commun 36:281–283
149. Darain F, Park DS, Park JS, Chang SC, Shim YB (2005) A separation-free amperometric immunosensor for vitellogenin based on screen-printed carbon arrays modified with a conductive polymer. Biosens Bioelectron 20:1780–1787
150. Drevenšek P, Zupančič T, Pihlar B, Jerala R, Kolitsch U, Plaper A, Turel I (2005) Mixed-valence Cu(II)/Cu(I) complex of quinolone ciprofloxacin isolated by a hydrothermal reaction in the presence of l-histidine: Comparison of biological activities of various copper-ciprofloxacin compounds. J Inorg Biochem 99:432–442
151. Lopez-Ortal P, Souza V, Bucio L, Gonzalez E, Gutierrez-Ruiz M (1999) DNA damage produced by cadmium in a human fetal hepatic cell line. Mutat Res Genet Toxicol Environ Mutagen 439:301–306
152. Marrazza G, Chianella I, Mascini M (1999) Disposable DNA electrochemical sensor for hybridization detection. Biosens Bioelectron 14:43–51

153. Doong RA, Shih HM, Lee SH (2005) Sol-gel-derived array DNA biosensor for the detection of polycyclic aromatic hydrocarbons in water and biological samples. Sens Actuators B Chem 111–112:323–330
154. Chiorcea Paquim AM, Diculescu VC, Oretskaya TS, Oliveira Brett AM (2004) AFM and electroanalytical studies of synthetic oligonucleotide hybridization. Biosens Bioelectron 20:933–944
155. Lazarides AA, Lance Kelly K, Jensen TR, Schatz GC (2000) Optical properties of metal nanoparticles and nanoparticle aggregates important in biosensors. J Mol Struct: Theochem 529:59–63
156. Mascini M, Palchetti I, Marrazza G (2001) DNA electrochemical biosensors. Anal Bioanal Chem 369:15–22
157. Tombelli S, Mascini M, Turner APF (2002) Improved procedures for immobilisation of oligonucleotides on gold-coated piezoelectric quartz crystals. Biosens Bioelectron 17: 929–936
158. Babkina SS, Ulakhovich NA (2005) Complexing of heavy metals with DNA and new bioaffinity method of their determination based on amperometric DNA-based biosensor. Anal Chem 77:5678–5685
159. Mugweru A, Wang B, Rusling J (2004) Voltammetric sensor for oxidized DNA using ultrathin films of osmium and ruthenium metallopolymers. Anal Chem 76:5557–5563
160. Lucarelli F, Authier L, Bagni G, Marrazza G, Baussant T, Aas E, Mascini M (2003) DNA biosensor investigations in fish bile for use as a biomonitoring tool. Anal Lett 36:1887–1901
161. Lu Y, Liu J, Li J, Bruesehoff PJ, Pavot CMB, Brown AK (2003) New highly sensitive and selective catalytic DNA biosensors for metal ions. Biosens Bioelectron 18:529–540
162. Wong ELS, Chow E, Justin Gooding J (2007) The electrochemical detection of cadmium using surface-immobilized DNA. Electrochem Commun 9:845–849
163. Oliveira SCB, Corduneanu O, Oliveira-Brett AM (2008) In situ evaluation of heavy metal-DNA interactions using an electrochemical DNA biosensor. Bioelectrochemistry 72:53–58
164. Chiti G, Marrazza G, Mascini M (2001) Electrochemical DNA biosensor for environmental monitoring. Anal Chim Acta 427:155–164
165. Lucarelli F, Palchetti I, Marrazza G, Mascini M (2002) Electrochemical DNA biosensor as a screening tool for the detection of toxicants in water and wastewater samples. Talanta 56:949–957
166. Oliveira-Brett AM, Da Silva LA (2002) A DNA-electrochemical biosensor for screening environmental damage caused by s-triazine derivatives. Anal Bioanal Chem 373:717–723
167. So M, Hvastkovs EG, Schenkman JB, Rusling JF (2007) Electrochemiluminescent/voltammetric toxicity screening sensor using enzyme-generated DNA damage. Biosens Bioelectron 23:492–498
168. Zhang GJ, Chua JH, Chee RE, Agarwal A, Wong SM, Buddharaju KD, Balasubramanian N (2008b) Highly sensitive measurements of PNA-DNA hybridization using oxide-etched silicon nanowire biosensors. Biosens Bioelectron 23:1701–1707
169. Feng Y, Yang T, Zhang W, Jiang C, Jiao K (2008) Enhanced sensitivity for deoxyribonucleic acid electrochemical impedance sensor: Gold nanoparticle/polyaniline nanotube membranes. Anal Chim Acta 616:144–151
170. Sorensen SJ, Burmolle M, Hansen LH (2006) Making bio-sense of toxicity: New developments in whole-cell biosensors. Curr Opin Biotechnol 17:11–16
171. Keane A, Phoenix P, Lau PCK, Ghoshal S (2002) Assessing the bioavailability of organic contaminants using a novel bioluminescent biosensor. Conference Proceedings - Joint 2002 CSCE/ASCE International Conference on Environmental Engineering - An International Perspective on Environmental Engineering, pp 1123–1127
172. Alexander DC, Costanzo MA, Guzzo J, Cai J, Charoensri N, Diorio C, Dubow MS (2000) Blazing towards the next millennium: Luciferase fusions to identify genes responsive to environmental stress. Water Air Soil Pollut 123:81–94
173. Kohler S, Belkin S, Schmid RD (2000) Reporter gene bioassays in environmental analysis. Fresenius' J Anal Chem 366:769–779

174. Lei Y, Chen W, Mulchandani A (2006) Microbial biosensors. Anal Chim Acta 568:200–210
175. Ertl P, Wagner M, Corton E, Mikkelsen SR (2003) Rapid identification of viable *Escherichia coli* subspecies with an electrochemical screen-printed biosensor array. Biosensors and Bioelectronics, 18:907–916
176. Tran MC (1993) Biosensors. Chapman-Hall, London
177. D'Souza SF (2001) Microbial biosensors. Biosens Bioelectron 16:337–353
178. D'Souza SF (2001) Immobilization and stabilization of biomaterials for biosensor applications. Appl Biochem Biotechnol A Enz Eng Biotechnol 96:225–238
179. Cosnier S (2007) Recent advances in biological sensors based on electrogenerated polymers: a review. Anal Lett 40:1260–1279
180. Martínez M, Hilding-Ohlsson A, Viale AA, Corton E (2007) Membrane entrapped *Saccharomyces cerevisiae* in a biosensor-like device as a generic rapid method to study cellular metabolism. J Biochem Biophys Methods 70:455–464
181. Nanduri V, Bhunia AK, Tu SI, Paoli GC, Brewster JD (2007) SPR biosensor for the detection of L. monocytogenes using phage-displayed antibody. Biosens Bioelectron 23: 248–252
182. Odaci D, Kiralp Kayahan S, Timur S, Toppare L (2008) Use of a thiophene-based conducting polymer in microbial biosensing. Electrochim Acta 53:4104–4108
183. Chan C, Lehmann M, Chan K, Chan P, Chan C, Gruendig B, Kunze G, Renneberg R (2000) Designing an amperometric thick-film microbial BOD sensor. Biosens Bioelectron 15:343–353
184. Chang IS, Jang JK, Gil GC, Kim M, Kim HJ, Cho BW, Kim BH (2004) Continuous determination of biochemical oxygen demand using microbial fuel cell type biosensor. Biosens Bioelectron 19:607–613
185. Riedel K, Lange KP, Stein HJ, Kuhn M, Ott Scheller PF (1990) A microbiol sensor for bod. Water Res 24:883–887
186. Lei Y, Mulchandani P, Chen W, Mulchandani A (2005) Direct determination of p-nitrophenyl substituent organophosphorus nerve agents using a recombinant *Pseudomonas putida* js444-modified clark oxygen electrode. J Agric Food Chem 53:524–527
187. Lei Y, Mulchandani P, Chen W, Mulchandani A (2007) Biosensor for direct determination of fenitrothion and epn using recombinant *Pseudomonas putida* JS444 with surface-expressed organophosphorous hydrolase. 2. Modified carbon paste electrode. Appl Biochem Biotechnol 136:243–250
188. Liu N, Cai X, Lei Y, Zhang Q, Chan-Park MB, Li C, Chen W, Mulchandani A (2007) Single-walled carbon nanotube based real-time organophosphate detector. Electroanalysis 19:616–619
189. Taranova L, Semenchuk I, Manolov T, Iliasov P, Reshetilov A (2002) Bacteria-degraders as the base of an amperometric biosensor for detection of anionic surfactants. Biosens Bioelectron 17:635–640
190. Taranova LA, Fesay AP, Ivashchenko GV, Reshetilov AN, Winther-Nielsen M, Emneus J (2004) Comamonas testosteroni strain TI as a potential base for a microbial sensor detecting surfactants. Appl Biochem Microbiol 40:404–408
191. Rajasekar S, Rajasekar R, Narasimhan KC (2000) *Acetobacter peroxydans* based electrochemical biosensor for hydrogen peroxide. Bull Electrochem 16:25–28
192. Farre M, Barcelo D (2001) Characterization of wastewater toxicity by means of a whole-cell bacterial biosensor, using *Pseudomonas putida*, in conjunction with chemical analysis. Anal Bioanal Chem 371:467–473
193. berlein S, Spener F, Zaborosch C (2000) Microbial and cytoplasmic membrane-based potentiometric biosensors for direct determination of organophosphorus insecticides. Appl Microbiol Biotechnol 54:652–658
194. Mulchandani A, Mulchandani P, Chauhan S, Kaneva I, Chen W (1998a) A potentiometric microbial biosensor for direct determination of organophosphate nerve agents. Electroanalysis 10:733–737
195. Mulchandani A, Mulchandani P, Kaneva I, Chen W (1998b) Biosensor for direct determination of organophosphate nerve agents using recombinant *Escherichia coli* with surface-expressed

organophosphorus hydrolase. 1. Potentiometric microbial electrode. Anal Chem 70:4140–4145
196. Bhatia R, Dilleen JW, Atkinson AL, Rawson DM (2003) Combined physico-chemical and biological sensing in environmental monitoring. Biosens Bioelectron 18:667–674
197. Kratasyuk VA, Esimbekova EN, Gladyshev MI, Khromichek EB, Kuznetsov AM, Ivanova EA (2001) The use of bioluminescent biotests for study of natural and laboratory aquatic ecosystems. Chemosphere 42:909–915
198. Girotti S, Bolelli L, Fini F, Monari M, Andreani G, Isani G, Carpené E (2006) Trace metals in arcid clam *Scapharca inaequivalvis*: Effects of molluscan extracts on bioluminescent bacteria. Chemosphere 65:627–633
199. Yin JQ, Li XZ, Zhou C, Zhang YH (2005) Luminescent bacterial sensors made from immobilized films of photobacterium phosphoreum. Chem Res Chin Univ 21:44–47
200. Stolper P, Fabel S, Weller MG, Knopp D, Niessner R (2008) Whole-cell luminescence-based flow-through biodetector for toxicity testing. Anal Bioanal Chem 390:1181–1187
201. Sun Y, Zhou T, Guo J, Li Y (2004) Dark variants of luminous bacteria whole cell bioluminescent optical fiber sensor to genotoxicants. J Huazhong Univ Sci Technol Med Sci 24:507–509
202. Yoo SK, Lee JH, Yun SS, Gu MB, Lee JH (2007) Fabrication of a bio-MEMS based cell-chip for toxicity monitoring. Biosens Bioelectron 22:1586–1592
203. Lee HJ, Villaume J, Cullen DC, Kim BC, Gu MB (2003) Monitoring and classification of PAH toxicity using an immobilized bioluminescent bacteria. Biosens Bioelectron 18:571–577
204. Troegl J, Ripp S, Kuncova G, Sayler GS, Churava A, Parik P, Demnerova K, Halova J, Kubicova L (2005) Selectivity of whole cell optical biosensor with immobilized bioreporter *Pseudomonas fluorescens* HK44. Sens Actuators B Chem 107:98–103
205. Gu MB, Choi SH (2001) Monitoring and classification of toxicity using recombinant bioluminescent bacteria. Water Sci Technol 43:147–154
206. Hwang ET, Lee JH, Chae YJ, Kim YS, Kim BC, Sang B-I, Gu MB (2008) Analysis of the toxic mode of action of silver nanoparticles using stress-specific bioluminescent bacteria. Small 4:746–750
207. Lee JH, Youn CH, Kim BC, Gu MB (2007) An oxidative stress-specific bacterial cell array chip for toxicity analysis. Biosens Bioelectron 22:2223–2229
208. Lee JH, Song CH, Kim BC, Gu MB (2006) Application of a muli-channel system for continuous monitoring and an early warning system. Water Sci Technol 53:341–346
209. Ge C, Xu M, Liu J, Lei J, Ju H (2008) Facile synthesis and application of highly luminescent CdTe quantum dots with an electrogenerated precursor. Chem Commun 8:450–452
210. Lee JH, Gu MB (2005) An integrated mini biosensor system for continuous water toxicity monitoring. Biosens Bioelectron 20:1744–1749
211. Bhattacharyya J, Read D, Amos S, Dooley S, Killham K, Paton GI (2005) Biosensor-based diagnostics of contaminated groundwater: assessment and remediation strategy. Environ Pollut 134:485–492
212. Kim BC, Gu MB (2003) A bioluminescent sensor for high throughput toxicity classification. Biosens Bioelectron 18:1015–1021
213. Pooley DT, Larsson J, Jones G, Rayner-Brandes MH, Lloyd D, Gibson C, Stewart WR (2004) Continuous culture of photobacterium. Biosens Bioelectron 19:1457–1463
214. Premkumar JR, Rosen R, Belkin S, Lev O (2002) Sol-gel luminescence biosensors: Encapsulation of recombinant *E coli* reporters in thick silicate films. Anal Chim Acta 462:11–23
215. Rosen R, Davidov Y, LaRossa RA, Belkin S (2000) Microbial sensors of ultraviolet radiation based on recA′::Lux fusions. Appl Biochem Biotechnol A Enz Eng Biotechnol 89:151–160
216. Polyak B, Bassis E, Novodvorets A, Belkin S, Marks RS (2000) Optical fiber bioluminescent whole-cell microbial biosensors to genotoxicants. Water Sci Technol 42:305–311
217. Taguchi K, Tanaka Y, Imaeda T, Hirai M, Mohri S, Yamada M, Inoue Y (2004) Development of a genotoxicity detection system using a biosensor. Environ Sci 11:293–302
218. Philp JC, Balmand S, Hajto E, Bailey MJ, Wiles S, Whiteley AS, Lilley AK, Hajto J, Dunbar SA (2003) Whole cell immobilised biosensors for toxicity assessment of a wastewater treatment plant treating phenolics-containing waste. Anal Chim Acta 487:61–74

219. Kim BC, Gu MB (2005) A multi-channel continuous water toxicity monitoring system: its evaluation and application to water discharged from a power plant. Environ Monitor Assess 109:123–133
220. Strachan G, Capel S, Maciel H, Porter AJR, Paton GI (2002) Application of cellular and immunological biosensor techniques to assess herbicide toxicity in soils. Eur J Soil Sci 53:37–44
221. Li YF, Li FY, Ho CL, Liao VHC (2008) Construction and comparison of fluorescence and bioluminescence bacterial biosensors for the detection of bioavailable toluene and related compounds. Environ Pollut 152:123–129
222. Bozeman J, Koopman B, Bitton G (1989) Toxicity testing using immobilized algae. Aquat Toxicol 14:345–352
223. Admiraal W, Blanck H, Buckert-De Jong M, Guasch H, Ivorra N, Lehmann V, Nyström BAH, Paulsson M, Sabater S (1999) Short-term toxicity of zinc to microbenthic algae and bacteria in a metal polluted stream. Water Res 33:1989–1996
224. Cassidy MB, Lee H, Trevors JT (1996) Environmental applications of immobilized microbial cells: a review. J Ind Microbiol 16:79–101
225. Awasthi M, Rai LC (2005) Toxicity of nickel, zinc, and cadmium to nitrate uptake in free and immobilized cells of *Scenedesmus quadricauda*. Ecotoxicol Environ Saf 61:268–272
226. Moreno-Garrido I, Lubian LM, Blasco J (2007) Sediment toxicity tests involving immobilized microalgae (*Phaeodactylum tricornutum* Bohlin). Environ Int 33:481–485
227. Twist H, Edwards AC, Codd GA (1997) A novel in-situ biomonitor using alginate immobilised algae (*Scenedesmus subspicatus*) for the assessment of eutrophication in flowing surface waters. Water Res 31:2066–2072
228. Moreira-Santos M, Soares AMVM, Ribeiro R (2004) An in situ bioassay for freshwater environments with the microalga *Pseudokirchneriella subcapitata*. Ecotoxicol Environ Saf 59:164–173
229. Moreira SM, Moreira-Santos M, Guilhermino L, Ribeiro R (2006) Immobilization of the marine microalga *Phaeodactylum tricornutum* in alginate for in situ experiments: Bead stability and suitability. Enz Microb Technol 38:135–141
230. Frense D, ller A, Beckmann D (1998) Detection of environmental pollutants using optical biosensor with immobilized algae cells. Sens Actuators B Chem 51:256–260
231. Naessens M, Leclerc JC, Tran-Minh C (2000) Fiber optic biosensor using *Chlorella vulgaris* for determination of toxic compounds. Ecotoxicol Environ Saf 46:181–185
232. Sanders CA, Rodriguez J, Greenbaum E (2001) Stand-off tissue-based biosensors for the detection of chemical warfare agents using photosynthetic fluorescence induction. Biosens Bioelectron 16:439–446
233. Shitanda I, Takada K, Sakai Y, Tatsuma T (2005) Compact amperometric algal biosensors for the evaluation of water toxicity. Anal Chim Acta 530:191–197
234. Chouteau C, Dzyadevych S, Chovelon JM, Durrieu C (2004) Development of novel conductometric biosensors based on immobilised whole cell *Chlorella vulgaris* microalgae. Biosens Bioelectron 19:1089–1096
235. De-Bashan LE, Hernandez JP, Morey T, Bashan Y (2004) Microalgae growth-promoting bacteria as "helpers" for microalgae: a novel approach for removing ammonium and phosphorus from municipal wastewater. Water Res 38:466–474
236. Weitz HJ, Campbell CD, Killham K (2002) Development of a novel, bioluminescence-based, fungal bioassay for toxicity testing. Environ Microbiol 4:422–429
237. Campanella L, Favero G, Mastrofini D, Tomassetti M (1996) Toxicity order of cholanic acids using an immobilised cell biosensor. J Pharm Biomed Anal 14:1007–1013
238. Billinton N, Barker MG, Michel CE, Knight AW, Heyer WD, Goddard NJ, Fielden PR, Walmsley RM (1998) Development of a green fluorescent protein reporter for a yeast genotoxicity biosensor. Biosens Bioelectron 13:831–838
239. Afanassiev V, Sefton M, Anantachaiyong T, Barker G, Walmsley R, Ifl S (2000) Application of yeast cells transformed with GFP expression constructs containing the RAD54 or RNR2 promoter as a test for the genotoxic potential of chemical substances. Mutat Res Genet Toxicol Environ Mutagen 464:297–308

240. Mitchell RJ, Gu MB (2006) Characterization and optimization of two methods in the immobilization of 12 bioluminescent strains. Biosens Bioelectron 22:192–199
241. Benton MG, Glasser NR, Palecek SP (2007) The utilization of a *Saccharomyces cerevisiae* HUG1P-GFP promoter-reporter construct for the selective detection of DNA damage. Mutat Res Genet Toxicol Environ Mutagen 633:21–34
242. Hollis RP, Killham K, Glover LA (2000) Design and application of a biosensor for monitoring toxicity of compounds to eukaryotes. Appl Environ Microbiol 66:1676–1679
243. Terziyska A, Waltschewa L, Venkov P (2000) A new sensitive test based on yeast cells for studying environmental pollution. Environ Pollut 109:43–52
244. Tucker CL, Fields S (2001) A yeast sensor of ligand binding. Nat Biotechnol 19:1042–1046
245. Sakaki T, Shinkyo R, Takita T, Ohta M, Inouye K (2002) Biodegradation of polychlorinated dibenzo-p-dioxins by recombinant yeast expressing rat CYP1A subfamily. Arch Biochem Biophys 401:91–98
246. Lehmann M, Riedel K, Adler K, Kunze G (2000) Amperometric measurement of copper ions with a deputy substrate using a novel *Saccharomyces cerevisiae* sensor. Biosens Bioelectron 15:211–219
247. Webb JS, Barratt SR, Sabev H, Nixon M, Eastwood IM, Greenhalgh M, Handley PS, Robson GD (2001) Green fluorescent protein as a novel indicator of antimicrobial susceptibility in *Aureobasidium pullulans*. Appl Environ Microbiol 67:5614–5620
248. Knight AW, Keenan PO, Goddard NJ, Fielden PR, Walmsley RM (2004) A yeast-based cytotoxicity and genotoxicity assay for environmental monitoring using novel portable instrumentation. J Environ Monitor 6:71–79
249. Polak ME, Rawson DM, Haggett BGD (1996) Redox mediated biosensors incorporating cultured fish cells for toxicity assessment. Biosens Bioelectron 11:1253–1257
250. Bentley A, Atkinson A, Jezek J, Rawson DM (2001) Whole cell biosensors - Electrochemical and optical approaches to ecotoxicity testing. Toxicol in Vitro 15:469–475
251. Zhang T, Rawson DM (2002) Studies on cryopreservation of Luc gene transfected bluegill sunfish fibroblast cell line. Cryo-Lett 23:191–196
252. Dierksen KP, Mojovic L, Caldwell BA, Preston RR, Upson R, Lawrence J, McFadden PN, Trempy JE (2004) Responses of fish chromatophore-based cytosensor to a broad range of biological agents. J Appl Toxicol 24:363–369
253. Mojovic L, Dierksen KP, Upson RH, Caldwell BA, Lawrence JR, Trempy JE, McFadden PN (2004) Blind and naive classification of toxicity by fish chromatophores. J Appl Toxicol 24:355–361
254. Gu MB, Kim BC, Cho J, Hansen PD (2001) The continuous monitoring of field water samples with a novel multi-channel two-stage mini-bioreactor system. Environ Monitor Assess 70:71–81
255. Diamond D (2008) Special section on wireless sensor networks. Talanta 75:605
256. Diamond D, Lau KT, Brady S, Cleary J (2008) Integration of analytical measurements and wireless communications-Current issues and future strategies. Talanta 75:606–612
257. Carrara S, Shumyantseva VV, Archakov AI, Samori B (2008) Screen-printed electrodes based on carbon nanotubes and cytochrome P450scc for highly sensitive cholesterol biosensors. Biosens Bioelectron 24:148–150
258. Chien JH, Yang CH, Chen PH, Yang CR, Lin CS, Wang H (2008) DNA detection using a radio frequency biosensor with gold nanoparticles. Front Biosci 13:4756–4764
259. Cui R, Huang H, Yin Z, Gao D, Zhu JJ (2008) Horseradish peroxidase-functionalized gold nanoparticle label for amplified immunoanalysis based on gold nanoparticles/carbon nanotubes hybrids modified biosensor. Biosens Bioelectron 23:1666–1673
260. Wang S, Tan Y, Zhao D, Liu G (2008) Amperometric tyrosinase biosensor based on Fe3O4 nanoparticles-chitosan nanocomposite. Biosens Bioelectron 23:1781–1787
261. Endo T, Yamamura S, Kerman K, Tamiya E (2008) Label-free cell-based assay using localized surface plasmon resonance biosensor. Anal Chim Acta 614:182–189
262. Cao Y, Jin R, Mirkin CA (2001) DNA-modified core-shell Ag/Au nanoparticles. J Am Chem Soc 123:7961–7962

263. Banerjee P, Lenz D, Robinson JP, Rickus JL, Bhunia AK (2008) A novel and simple cell-based detection system with a collagen-encapsulated B-lymphocyte cell line as a biosensor for rapid detection of pathogens and toxins. Lab Investig 88:196–206
264. Souza GR, Christianson DR, Staquicini FI, Ozawa MG, Snyder EY, Sidman RL, Miller JH, Arap W, Pasqualini R (2006) Networks of gold nanoparticles and bacteriophage as biological sensors and cell-targeting agents. Proc Natl Acad Sci U S A 103:1215–1220
265. Boldt TS, S(rensen J, Karlson U, Molin S, Ramos C (2004) Combined use of different Gfp reporters for monitoring single-cell activities of a genetically modified PCB degrader in the rhizosphere of alfalfa. FEMS Microbiol Ecol 48:139–148
266. Lattier DL, Gordon DA, Burks DJ, Toth GP (2001) Vitellogenin gene transcription: a relative quantitative exposure indicator of environmental estrogens. Environ Toxicol Chem 20:1979–1985
267. Malhotra BD, Chaubey A, Singh SP (2006) Prospects of conducting polymers in biosensors. Anal Chim Acta 578:59–74
268. Nemecek S, Marisch K, Juric R, Bayer K (2008) Design of transcriptional fusions of stress sensitive promoters and GFP to monitor the overburden of *Escherichia coli* hosts during recombinant protein production. Bioproc Biosyst Eng 31:47–53

Hdb Env Chem (2009) 5J: 161–186
DOI:10.1007/698_2008_12

Genetically Engineered Bacteria for Genotoxicity Assessment

Alva Biran, Rami Pedahzur, Sebastian Buchinger, Georg Reifferscheid, and Shimshon Belkin

Abstract Ever since the introduction of the *Salmonella typhimurium* mammalian microsome mutagenicity assay (the *Ames test*) over three decades ago, there has been a constant development of additional genotoxicity assays based upon the use of genetically engineered microorganisms. Such assays rely either on reversion principles similar to those of the Ames test or on promoter–reporter fusions that generate a quantifiable dose–dependent signal in the presence of potential DNA damaging compounds and the induction of repair mechanisms. While some of these assays were only briefly described in the scientific literature, others have been developed all the way to commercial products. Out of these, only one, the *umu*-test has been fully validated and ISO- and OECD-standardized. Herein we review the main directions undertaken in the construction and testing of bacterial-based genotoxicity bioassays, including the attempts to incorporate at least a partial metabolic activation capacity into the molecular design. We list the genetic modifications introduced into the tester strains, compare the performance of the different assays, and briefly describe the first attempts to incorporate such bacterial reporters into actual genotoxicity testing devices.

Keywords Biosensors, Genetically engineered bacteria, Genotoxicity, Promoter-reporter fusions, Reversion assays

---

A. Biran, R. Pedahzur, and S. Belkin (✉)
Institute of Life Sciences, The Hebrew University of Jerusalem, Jerusalem, 91904, Israel
e-mail: shimshon@vms.huji.ac.il

S.Buchinger and G. Reifferscheid
Division of Qualitative Hydrology, Federal Institute of Hydrology (BfG),
Koblenz, 56068, Germany

Contents

1 Introduction

The increasing need to assay and monitor the potential genotoxic effects of an ever growing number of chemicals and environmental samples is countered by the logistic, economical, and ethical constraints imposed by the use of animal-based test systems. Consequently, ever since the introduction of the revolutionary *Salmonella typhimurium* mammalian microsome mutagenicity assay (the *Ames test*) over three decades ago [1], continuous efforts have been directed towards the development, improvement, and implementation of additional bacterial-based genotoxicity assays. Several such assays employ genetically engineered microorganisms, "tailored" to generate a quantifiable signal that reflects the genotoxic potency of the tested sample. Such assays share several significant advantages including rapid response times, high reproducibility, facility of use, and low operational cost. Yet, bacterial-based assays can not carry out the complex biochemical reactions collectively known as "metabolic activation" which take place mainly in mammalian liver cells, in which many xenobiotics are transformed into genotoxic forms. Herein we review the main directions undertaken in the construction and testing of bacterial-based genotoxicity bioassays, including the attempts to incorporate at least a partial metabolic activation capacity into the molecular design.

2 The Promoter–Reporter Concept

As will be discussed below, several genetic engineering approaches have been employed over the years in the construction of bacterial reporter strains that respond to the presence of genotoxic compounds. Many of these share the same basic principle: the fusion of a gene promoter, known to be activated by the presence of genotoxic chemicals, to a gene or a group of genes the activity of which can be monitored quantitatively, preferably in real time [2]. The gene promoter acts as the sensing element, which – upon activation – drives the transcription of the downstream reporter gene(s). Consequently, the gene promoter will dictate the response spectrum of the construct and, to some extent, its sensitivity. The reporter genes determine the nature of the generated signal (bioluminescence, fluorescence, etc.) and thus also the instrumentation required for its acquisition. The host cell, the third major component in the construction of a genotoxicity reporter strain, is selected for ease of genetic manipulation, for its relevance, and – most importantly – for its effects on detection sensitivity and threshold.

3 Sensing Elements

In the selection of sensing elements to be used for the construction of genotoxicity reporter systems, the most promising candidates are promoters of genes involved in DNA repair. Such genes, induced in response to either actual DNA damage or to the presence of DNA damaging agents, are mainly part of two inducible systems, the *recA*-dependent, *lexA*-controlled SOS response and the *recA*-independent, *ada*-controlled adaptive system induced in response to alkylation damage of DNA. The latter system responds specifically to the presence of methylated phosphotriesters generated by DNA alkylation that activate the *ada* gene product which, in turn, triggers the transcription of genes such as *ada*, *alkA*, *alkB*, and *aid* [3, 4].

The SOS response is under the control of the LexA protein that binds to the SOS box in the promoter region of the regulon genes, repressing their expression. Derepression occurs when the RecA protein binds to single-stranded DNA at replication forks that are blocked by DNA damage, forming RecA-ssDNA nucleoprotein filaments [5]. Once bound to DNA, the RecA protein changes conformation and acts as a coprotease in the cleavage of LexA, thus allowing transcription of the SOS genes [6–8]. Among these are genes such as *uvrA*, *recA*, *recN*, or *umuCD*, responsible for DNA repair, and others such as *sulA*, that couple DNA damage to cell division [7, 9]. Expression of a given SOS gene depends on the specific LexA-binding properties of its promoter, determined by the sequence of the LexA-binding sites (SOS boxes), their number and arrangement [10–12].

3.1 SOS Promoters

Several gene promoters from over 30 known SOS regulon genes induced in times of DNA damage were used for the construction of genotoxicity sensors, as is briefly described below.

3.1.1 *umuDC*

The two proteins coded by this operon, UmuD and UmuC, are induced under DNA damage conditions by the LexA- and RecA-dependent transcriptional upregulation of the SOS regulon operon. They first form a $UmuD_2C$ complex, which acts as a checkpoint inhibitor of cell division until repair can address the original inducing DNA damage signal. After RecA–ssDNA mediated UmuD cleavage, these proteins form a $UmuD'_2C$ complex (DNA polymerase V), which carries out the error-prone replication of damaged DNA (SOS mutagenesis; for a review see [13]).

The first description of a *umuC'–lacZ* fusion coded on plasmid pSK1002 in *S. typhimurium* for the detection of genotoxic agents was published by Oda et al. [14], and has since been recognized as the "*umu* - test." The *S. typhimurium* strain carrying this fusion (TA1535) has undergone several modifications including excision repair deficiency (*uvrB*), an *rfa* deletion which increases permeability to many chemicals, and a deletion of the natural *lac* operon. The *umu* test was standardized according to DIN (DIN 38415-3 [15]) and the International Standardization Organization (ISO) (ISO/CD 13829 [16]). It is now a part of the set of tools available to authorities and researchers for the investigation and monitoring of genotoxicity of environmental samples. The system was adapted to a 96-well microtiter plate format [17] and has been used, for example, to detect a wide range of carcinogenic mutagens [14, 18–21], as well as genotoxic activity in disinfectants [22], complex mixtures [23, 24], environmental pollutants [25], river waters and industrial wastewaters [17, 26, 27].

3.1.2 *sulA* (sfiA)

The SulA protein, produced in large amounts during the SOS response, halts cell division in *E. coli* by binding to the tubulin-like GTPase, FtsZ [28]. It has been used as a sensing element in genotoxicity detection in several cases, most notably in the colorimetric SOS-chromotest [29], commercialized in 1984. The *E. coli* PQ37 tester strain used in the SOS-chromotest harbors a *sfiA'::lacZ* fusion and carries a deletion of the normal *lac* region, so that β-galactosidase activity is strictly dependent on *sfiA* expression. Similarly to the *umu*-test bacterium it is mutated in *uvrA* to hinder DNA repair, and in *rfa* to increase permeability [30]. A different colorimetric assay based on a plasmid-borne *sulA'::lacZ* fusion in *S. typhimurium* TA1538 was proposed by El Mzibri et al. [31]; a procedure that includes metabolic activation based on S9-mix has been described as well.

3.1.3 *recN*

Another *E. coli* SOS gene promoter fusion that has been developed into a commercial product (VITOTOX®) is based on *recN*, coding for a protein that is involved in double-stranded DNA break repair. *E. coli* and several *S. typhimurium* strains (TA98, TA100,

TA104) are used as bacterial hosts. A multicopy plasmid harboring a fusion of the *recN* promoter to the *Vibrio fischeri luxCDABE* genes drives the emission of light in response to the presence of DNA damaging agents [32], allowing real-time monitoring of the bacterial response. The VITOTOX® test strains were tested with a variety of chemicals [32, 33] as well as river waters [34], ground water [35], and air samples.

3.1.4 *recA*

The gene of the RecA recombinase, which plays a key role in the SOS response by its coprotease activity on the LexA repressor, has been the basis of several attempts for the construction of genotoxicity sensors. Nunoshiba and Nishioka [36, 37] described the *E. coli* colorimetric "Rec–lac test," based on the GE94 strain that carries the *recA–lacZ* fusion gene and its DNA repair-deficient derivatives such as KY946 (*uvrA*), KY945 (*recA*), and KY943 (*lexA*). The system was tested against 4-nitroquinoline-*N*-oxid (4-NQO), *N*-methyl-*N'*-nitro-*N*-nitrosoguanidine (MNNG), mitomycin C (MMC) and UV radiation, as well as with hydrogen peroxide (H_2O_2), formaldehyde (CH_2O), *tert-butyl* hydroperoxide, cumene hydroperoxide, and streptonigrin.

A different system, the *V. fischeri luxCDABE* genes, was used by Vollmer et al. [38] to generate several *E. coli* reporter strains, one of them (DPD2794) carrying a *recA'::luxCDABE* fusion on the multicopy plasmid pUCD615. These bioluminescent fusions allow real-time visualization of the transcriptional responses induced by DNA damage, without the need for cell-free enzyme assays or the exogenous addition of luciferase substrates. To make full use of these advantages, Polyak et al. [39] have alginate-immobilized a similar *recA'::lux* harboring strain to the tip of an optic fiber, the other end of which was connected to a photon counter. The instrument allowed a real-time determination of genotoxicity by dipping the bacteria-clad end into a sample.

3.1.5 *Cda*

The colicin D gene *cda*, a constituent of the ColD plasmid [40], also served as a basis for a bioluminescent genotoxicity sensor using *Photobacterium leiognathi luxCDABE* as a reporter. This "SOS-*lux*" test responded sensitively to diverse genotoxins, such as MMC, MNNG, nalidixic acid (NA), dimethylsulfate (DMS), H_2O_2, CH_2O, and UV and γ radiation [41]. This assay was later combined with the GFPuv-based Lac-Fluoro test to generate a combined toxicity–genotoxicity sensor [42].

Four different SOS promoters (*recA*, *umuCD*, *sulA*, and *cda*) were compared by Norman et al. [12] using the same fluorescent reporter (*gfp* mut3*, in plasmid pANO1). The differences between the constructs were evaluated after the exposure of host cells harboring the fusion plasmids (MG1655/pANO1::SOS promoter) to the known genotoxicant *N*-methyl-*N'*-nitro-*N*-nitrosoguanidine (MNNG). A *tolC* mutation enhanced the sensitivity to this agent; no other chemicals were tested in this study. Performance of the *cda*-based sensor in response to MNNG clearly surpassed the other three with respect to the SOS-induction factor, as a result of high rates of gene expression combined with a low background activity of the *cda* promoter. Thus, the *cda* promoter was selected for the further development of the GenoTox test [43].

3.2 Non SOS Promoters

3.2.1 *alkA*

Several bacterial DNA protection and repair systems that are independent of the SOS regulon have been described, one of which, most efficiently induced by alkylating agents, has been generally termed the "adaptive response." Several genes of this system have been characterized, including *alkA*, which encodes a repair glycosylase (*N*3-methyladenine DNA glycosylase II; [3]). A promoter of this gene has been fused by Vollmer et al. [38] to the *V. fischeri luxCDABE* genes. The construct displayed a very strong response to the alkylating agent MNNG, the magnitude of which was attributed to a very low background bioluminescence; the responses of an equivalent *lacZ* fusion were much more moderate.

3.2.2 *nrdA*

The expression of the *E. coli nrdA* gene, which encodes for a ribonucleoside diphosphate reductase, is strongly affected by DNA damage, induced for example by an exposure to UV irradiation, but is independent of LexA [44]. The *nrdA* promoter was fused by Hwang et al. [45] to *Photorhabdus luminescens luxCDABE* genes. *E. coli* strain BBTNrdA carrying this plasmid-borne fusion responded to the DNA damaging agents NA, MMC, MNNG, and 4-NQO and hydrogen peroxide but not to other oxidants or phenolic compounds [45].

4 Reporter Systems

The spectrum of reporter systems available for monitoring gene expression by transcriptional fusions is continuously expanding, as is the instrumentation for signal detection and quantification. Colorimetric, fluorescent, bioluminescent and (to a small extent) electrochemical detection of genotoxicity have been described, and will be briefly outlined below.

4.1 Colorimetric and Electrochemical (lacZ, phoA)

The β-galactosidase gene, *lacZ*, has been used as a gene expression reporter for several decades. The most common substrates employed for assaying the activity of this enzyme are *o*-nitrophenyl *β*-D-galactopyranoside (ONPG) and 5-bromo-4-

chloro-3-indolyl β-D-galactoside (X-gal) for colorimetric detection, 4-methylumbelliferyl-β-D-galactopyranoside (MUG) for fluorimetry, 1,2-dioxetane substrates for luminescence, and *p*-aminophenyl-β-D-galactopyranoside (PAPG) for electrochemical analysis. The advantages of colorimetric assays lie in their simplicity and rapidity, but the need for improved sensitivity, faster response times, a broader dynamic range and the capability of real-time monitoring has led to a continuous search for alternatives [46]. The *umu*-test [14], SOS-chromotest [29], *sulA*-test [31] and Rec–lac test [37] were all developed using *lacZ* as the reporter and ONPG as the substrate. Several remedies have been proposed to overcome interferences by colored samples, such as the inclusion of a washing step after the exposure of the bacteria to the samples [19, 47], or a post-treatment dilution and reincubation [17, 18]. The latter procedure was reported to enhance the sensitivity of the *umu*-test to genotoxicants in environmental samples in a high-throughput microtiter plate system [23]. Oda et al. [48] achieved higher sensitivity by using a different substrate, chlorophenol red-β-D-galactopyranoside (CRPG). The red reaction product after cleavage by the β-galactosidase has a longer life time than *o*-nitrophenol, the ONPG reaction product.

Similar modifications were also introduced to the SOS-chromotest [49]. The original colorimetric procedure of the assay [29] was successfully changed to a fluorimetric one by using the fluorescent substrate 4-methylumbelliferyl-β-D-galactopyranoside (MUG) [50].

A different approach was proposed by Matsui et al. [51], who provided the *umu*-test bacteria (TA1535/pSK1002) with the substrate PAPG, the end product of which (*p*-aminophenol) can be monitored electrochemically. This approach, utilized earlier for other bacterial sensor systems [52–55], requires the addition of an external substrate but does not involve lysis or permeabilization of the cells. Similarly to bioluminescence, it is thus suitable for continuous online measurement of enzymatic activity, even in turbid solutions and under anaerobic conditions. Matsui et al. [51] have demonstrated this by scanning electrochemical microscopy (SECM) in a specialized glass biochip configuration, using 5 nl cell aliquots immobilized in collagen gel. Overall, lower limits of detection of 2-aminoflouren (2-AF), MMC and 2-aminoanthracene (2-AA; +S9-mix) were obtained by the microbial chip as compared to the conventional *umu* tests, but it should be noted that the definition of the detection limit was different and exposure times were longer.

4.2 Bioluminescence (*lux*, *luc*)

The reaction by which photons are released by living organisms, bioluminescence, appears in numerous groups of organisms including bacteria, protozoa, fungi, insects, and fish. In all cases the reaction is catalyzed by an enzyme termed luciferase that oxidizes a substrate known as luciferin, but the chemical and enzymatic nature of both entities depends on the organism from which the system is derived. The two bioluminescent systems most commonly used as reporters of gene activation are of bacterial and insect (firefly) origin. Firefly luciferase, coded by the *luc* gene, is a 62-kDa monomeric protein, and its activity is oxygen- and ATP-dependant. Its luciferin, benzothiazoyl-thiazole, has to be added externally when *luc* is used as a reporter gene. Bacterial luciferase catalyzes the oxidation of a reduced flavin mononucleotide ($FMNH_2$) by a long-chain fatty aldehyde to FMN and the corresponding fatty acid in the presence of molecular oxygen. All bacterial luciferases are heterodimeric proteins composed of two subunits, α (40 kDa) and β (37 kDa), encoded by the *luxA* and *luxB* genes of the *lux* operon. The three other genes in this operon (*luxCDE*) encode the synthesis and recycling enzymes of the fatty acid aldehyde [56]. Constructs carrying just *luxA* and *luxB* are sufficient to generate a bioluminescent signal, but necessitate the external addition of the aldehyde substrate. The commonly used luciferases of *V. fischeri* and *Vibrio harveyi* have limited upper temperatures of 30°C or 37°C, respectively. In recent years, *Photorhabdus luminescens lux* genes have thus often been used due to the higher upper temperature limit (45°C) of their gene products [57]. The noninvasive protocol using *lux* fusions allows real-time reporting of the transcriptional activation of the monitored gene promoters. As described above, the VITOTOX® test uses the *V. fischeri luxCDABE* operon under the control of the *recN* promoter [32, 33], and the SOS-*lux* test employs the *lux* operon (*luxCDABFE*) of *P. leiognathi* under control of the *cda* gene promoter of the plasmid ColD [41]. Vollmer et al. [38] have fused the *E. coli recA*, *uvrA*, and *alkA* promoters to the *V. fischeri luxCDABE* operon. Further modifications to the same system [58, 59] included integration of the *recA'::lux* fusion into the *E. coli* chromosome, a change of the reporter system to *P. luminescens lux*, and the use of either *S. typhimurium* or a *tolC E. coli* mutant as alternative hosts. Application of the *P. luminescens* reporter, which allowed a working temperature of 37°C, resulted in a more rapid response to various genotoxic chemicals and UV.

The *luxCDABE* genes of *V. fischeri* were also fused to the *recA* promoter of *Pseudomonas aeruginosa* [60]. As a soil and freshwater bacterium, *P. aeruginosa* was presented as a good candidate to serve as a sensor for the state of natural bacterial communities of both pristine and polluted habitats. Light production in response to UV exposure was monitored in this strain as part of a study of UV effects on natural bacterial populations.

To increase the sensitivity of the *umu*-test and to expand its detection capabilities, two groups independently replaced its β-galactosidase reporting gene by either bacterial [61] or insect [62] luciferase. In both cases, improvements in performance were reported, including enhanced sensitivity, improved signal-to-noise ratios, stronger signals, and a better neutralization of color interferences.

4.3 Fluorescent Protein Genes

The highly stable green fluorescent protein, GFP, of the jellyfish *Aequorea victoria* was the first fluorescent protein the gene of which was utilized as a molecular reporter [63]. It was soon followed by additional fluorescent protein genes isolated from various marine organisms as well as by mutated forms with improved performance [64–70]. The GFP protein has a high quantum yield and can be expressed in both prokaryotic and eukaryotic systems with no need of a substrate or cofactor [71]. To increase the sensitivity of assays based on the GFP reporter system, several green fluorescent protein mutants were constructed [65, 72–74].

Arai et al. [75] have modified the *umu*-test by replacing the *lacZ* gene with a DNA fragment encoding for EGFP (enhanced green fluorescent protein). This construct was tested in *E. coli* strain KY706 with 3 μg/ml 4-NQO, a concentration that strongly induced β-galactosidase activity in the *umu*-test. The GFP reporter system was equivalent to the β-galactosidase reporting system with respect to its detection sensitivity only after inserting additional modifications to the plasmid. These modifications include utilization of tandem *lacUV5* and chimeric *trp/umu* promoters, and coexpression of the *E. coli recA5327* mutant. An additional construct that harbors the fusion of the *umuCD* promoter to the *gfp* gene was generated by Justus and Thomas [76], who reported an overall poorer performance compared to the *lacZ*-based assay.

Other fluorescent genotoxicity sensors were constructed by fusing the *gfp* and *gfp*mut3 reporting genes to the *recA* promoter [77]. GFPmut3 is a mutant that is approximately 20-times more fluorescent than wild-type GFP and only weakly excited by UV-light [72]. The use of the wild-type *gfp* yielded dose-dependent but weak results, while with *gfp*mut3 the detection thresholds for MMC, MNNG, NA, hydrogen peroxide, and formaldehyde were comparable to the SOS-chromotest [29], the *umu*-test [19], and the SOS-*lux* test [41].

The three fluorescent protein genes coding for EGFP, GFPuv, and DsRed (fluorescent protein derived from the sea anemone *Discosoma* sp. [66]) were similarly fused to the *recA* promoter [78], and the responses to nalidixic acid were compared to a luminescent *recA'::lux* strain. Performance was usually poorer compared to bioluminescent *recA*-based reporters: lag times were longer and detection thresholds were higher, unless incubation times were very long. The *recA'::DsRed* plasmid, hosted in *E. coli* UTL2, was used in order to monitor antigenotoxic activity of plant extracts and exhibited some protection against MMC, NA, and hydrogen peroxide [79].

The use of fluorescent proteins as reporters has been characterized as superior in terms of stability but inferior to enzyme-based reporters in terms of sensitivity and response kinetics [78, 80]. Norman et al. [81] have demonstrated that these drawbacks can be circumvented by the use of flow cytometry, showing that the response threshold of strain *cda'::gfp*mut3 to MNNG was 5 nM, tenfold lower than the minimal detectable concentration (MDC) of the *umu*-test [17]. Moreover, the experimental procedure enabled the detection of MMC in spiked soil.

5 Cytotoxicity Controls

As samples or chemicals suspected of DNA damaging activity are likely to also be cytotoxic, genotoxicity assays often incorporate suitable controls to neutralize or correct for the effects cell damage and death may have on assay results. One simple measure is an optical determination of cell growth in parallel to assaying reporter gene activity [82]. However, this solution is limited, since optical density does not necessarily reflect the viability status of a cell suspension. A different approach is based on the inclusion of an additional, constitutive reporting strain or enzyme which serves as a "light off" sensor: a decrease in its signal indicates a toxic effect of the sample. This approach, for example, was adopted in the VITOTOX® test that introduced a constitutive light-producing strain with a *lux* operon under the control of the strong promoter, *pr1* [33].

The SOS-*lux* test similarly incorporated a cytotoxicity reporting strain harboring a constitutive *lac*-GFPuv plasmid in the same *S. typhimurium* host strain [42]. In a further development of this system a *SWITCH* plasmid was added, combining the SOS *lux* plasmid pPLS-1 and the LAC-*Fluoro* plasmid pGFPuv [83].

Using a different approach, the tester strain in the SOS-chromotest was made constitutive for alkaline phosphatase synthesis [84]. This enzyme, noninducible by DNA-damaging agents, is assayed in parallel to β-galactosidase and the ratio of the two activities is taken as a measure of the specific activity of β-galactosidase [29].

The toxicity of a sample can also be evaluated with promoters that are induced by a broad spectrum of environmental insults and are thus good indicators of toxic cellular stress, such as the *grpE* promoter, a component of the functionally cooperating chaperone network in *E. coli* [85, 86]. The use of two strains, one harboring the plasmid *recA′*::GFPuv and the other *grpE′::lux* allowed an assessment of the toxicity of the sample along with its genotoxicity [78]. A dual-function toxicity/genotoxicity bioreporter system was reported by Hever and Belkin [87] who described a plasmid containing both *recA′*::EGFP and *grpE′*::DsRed fusions. A somewhat different double reporter concept was demonstrated by Mitchell and Gu [88], who presented a strain containing a fluorescent genotoxicity reporter fusion (*recA′::GFPuv4*) and a bioluminescent oxidative stress reporter (*katG′::luxCDABE*).

6 Genotoxicants Detection Performance

As described in detail above, numerous genotoxicity bioassays based on genetically engineered bacteria have been presented over the years. While some of them, such as the *unu*-test and the SOS-chromotest, have undergone intensive validation and have been tested against hundreds of compounds, others have only been briefly described along with their responses to a very limited range of chemicals. Quite clearly, therefore, pending further validation of the latter group, the validated tests are of a much higher value for routine testing and their results receive higher credibility. Detailed reports of an extensive testing of these assays and their comparison to the Ames test can be found in Nakamura et al. [19], Reifferscheid and Heil [20], and Quillardet and Hofnung [89]. The *umu*-test has been standardized and accepted as an ISO and a DIN test for waste water quality (ISO/CD 13829, DIN 38423-5) [15, 16].

Table 1 summarizes the detection thresholds to selected genotoxicants exhibited by many of the assays described in the present review. Possibly the major factor which stands out from even a brief look at Table 1 is that only two systems, the *umu*-test and the SOS-chromotest, were challenged with the required spectrum of genotoxic chemicals necessary to demonstrate their applicability to environmental testing. All the others were only preliminarily challenged with a very limited number of compounds. In fact, this Table only lists chemicals that have been tested by at least one bioassay in addition to the *umu*-test and the SOS-chromotest; it thus does not contain the detection thresholds for hundreds of other compounds that have been reported for these two tests [14, 19, 20, 29, 49, 89, 94, 99]. On the basis of this limited comparison it may also be observed that detection thresholds vary greatly between the different assays, sometimes by several orders of magnitude. Other factors such as response times, detection spectra or facility of use, which have not been compared in Table 1, confer additional advantages to some of the reporter strains.

A further comparison of the performance of some of these bioassays has been performed in several hands-on workshops conducted in Belgium (Mol TECHNOTOX; [90, 100]) and in the USA (Eilatox-Oregon; [101-104]). In addition to highlighting differences in response characteristics, such workshops help to emphasize the difficulties encountered when taking a newly developed test out of the lab and into the field.

7 Sensitivity Enhancement and Expansion of the Response Spectrum

Very early in the short history of genetically engineered bacterial reporters it became apparent that simple promoter–reporter fusions may be sufficient to demonstrate the applicability of the concept for genotoxicity testing, but that additional molecular manipulations are required in order to turn them into efficient tools for routine use. Such manipulations have taken several forms including modification of the sensing elements, introducing mutations into the reporter strains for enhanced sensitivity and permeability, and the incorporation of metabolic activation capabilities.

Table 2 lists some of the genetic manipulations introduced into the *E. coli* or *S. typhimurium* host strains and their reported effects. The modifications can be divided into two classes: deficiencies that reduce the ability of the cells to defend against DNA damaging agents and new or enhanced abilities of bacterial cells to metabolically activate progenotoxic compounds, thus at least partially mimicking the metabolic pathways such compounds may undergo in mammalian systems.

Table 1 Published detection thresholds of selected chemicals by genetically engineered bacterial genotoxicity reporters

	umu-test	SOS-chromotest	SOS-*lux* test [41, 42, 90, 91]	VITOTOX® [32, 33]	sulA-test [31]	*recA'::lux* [38, 92]	*nrdA'::lux* [45]	*recA'::gfp* [77]	*cda':: gfp* [12]	*recA'::DsRed* [79]
Promoter	*umuCD*	*sulA (sfiA)*	*Cda (ColD)*	*recN*	*sulA*	*recA*	*nrdA*	*recA*	*Cda (ColD)*	*recA*
Reporter gene(s)	*lacZ*	*lacZ*	*luxCDABFE P. leiognathi*	*luxCDABE V. fischeri*	*lacZ*	*luxCDABE V. fischeri*	*luxCDABE P. luminescens*	*gfp*mut3	*gfp*mut3	*DsRed2*
Host strain	*S. typhimurium* TA1535	*E. coli* PQ37	*S. typhimurium* TA1535	*S. typhimurium* TA104	*S. typhimurium* TA1535	*E. coli* RFM443	*E. coli* RFM443	*E.coli* C600	*S. typhimurium* TA1535	*E. coli* UTL2
Compound	MDC- Minimal detectable concentration [μM]									
Fungal toxins and antibiotics										
MMC	0.01 [19]	0.016 [29]	4.3×10^{-3}	0.046	5.3×10^{-3}	2.9×10^{-4}	0.93	0.012	9.1×10^{-3}	0.011
Doxorubicin	1.1 [19]	0.41 [93]	0.9		0.43					
NA	2.4 [19]		4.6	0.69			10.77	3.57	1.02	3.01
Bleomycin	0.04 [19]	22.5 [93]	0.35							
Esters										
MMS	150 [17]	63 [29]		73	117					
EMS	1.8×10^{3} [19]	10^{4} [94]		2,061	3,414					
DMS	300 [19]	6.7 [29]	7.5							
Nitroso-, nitro-										
DEN	5×10^{4} [19]	199 [29]			2,349	1.5×10^{4}				
MNNG	3.0 [23]	0.45 [94]		0.6	0.40	1	0.33	1.06	0.763	0.16
NPAHs										
4-NOPD	32 [14]	ND [49]			10.4					
3-NFA	0.04 [95]	0.25 [96]			0.06					
4-NQO	0.1 [19]	0.02 [29]		0.042	0.004	0.018		13.1		
Furazolidone	0.22 [47]	<0.44 [49]			0.002					
HAs										
2-AA	1 [17]	3 [94]		1.08		0.037				0.08
2-AF	0.4 [17]	< 274 [94]			1.1					

PAHs										
B[a]P	4 [19]	2.33 [29]		0.79	0.58	5×10^{-5}				
Fluoranthene	ND [19]	ND [97]		15.3						
Chrysene	65 [19]	<87 [49]		21.9						
Acridine										
Acridine orange	12 [19]	75 [29]			0.83					
ICI 191	0.5 [19]	0.7 [94]		0.68	0.3					
9-aminoacridine	46 [19]	ND [49]			5					
Miscellaneous										
H_2O_2	1,322 [19]	1,000 [94]	50	59	84	23.5	1,470	286	24.7	588
CH_2O	623 [19]	>1,000 [94]	6,500					305	141	
Epi	649 [19]	3,300 [94]		1,383						
EtBr	127 [19]	1,165 [30]				0.32				
Metals										
$K_2Cr_2O_7$	258 [19]	68 [98]	0.68	14						
$CdCl_2$	ND [20]	ND [30]		ND		0.99				

MDC The lowest concentration at which the response is systematically over twice the background; *ND* Not detected; *NPAH* Nitrated polycyclic aromatic hydrocarbons; *HA* Heterocyclic amines; *PAH* Polycyclic aromatic hydrocarbons; *MMC* Mitomycin C; *NA* Nalidixic acid; *MMS* Methyl methanesulfonate; *EMS* Ethyl methanesulfonate; *DMS* Dimethylsulfate; *DEN* Diethylnitrosamine; *MNNG N*-methyl-*N'*-nitro-*N*-nitrosoguanidine; *4-NOPD* 4-Nitro-*o*-phenylenediamine; 2-NFA 2-Nitrofluoranthene; *4-NQO* 4-Nitroquinoline-N-oxide; *2-AA* 2-Aminoanthracene; *2-AF* 2-Aminoflouren; *B[a]P* Benzo[a]pyrene; H_2O_2 Hydrogen peroxide; CH_2O Formaldehyde; *Epi* Epichlorohydrin; *EtBr* Ethidium bromide

7.1 Introduction of Host Strain Mutations for Enhanced Sensitivity

As listed in Table 2, several mutations have been introduced into genotoxicity reporter strains to enhance their sensitivity and thus lower their detection thresholds. While some mutations, such as *tag* or *oxyR* have only been reported once (for the SOS-chromotest, [89]), others have become almost a prerequisite in microbial genotoxicity reporters. Most notable in the latter group are *uvrAB* mutants deficient in excision repair, and *rfa* mutations that, by increasing membrane permeability, allow higher intracellular concentrations of the tested chemicals [111, 112].

7.2 Altering the Sensing Element: Manipulation in Regulatory Sequences

Molecular manipulations of the DNA fragment harboring the sensing promoter element in order to improve the bacterial response to the target chemicals have also been described. The most notable example is the VITOTOX® strain [32]. In addition to the wild-type *recN* promoter, two different promoter mutants were constructed and tested singly and in combination: a deletion of one of the LexA binding sites, and a "promoter up" mutation where a consensus nucleotide was introduced in the -35 region. Each of the single mutants was superior to the wild type in at least one respect, but the double mutation resulted in poorer performance compared to the wild type. Another successful effort was done by Arai et al. [75], who improved the performance of a *umuCD'::gfp* construct by replacing the wild-type -35 promoter sequence with the -35 sequence of the highly active *trp* gene promoter.

7.3 Introduction of Metabolic Activation Enzymes

To at least partially alleviate the lack of metabolic activation potential in prokaryotes and correspondingly reduce the dependency on external metabolic activation by rodent-derived cytochrome P450 (S9) preparations [1, 113], several attempts have been made to genetically engineer bacterial cells to incorporate some of the enzymatic activities involved in the activation process of xenobiotics. Some of these efforts are summarized in Table 2, clearly demonstrating the viability and potential of this approach. Nevertheless, while the study of single enzymes is a necessary step in improving our understanding of the activation of promutagens to mutagens [114], it should be remembered that the cyt P-450 complex is composed of a plurality of activities that are not very likely to be engineered into a single reporter strain in the near future.

Table 2 Molecular modifications introduced into genotoxicity reporter strains to enhance sensitivity, expand the response spectrum and incorporate metabolic activation capabilities

Manipulation	Test	Strain	Modified capabilities	Effect	Reference
uvrAB mutation	SOS-chromotest	PQ37/*sfiA::Mud(Ap lac) cts*	Deficiency in nucleotide excision repair (NER)	Increased sensitivity toward certain genotoxicants	[1, 14, 29, 31, 37, 42, 43]
	umu-test	TA1535/*umuCD'::lacZ*			
	GenoTox	TA1535/*cda'::gfp*			
	SOS-*lux*	TA1538/*cda'::lux*			
	VITOTOX®	TA104/*recN2-4::lux*			
	sulA-test	TA1538/*sulA'::lacZ*			
	Rec-lac test	KY946 φ(*recA-lacZ)*			
tag mutation	SOS-chromotest	PQ243/*sfiA::Mud(Ap lac) cts*	Inactivation of the constitutive 3-methyl-adenine DNA glycosylase I	Response to lower concentrations of alkylating agent as MNNG, MMS, etc.	[89, 105]
oxyR mutation	SOS-chromotest	PQ300/*sfiA::Mud(Ap lac) cts*	Depletes the oxidative stress responses under the control of OxyR transcription regulator	More sensitive to various classes of peroxides and compounds generating peroxides	[89, 93]
rfa mutation	SOS-chromotest	PQ37/*sfiA::Mud(Ap lac) cts*	Mutation in the core enzymes of the Lypopolysaccharide (LPS) biosynthesis. Incomplete LPS composed of the ketodeoxy-octanoate-lipid core	Higher permeability to substances, especially important with larger hydrophobic genotoxins	[1, 14, 30, 31, 33, 43, 106]
	umu-test	TA1535/*umuCD'::lacZ*			
	GenoTox	TA1535/*cda'::gfp*			
	SOS-*lux*	TA1538/*cda'::lux*			
	VITOTOX®	TA104/*recN2-4::lux*			
	sulA-test	TA1538/*sulA'::lacZ*			
tolC mutation	GenoTox	N43/*cda'::gfp*mut3	Inactivation of the efflux, outer membrane transporter- TolC	Limited efflux capability, increases sensitivity to genotoxins	[12, 58, 106]
	SOS-*lux*	PB3/*cda'::lux*			
	recA'::lux	DE112/*recA'::lux*			
S. typhimurium NR overexpression	*umu*-test	NM1011/*umuCD'::lacZ*	High nitroreductase activity	Highly sensitive towards many nitroarenes as 2-NF, 1-NP, etc.	[95]

(continued)

Table 2 (continued)

Manipulation	Test	Strain	Modified capabilities	Effect	Reference
S. typhimurium O-AT overexpression	*umu*-test	NM2009/*umuCD′::lacZ*	13-fold higher isoniazid-N-acetyltransferase activity	High sensitivity toward nitro- and dinitro- containing compounds, as well as arylamines, aminoanzo and HAs	[107, 108]
S. typhimurium O-AT and NR overexpression	*umu*-test GenoTox	NM3009/*umuCD′::lacZ* TGO2/*cda′::gfp*	High O-AT and NR activity	Increased sensitivity to aromatic amines and nitroarenes with/without external MA (S-9)	[43, 48, 108]
Human CYP1A2 and NADPH-P450 reductase expression in *S. typhimurium*	*umu*-test	OY1001/*umuCD′::lacZ*	7-Ethoxyresorufin O-deethylation and NADPH-cytochrome c reductase activity	Detection of some carcinogenic HAs, without the addition of metabolic activation system (S-9)	[109]
Human CYP1A2 and NADPH-P450 reductase with O-AT expression in *S. typhimurium*	*umu*-test	OY1002/*umuCD′::lacZ*	7-Ethoxyresorufin O-deethylation and NADPH-cytochrome c reductase activity, joint with high O-AT activity	More sensitive to HAs than the previous strain, without external MA. Detects the mutagens APNH and APH	[48, 110]

CYP1A2 Cytochrome P450 1A2; *O-AT O-* acetyltransferase; *NR* nitroreductase; *HA* heterocyclic amines; *MA* metabolic activation; *APNH* aminophenylnorharman; *APH* aminophenylharman

8 Tests that Do not Involve Promoter–Reporter Fusions: Reversion Assays

The second major group of bacterial test systems for the assessment of genotoxicity is based on the reversion of auxotrophic bacterial strains to prototrophic strains by mutation events that occur more frequently in the presence of genotoxic compounds. This basic concept was described by Ames and Hartmann already in the 1970s [1, 111]. In this system the target gene sequence for mutagenesis is the histidine operon (*hisLGDCBHAFI*). The *Salmonella typhimurium* tester strains do not proliferate in the absence of histidine due to mutations in the *his* operon but can recover histidine synthesis by remutations. Different strains are available allowing a determination of the type of mutation that is caused by the genotoxicant. Mostly used are the *S. typhimurium* strains TA98, sensitive to +1 and –2 frameshift mutations, and TA100, responsive to base substitutions. The sensitivity of these strains to genotoxic compounds is enhanced by *uvrB and rfa mutations* (see above). Further milestones were the addition of S9-preparations to the assay in order to mimic the metabolism of xenobiotics in eukaryotes and the introduction of the *mucAB* genes via the plasmid pKM101 [115].

Many attempts have since been reported to further improve the performance of the Ames test, such as the already discussed introduction of genes coding for enzymes involved in the activation of xenobiotics in order to enhance the metabolic competence of the bacterial cells. An interesting approach was described by Aubrecht et al. [116] with the cloning of the *P. luminescens luxCDABE* genes into strains TA98 and TA100. Bioluminescence, which is coupled to the metabolic state of the cell, is used as a sensitive sensing element allowing the identification of histidine-independent revertants in a high-throughput fashion.

The basic concept of growth recovery after reversion was adopted to rationally designed *lacZ* mutants of *E. coli* that are sensitive for defined mutation events [117, 118]. After reversion the test strains grow on lactose minimal medium because of the regained enzymatic activity of β-galactosidase [119, 120].

9 Reversion Tests Using Reporting Genes/Selection Markers

An attractive alternative to the traditional reversion test is its combination with reporter genes and/or selection markers. Ulitzur et al. isolated a spontaneous dark variant of *P. leiognathi* and showed that its reversion frequency to the luminescent wild type is increased by the presence of mutagenic compounds [121]. The Mutatox test was developed using this approach [122]. A further example for this strategy is the usage of a dim *luxE* mutant of the marine bacterium *V. harveyi* [123–127].

Other approaches utilize the gain of resistance to antibiotics induced by genotoxicants for the assessment of the mutagenic potential of a compound. The "MutaGen" assay uses a reversibly knocked-out beta-lactamase gene coding for ampicillin resistance. Revertants are able to grow in the presence of ampicillin whereas nonreverted bacteria disappear due to lytic death. The *lacZ* gene under the stringent control of the tetracycline repressor provides a sensitive detection system for revertants [128, 129].

Another approach applying an antibiotic selection marker quantifies the appearance of neomycin-resistant mutants of the marine bacterium *V. harveyi* [130, 131]. To improve performance, a mutant with enhanced UV-sensitivity was generated by transposon mutagenesis and a plasmid containing the *mucAB* genes was added.

10 Devices Incorporating Bacterial Genotoxicity Reporters

Several reports describe genetically engineered bacteria incorporated into specially designed hardware to generate dedicated genotoxicity biosensors. Polyak et al. [39] have immobilized the *E. coli* strain DPD1718 [58, 59] that contains a chromosomally integrated *recA'::lux* fusion in sodium alginate onto the tip of an optical fiber. The luminescent signal induced in the bacteria by the presence of genotoxicants was collected by the fiber and electronically amplified. Sensor strains embedded in alginate retained their sensitivity following a 2-month incubation, but at the cost of a significantly delayed response [58]. A biosensor composed of a high-density living bacterial cell array was fabricated by depositing single *E. coli* cells carrying a *recA'::gfp*mut2 into a microwell array formed on one end of an imaging fiber bundle [132]. Each fiber in the array had its own distinct light pathway, enabling thousands of individual cell responses to be monitored simultaneously with both spatial and temporal resolution. The biosensors demonstrated an active sensing lifetime of more than 6 h and a shelf lifetime of 2 weeks. An on-chip whole-cell genotoxicity bioassay was developed [133], using a three-dimensional microfluidic network system composed of one perforated microwell chip bound in two microchannel chips, in which the sensor cells were immobilized in agarose. The bioluminescent responses to MMC of the sensor strain, SOSluc, in a wild-type *E. coli* or its *tolC* deficient derivative were measured by a charge-coupled camera (CCD). The immobilized cells were stable for at least 1 week at 4°C.

An electrochemical mutagen screening based on the *umu*-test was performed on a microbial chip combined with a scanning electrochemical microscopy (SECM) device [51]. The microbial chip was fabricated by embedding 5 nl of collagen-immobilized genetically engineered *S. typhimurium* strain (TA1535/pSK1002) in a microcavity on a glass substrate. β-Galactosidase activity was monitored by electrochemical determination of the concentration of *p*-aminophenol (PAP), the enzymatic hydrolosis product of PAPG. Although such biosensors open potential horizons for both field applications and laboratory high-throughput screening systems, their current status requires much additional development before they become available for routine use. Such future development should take into account the necessity of including metabolic activation as an integral part of the process.

11 Summary and Outlook

Genetically engineered bacterial sensor systems display a central element in effect-directed analysis of environmental contaminants and the assessment of the DNA damaging potential of chemicals. The basis for a successful construction of bacterial sensors consist of three elements: (a) a bacterial tester strain that offers an appropriate genetic background facilitating high permeability for chemicals, an appropriate DNA-repair capacity and negligible background activity of the reporter gene; (b) a sensitive promoter that offers well-balanced repressor binding properties for sensitive induction of the reporter gene; and (c) a sensitive, fast responding reporter system with a broad dynamic range and preferably the capability of real-time monitoring.

The discovery and the understanding of the bacterial SOS-system, a regulatory pathway that is mainly responsible for inducible DNA repair and induced mutagenesis in bacteria, opened a wide variety of possibilities to genetically tailor bacteria for the specific, sensitive, and fast detection of genotoxic contaminants. The fusion of SOS gene promoters to reporter genes that generate quantifiable signals allows one to easily detect the genotoxic potency of a sample. Herein we have described the current stage of development of the sensing systems by discussing the molecular, biochemical, and physico-chemical characteristics of the different promoters and suitable reporter genes based on colorimetric, luminometric, fluorimetric, and electrochemical detection. The advantages compared with classical test systems were discussed.

The potential of future sensor developments with enhanced external and/or internal metabolic competence is immense. Against the background of a worldwide increasing freshwater demand and the need for reclamation of process water as a drinking water resource, bacterial sensors can play a crucial role in risk minimization. To fully reach this objective, additional progress needs to be made in several directions including enhancement of sensitivity, expansion of the response spectrum, stabilization of the more sensitive reagents and, most importantly, the introduction of broad-spectrum metabolic activation capabilities into the sensor strains.

Acknowledgements The authors are grateful for funding provided by the German BMBF and the Israeli MOST in the framework of project WT601 ("DipChip") of the binational Water Technology Program.

References

1. Ames BN, Durston WE, Yamasaki E, Lee FD (1973) Carcinogens are mutagens: a simple test system combining liver homogenates for activation and bacteria for detection. Proc Natl Acad Sci U S A 70:2281–2285
2. Belkin S (2003) Microbial whole-cell sensing systems of environmental pollutants. Curr Opin Microbiol 6:206–212

3. Volkert MR (1988) Adaptive response of *Escherichia coli* to alkylation damage. Environ Mol Mutagen 11:241–255
4. Volkert MR, Gately FH, Hajec LI (1989) Expression of DNA damage-inducible genes of *Escherichia coli* upon treatment with methylating, ethylating and propylating agents. Mutat Res DNA Repair 217:109–115
5. Courcelle J, Hanawalt PC (2003) RecA-dependent recovery of arrested DNA replication fork. Annu Rev Genet 37:611–646
6. Giese KC, Michalowski CB, Little JW (2008) RecA-dependent cleavage of LexA dimers. J Mol Biol 377:148–161
7. Janion C (2001) Some aspects of the SOS response system – a critical survey. Acta Biochim Pol 48:599–610
8. Little JW (1991) Mechanism of specific LexA cleavage: Autodigestion and the role of RecA coprotease. Biochimie 73:411–421
9. D'Ari R (1985) The SOS system. Biochimie 67:343–347
10. Fernández de Henestrosa AR, Ogi T, Aoyagi S, Chafin D, Hayes JJ, Ohmori H, Woodgate R (2000) Identification of additional genes belonging to the LexA regulon in *Escherichia coli*. Mol Microbiol 35:1560–1572
11. Lewis LK, Harlow GR, Gregg-Jolly LA, Mount DW (1994) Identification of high affinity binding sites for LexA which define new DNA damage-inducible genes in *Escherichia coli*. J Mol Biol 241:507–523
12. Norman A, Hansen LH, Sørensen SJ (2005) Construction of a ColD *cda* promoter-based SOS-green fluorescent protein whole-cell biosensor with higher sensitivity toward genotoxic compounds than constructs based on *recA*, *umuDC*, or *sulA* promoters. Appl Environ Microbiol 71:2338–2346
13. Sutton MD, Smith BT, Godoy VG, Walker GC (2000) The SOS Response: recent insights into *umu-DC*-dependent mutagenesis and DNA damage tolerance. Annu Rev Genet 34:479–497
14. Oda Y, Nakamura S-i , Oki I, Kato T, Shinagawa H (1985) Evaluation of the new system (umu-test) for the detection of environmental mutagens and carcinogens. Mutat Res Environ Mutagen Relat Subj 147:219–229
15. DIN 38415-3- German standard methods for the examination of water, waste water and sludge – Sub-animal testing (group T) – Part 3: Determination of the genotoxic potential of water with the umu-test (T 3)
16. ISO/CD13829 Water quality – Determination of the genotoxicity of water and waste water using the umu-test
17. Reifferscheid G, Heil J, Oda Y, Zahn RK (1991) A microplate version of the SOS/*umu*-test for rapid detection of genotoxins and genotoxic potentials of environmental samples. Mutat Res Environ Mutagen Relat Subj 253:215–222
18. McDaniels AE, Reyes AL, Wymer LJ, Rankin CC, Stelma GN Jr (1990) Comparison of the *Salmonella* (Ames) test, *Umu* tests, and the SOS chromotests for detecting genotoxins. Environ Mol Mutagen 16:204–215
19. Nakamura S-i , Oda Y, Shimada T, Oki I, Sugimoto K (1987) SOS-inducing activity of chemical carcinogens and mutagens in *Salmonella typhimurium* TA1535/pSK1002: examination with 151 chemicals. Mutat Res Lett 192:239–246
20. Reifferscheid G, Heil J (1996) Validation of the SOS/*umu*-test using test results of 486 chemicals and comparison with the Ames test and carcinogenicity data. Mutat Res Genet Toxicol 369:129–145
21. Shimada T, Nakamura S-i (1987) Cytochrome P-450-mediated activation of pro-carcinogens and promutagens to DNA-damaging products by measuring expression of *umu* gene in *Salmonella typhimurium* TA1535/pSK1002. Biochem Pharmacol 36:1979–1987
22. Sakagami Y, Yamazaki H, Ogasawara N, Yokoyama H, Ose Y, Sato T (1988) The evaluation of genotoxic activities of disinfectants and their metabolites by *umu*-test. Mutat Res Lett 209:155–160

23. Hamer B, Bihari N, Reifferscheid G, Zahn RK, Müller WEG, Batel R (2000) Evaluation of the SOS/*umu*-test post-treatment assay for the detection of genotoxic activities of pure compounds and complex environmental mixtures. Mutat Res Genet Toxicol Environ Mutagen 466:161–171
24. Whong W-Z, Wen Y-F, Stewart J, Ong T-m (1986) Validation of the SOS/*Umu* test with mutagenic complex mixtures. Mutat Res Lett 175:139–144
25. Bihari N, Vukmirovic M, Batel R, Zahn R (1990) Application of the SOS *umu*-test in detection of pollution using fish liver S9 fraction. Comp Biochem Physiol 95:15–18
26. Dizer H, Wittekindt E, Fischer B, Hansen P (2002) The cytotoxic and genotoxic potential of surface water and wastewater effluents as determined by bioluminescence, *umu*-assays and selected biomarkers. Chemosphere 46:225–233
27. Ehrlichmann H, Dott W, Eisentraeger A (2000) Assessment of the water-extractable genotoxic potential of soil samples from contaminated sites. Ecotoxicol Environ Saf 46:73–80
28. Higashitani A, Higashitani N, Horiuchi K (1995) A cell division inhibitor SulA of *Escherichia coli* directly interacts with FtsZ through GTP hydrolysis. Biochem Biophys Res Commun 209:198–204
29. Quillardet P, Huisman O, D'Ari R, Hofnung M (1982) SOS chromotest, a direct assay of induction of an SOS function in *Escherichia coli* K-12 to measure genotoxicity. Proc Natl Acad Sci U S A 79:5971–5975
30. Quillardet P, Hofnung M (1985) The SOS chromotest, a colorimetric bacterial assay for genotoxins: procedures. Mutat Res Environ Mutagen Relat Subj 147:65–78
31. El Mzibri M, De Méo MP, Laget M, Guiraud H, Séree E, Barra Y, Duménil G (1996) The Salmonella sulA-test: a new in vitro system to detect genotoxins. Mutat Res Genet Toxicol 369:195–208
32. van der Lelie D, Regniers L, Borremans B, Provoost A, Verschaeve L (1997) The VITOTOX® test, an SOS bioluminescence *Salmonella typhimurium* test to measure genotoxicity kinetics. Mutat Res Genet Toxicol Environ Mutagen 389:279–290
33. Verschaeve L, Van Gompel J, Thilemans L, Regniers L, Vanparys P, van der Lelie D (1999) VITOTOX® bacterial genotoxicity and toxicity test for the rapid screening of chemicals. Environ Mol Mutagen 33:240–248
34. Vijayashree B, Ahuja Y, Regniers L, Rao V, Verschaeve L (2005) Genotoxicity of the Musi river (Hyderabad, India) investigated with the VITOTOX test. Folia Biol (Praha) 51:133–139
35. Verschaeve L (2002) Genotoxicity studies in groundwater, surface waters, and contaminated soil. Sci World J 2:1247–1253
36. Nunoshiba T, Nishioka H (1989) Genotoxicity of quinoxaline 1,4-dioxide derivatives in *Escherichia coli* and *Salmonella typhimurium*. Mutat Res DNA Repair 217:203–209
37. Nunoshiba T, Nishioka H (1991) 'Rec-lac test' for detecting SOS-inducing activity of environmental genotoxic substances. Mutat Res DNA Repair 254:71–77
38. Vollmer AC, Belkin S, Smulski DR, Van Dyk TK, LaRossa RA (1997) Detection of DNA damage by use of *Escherichia coli* carrying *recA'::lux*, *uvrA'::lux*, or *alkA'::lux* reporter plasmids. Appl Environ Microbiol 63:2566–2571
39. Polyak B, Bassis E, Novodvorets A, Belkin S, Marks RS (2000) Optical fiber bioluminescent whole-cell microbial biosensors to genotoxicants. Water Sci Technol 42:305–311
40. Frey J, Ghersa P, Palacios PG, Belet M (1986) Physical and genetic analysis of the ColD plasmid. J Bacteriol 166:15–19
41. Ptitsyn LR, Horneck G, Komova O, Kozubek S, Krasavin EA, Bonev M, Rettberg P (1997) A biosensor for environmental genotoxin screening based on an SOS lux assay in recombinant *Escherichia coli* cells. Appl Environ Microbiol 63:4377–4384
42. Baumstark-Khan C, Rode A, Rettberg P, Horneck G (2001) Application of the Lux-Fluoro test as bioassay for combined genotoxicity and cytotoxicity measurements by means of recombinant *Salmonella typhimurium* TA1535 cells. Anal Chim Acta 437:23–30
43. Østergaard TG, Hansen LH, Binderup M-L, Norman A, Sørensen SJ (2007) The *cda* GenoTox assay: A new and sensitive method for detection of environmental genotoxins, including nitroarenes and aromatic amines. Mutat Res Genet Toxicol Environ Mutagen 631:77–84

44. Courcelle J, Khodursky A, Peter B, Brown PO, Hanawalt PC (2001) Comparative gene expression profiles following UV exposure in wild-type and SOS-deficient *Escherichia coli*. Genetics 158:41–64
45. Hwang ET, Ahn J-M, Kim BC, Gu MB (2008) Construction of a *nrdA::luxCDABE* fusion and its use in *Escherichia coli* as DNA damage biosensor. Sensors 8:1297–1307
46. Jain VK, Magrath IT (1991) A chemiluminescent assay for quantitation of β-galactosidase in the femtogram range: application to quantitation of β-galactosidase in lacZ-transfected cells. Anal Biochem 199:119–124
47. Pal AK, Rahman MS, Chatterjee SN (1992) On the induction of *Umu* gene expression in *Salmonella typhimurium* strain TA1535/pSK1002 by some nitrofurans. Mutat Res Genet Toxicol 280:67–71
48. Oda Y, Kunihiro F, Masaaki K, Akihiko N, Taro Y (2004) Use of a high-throughput *umu*-microplate test system for rapid detection of genotoxicity produced by mutagenic carcinogens and airborne particulate matter. Environ Mol Mutagen 43:10–19
49. Ohta T, Nakamura N, Moriya M, Shirasu Y, Kada T (1984) The SOS-function-inducing activity of chemical mutagens in *Escherichia coli*. Mutat Res DNA Repair Rep 131:101–109
50. Fuentes JL, Alonso A, Cuétara E, Vernhe M, Alvarez N, Sénchez-Lamar A, Llagostera M (2006) Usefulness of the SOS chromotest in the study of medicinal plants as radioprotectors. Int J Radiat Biol 82:323–329
51. Matsui N, Kaya T, Nagamine K, Yasukawa T, Shiku H, Matsue T (2006) Electrochemical mutagen screening using microbial chip. Biosens Bioelectron 21:1202–1209
52. Biran I, Babai R, Levcov K, Rishpon J, Ron EZ (2000) Online and *in situ* monitoring of environmental pollutants: electrochemical biosensing of cadmium. Environ Microbiol 2:285–290
53. Biran I, Klimentiy L, Hengge-Aronis R, Ron EZ, Rishpon J (1999) On-line monitoring of gene expression. Microbiology 145:2129–2133
54. Paitan Y, Biran D, Biran I, Shechter N, Babai R, Rishpon J, Ron EZ (2003) On-line and in situ biosensors for monitoring environmental pollution. Biotechnol Adv 22:27–33
55. Schwartz-Mittelmann A, Neufeld T, Biran D, Rishpon J (2003) Electrochemical detection of protein-protein interactions using a yeast two hybrid: 17-β-Estradiol as a model. Anal Biochem 317:34–39
56. Meighen EA (1993) Bacterial bioluminescence: organization, regulation, and application of the lux genes. FASEB J 7:1016–1022
57. Meighen EA, Szittner RB (1992) Multiple repetitive elements and organization of the lux operons of luminescent terrestrial bacteria. J Bacteriol 174:5371–5381
58. Davidov Y, Rozen R, Smulski DR, Van Dyk TK, Vollmer AC, Elsemore DA, LaRossa RA, Belkin S (2000) Improved bacterial SOS promoter::lux fusions for genotoxicity detection. Mutat Res Genet Toxicol Environ Mutagen 466:97–107
59. Rozen R, Davidov Y, LaRossa RA, Belkin S (2000) Microbial sensors of ultraviolet radiation based on *recA'::lux* fusions. Appl Biochem Biotechnol 89:151–160
60. Elasri M, Miller R (1998) A *Pseudomonas aeruginosa* biosensor responds to exposure to ultraviolet radiation. Appl Microbiol Biotechnol 50:455–458
61. Justus T, Thomas SM (1998) Construction of a *umuC'-luxAB* plasmid for the detection of mutagenic DNA repair via luminescence. Mutat Res Fund Mol Mech Mutagen 398:131–141
62. Schmid C, Reifferscheid G, Zahn RK, Bachmann M (1997) Increase of sensitivity and validity of the SOS/*umu*-test after replacement of the β-galactosidase reporter gene with luciferase. Mutat Res Genet Toxicol Environ Mutagen 394:9–16
63. Chalfie M, Tu G, Euskirchen G, Ward WW, Prasher DC (1994) Green fluorescent protein as a marker for gene expression. Science 263:802–804
64. Fradkov AF, Chen Y, Ding L, Barsova EV, Matz MV, Lukyanov SA (2000) Novel fluorescent protein from Discosoma coral and its mutants possesses a unique far-red fluorescence. FEBS Lett 479:127–130
65. Crameri A, Whitehorn EA, Tate E, Stemmer WPC (1996) Improved green fluorescent protein by molecular evolution using DNA shuffling. Nat Biotechnol 14:315–319

66. Matz MV, Fradkov AF, Labas YA, Savitsky AP, Zaraisky AG, Markelov ML, Lukyanov SA (1999) Fluorescent proteins from nonbioluminescent Anthozoa species. Nat Biotechnol 17:969–973
67. Nagai T, Ibata K, Park ES, Kubota M, Mikoshiba K, Miyawaki A (2002) A variant of yellow fluorescent protein with fast and efficient maturation for cell-biological applications. Nat Biotechnol 20:87–90
68. Shaner NC, Patterson GH, Davidson MW (2007) Advances in fluorescent protein technology. J Cell Sci 120:4247–4260
69. Tsien RY (1998) The green fluorescent protein. Annu Rev Biochem 67:509–544
70. Wiedenmann J, Ivanchenko S, Oswald F, Schmitt F, Röcker C, Salih A, Spindler K-D, Nienhaus GU (2004) EosFP, a fluorescent marker protein with UV-inducible green-to-red fluorescence conversion. Proc Natl Acad Sci U S A 101:15905–15910
71. Kain SR, Kitts P (1997) Expression and detection of green fluorescent protein (GFP). In: Recombinant protein protocols: detection and isolation, p 305–324
72. Cormack BP, Valdivia RH, Falkow S (1996) FACS-optimized mutants of the green fluorescent protein (GFP). Gene 173:33–38
73. Heim R, Tsien RY (1996) Engineering green fluorescent protein for improved brightness, longer wavelengths and fluorescence resonance energy transfer. Curr Biol 6:178–182
74. Welsh S, Kay SA (1997) Reporter gene expression for monitoring gene transfer. Curr Opin Biotechnol 8:617–622
75. Arai R, Makita Y, Oda Y, Nagamune T (2001) Construction of green fluorescent protein reporter genes for genotoxicity test (SOS/*umu*-test) and improvement of mutagen-sensitivity. J Biosci Bioeng 92:301–304
76. Justus T, Thomas SM (1999) Evaluation of transcriptional fusions with green fluorescent protein versus luciferase as reporters in bacterial mutagenicity tests. Mutagenesis 14:351–356
77. Kostrzynska M, Leung KT, Lee H, Trevors JT (2002) Green fluorescent protein-based biosensor for detecting SOS-inducing activity of genotoxic compounds. J Microbiol Methods 48:43–51
78. Sagi E, Hever N, Rosen R, Bartolome AJ, Rajan Premkumar J, Ulber R, Lev O, Scheper T, Belkin S (2003) Fluorescence and bioluminescence reporter functions in genetically modified bacterial sensor strains. Sens Actuator B Chem 90:2–8
79. Bartolome A, Mandap K, David KJ, Sevilla Iii F, Villanueva J (2006) SOS-red fluorescent protein (RFP) bioassay system for monitoring of antigenotoxic activity in plant extracts. Biosens Bioelectron 21:2114–2120
80. Hakkila K, Maksimow M, Karp M, Virta M (2002) Reporter genes *lucFF*, *luxCDABE, gfp*, and *dsred* have different characteristics in whole-cell bacterial sensors. Anal Biochem 301:235–242
81. Norman A, Hansen LH, Sørensen SJ (2006) A flow cytometry-optimized assay using an SOS-green fluorescent protein (SOS-GFP) whole-cell biosensor for the detection of genotoxins in complex environments. Mutat Res Genet Toxicol Environ Mutagen 603:164–172
82. Baun A, Andersen JS, Nyholm N (1999) Correcting for toxic inhibition in quantification of genotoxic response in the *umuC* test. Mutat Res Genet Toxicol Environ Mutagen 441:171–180
83. Baumstark-Khan C, Cioara K, Rettberg P, Horneck G (2005) Determination of geno- and cytotoxicity of groundwater and sediments using the recombinant SWITCH test. J Environ Sci Health A Tox Hazard Subst Environ Eng 40:245–263
84. Torriani A, Rothman F (1961) Mutants of *Escherichia coli* constitutive for alkaline phosphatase. J Bacteriol 81:835–836
85. de Marco A, Deuerling E, Mogk A, Tomoyasu T, Bukau B (2007) Chaperone-based procedure to increase yields of soluble recombinant proteins produced in *E. coli*. BMC Biotechnol 7:32
86. Van Dyk TK, Majarian WR, Konstantinov KB, Young RM, Dhurjati PS, LaRossa RA (1994) Rapid and sensitive pollutant detection by induction of heat shock gene-bioluminescence gene fusions. Appl Environ Microbiol 60:1414–1420

87. Hever N, Belkin S (2006) A dual-color bacterial reporter strain for the detection of toxic and genotoxic effects. Eng Life Sci 6:319–323
88. Mitchell RJ, Gu MB (2004) Construction and characterization of novel dual stress-responsive bacterial biosensors. Biosens Bioelectron 19:977–985
89. Quillardet P, Hofnung M (1993) The SOS chromotest: a review. Mutat Res Rev Genet Toxicol 297:235–279
90. Baumstark-Khan C, Rabbow E, Rettberg P, Horneck G (2007) The combined bacterial Lux-Fluoro test for the detection and quantification of genotoxic and cytotoxic agents in surface water: Results from the "technical workshop on genotoxicity biosensing". Aqua Toxicol 85:209–218
91. Rabbow E, Rettberg P, Baumstark-Khan C, Horneck G (2002) SOS-LUX- and LAC-FLUORO-TEST for the quantification of genotoxic and/or cytotoxic effects of heavy metal salts. Anal Chim Acta 456:31–39
92. Min J, Kim EJ, LaRossa RA, Gu MB (1999) Distinct responses of a recA::luxCDABE Escherichia coli strain to direct and indirect DNA damaging agents. Mutat Res Genet Toxicol Environ Mutagen 442:61–68
93. Müller J, Janz S (1992) Assessment of oxidative DNA damage in the oxyR-deficient SOS chromotest strain Escherichia coli PQ300. Environ Mol Mutagen 20:297–306
94. von der Hude W, Behm C, Gürtler R, Basler A (1988) Evaluation of the SOS chromotest. Mutat Res Environ Mutagen Relat Subj 203:81–94
95. Oda Y, Shimada T, Watanabe M, Ishidate M, Nohmi T (1992) A sensitive *umu* test system for the detection of mutagenic nitroarenes in Salmonella typhimurium NM1011 having a high nitroreductase activity. Mutat Res Environ Mutagen Relat Subj 272:91–99
96. Mersch-Sundermann V, Kern S, Wintermann F (1991) Genotoxicity of nitrated polycyclic aromatic hydrocarbons and related structures on Escherichia coli PQ37 (SOS chromotest). Environ Manage 18:41–50
97. Mersch-Sundermann V, Mochayedi S, Kevekordes S (1992) Genotoxicity of polycyclic aromatic hydrocarbons in Escherichia coli PQ37. Mutat Res Genet Toxicol 278:1–9
98. Venier P, Montini R, Zordan M, Clonfero E, Paleologo M, Levis AG (1989) Induction of SOS response in *Escherichia coli* strain PQ37 by 16 chemical compounds and human urine extracts. Mutagenesis 4:51–57
99. Quillardet P, de Bellecombe C, Hofnung M (1985) The SOS chromotest, a colorimetric bacterial assay for genotoxins: validation study with 83 compounds. Mutat Res Environ Mutagen Relat Subj 147:79–95
100. Corbisier P, Hansen P-D, Barcelo D (2000) Proceedings of the BIOSET Technical Workshop on Genotoxicity Biosensing TECHNOTOX 2000, http://wwwa.vito.be/english/environment/pdf/technotox/Proceedings4_00.PDF.
101. Hakkila K, Lappalainen J, Virta M (2004) Toxicity detection from EILATox-Oregon Workshop samples by using kinetic photobacteria measurement: the flash method. J Appl Toxicol 24:349–353
102. Meriläinen J, Lampinen J (2004) EILATox-Oregon Workshop: blind study evaluation of vitotox test with genotoxic and cytotoxic sample library. J Appl Toxicol 24:327–332
103. Pancrazio JJ, McFadden PN, Belkin S, Marks RS (2004) EILATox-Oregon Biomonitoring Workshop: summary and observations. J Appl Toxicol 24:317–321
104. Pedahzur R, Polyak B, Marks RS, Belkin S (2004) Water toxicity detection by a panel of stress-responsive luminescent bacteria. J Appl Toxicol 24:343–348
105. Costa de Oliveira R, Laval J, Boiteux S (1987) Induction of SOS and adaptive responses by alkylating agents in *Escherichia coli* mutants deficient in 3-methyladenine-DNA glycosylase activities. Mutat Res DNA Repair Rep 183:11–20
106. Rettberg P, Bandel K, Baumstark-Khan C, Horneck G (2001) Increased sensitivity of the SOS-LUX-Test for the detection of hydrophobic genotoxic substances with *Salmonella typhimurium* TA1535 as host strain. Anal Chim Acta 426:167–173
107. Oda Y, Yamazaki H, Watanabe M, Nohmi T, Shimada T (1993) Highly sensitive *umu* test system for the detection of mutagenic nitroarenes in *Salmonella typhimurium* NM3009 having high O-acetyltransferase and nitroreductase activities. Environ Mol Mutagen 21:357–364

108. Oda Y, Yamazaki H, Watanabe M, Nohmi T, Shimada T (1995) Development of high sensitive *umu* test system: rapid detection of genotoxicity of promutagenic aromatic amines by *Salmonella typhimurium* strain NM2009 possessing high O-acetyltransferase activity. Mutat Res Environ Mutagen Relat Subj 334:145–156
109. Aryal P, Yoshikawa K, Terashita T, Guengerich FP, Shimada T, Oda Y (1999) Development of a new genotoxicity test system with *Salmonella typhimurium* OY1001/1A2 expressing human CYP1A2 and NADPH-P450 reductase. Mutat Res Genet Toxicol Environ Mutagen 442:113–120
110. Aryal P, Terashita T, Guengerich FP, Shimada T, Oda Y (2000) Use of genetically engineered *Salmonella typhimurium* OY1002/1A2 strain coexpressing human cytochrome P450 1A2 and NADPH-cytochrome P450 reductase and bacterial *O*-acetyltransferase in SOS/*umu* assay. Environ Mol Mutagen 36:121–126
111. Ames BN, McCann J, Yamasaki E (1975) Methods for detecting carcinogens and mutagens with the salmonella/mammalian-microsome mutagenicity test. Mutat Res Environ Mutagen Relat Subj 31:347–363
112. Makela PH, Mayer H, Whang HY, Neter E (1974) Participation of lipopolysaccharide genes in the determination of the enterobacterial common antigen: Analysis of R mutants of *Salmonella minnesota*. J Bacteriol 119:760–764
113. Malling HV (1971) Dimethylnitrosamine: formation of mutagenic compounds by interaction with mouse liver microsomes. Mutat Res Fund Mol Mech Mutagen 13:425–429
114. Muckel E, Frandsen H, Glatt HR (2002) Heterologous expression of human *N*-acetyltransferases 1 and 2 and sulfotransferase 1A1 in *Salmonella typhimurium* for mutagenicity testing of heterocyclic amines. Food Chem Toxicol 40:1063–1068
115. McCann J, Spingarn NE, Kobori J, Ames BN (1975) Detection of carcinogens as mutagens: bacterial tester strains with R factor plasmids. Proc Natl Acad Sci U S A 72:979–983
116. Aubrecht J, Osowski JJ, Persaud P, Cheung JR, Ackerman J, Lopes SH, Ku WW (2007) Bioluminescent Salmonella reverse mutation assay: a screen for detecting mutagenicity with high throughput attributes. Mutagenesis 22:335–342
117. Cupples CG, Cabrera M, Cruz C, Miller JH (1990) A set of *lacZ* mutations in *Escherichia coli* that allow rapid detection of specific frameshift mutations. Genetics 125:275–280
118. Cupples CG, Miller JH (1989) A set of *lacZ* mutations in *Escherichia coli* that allow rapid detection of each of the six base substitutions. Proc Natl Acad Sci U S A 86:5345–5349
119. Josephy PD (2000) The *Escherichia coli* lacZ reversion mutagenicity assay. Mutat Res Fund Mol Mech Mutagen 455:71–80
120. Ohta T, Watanabe-Akanuma M, Yamagata H (2000) A comparison of mutation spectra detected by the *Escherichia coli* Lac+ reversion assay and the *Salmonella typhimurium* His+ reversion assay. Mutagenesis 15:317–323
121. Ulitzur S, Weiser I, Yannai S (1980) A new, sensitive and simple bioluminescence test for mutagenic compounds. Mutat Res Environ Mutagen Relat Subj 74:113–124
122. Kwan KK, Dutka BJ, Rao SS, Liu D (1990) Mutatox test: A new test for monitoring environmental genotoxic agents. Environ Pollut 65:323–332
123. Chec E, Podgórska B, Wegrzyn G (2006) Direct addition of cultures of tester bacteria into semi-permeable membrane devices (SPMDs) as a modified procedure for preliminary detection of mutagenic pollution of the marine environment by use of microbiological mutagenicity assays. Mutat Res Genet Toxicol Environ Mutagen 611:17–24
124. Podgórska B, Królicka A, Lojkowska E, Wegrzyn G (2008) Rapid detection of mutagens accumulated in plant tissues using a novel *Vibrio harveyi* mutagenicity assay. Ecotoxicol Environ Saf 70:231–235
125. Podgórska B, Pazdro K, Pempkowiak J, Wegrzyn G (2007b) The use of a novel *Vibrio harveyi* luminescence mutagenicity assay in testing marine water for the presence of mutagenic pollution. Mar Pollut Bull 54:808–814
126. Podgórska B, Pazdro K, Wegrzyn G (2007a) The use of the *Vibrio harveyi* luminescence mutagenicity assay as a rapid test for preliminary assessment of mutagenic pollution of marine sediments. J Appl Genet 48:409–412

127. Podgórska B, Wegrzyn G (2006) A modified *Vibrio harveyi* mutagenicity assay based on bioluminescence induction. Lett Appl Microbiol 42:578–582
128. Reifferscheid G, Arndt C, Schmid C (2005) Further development of the beta-lactamase MutaGen assay and evaluation by comparison with Ames fluctuation tests and the *umu* test. Environ Mol Mutagen 46:126–139
129. Schmid C, Arndt C, Reifferscheid G (2003) Mutagenicity test system based on a reporter gene assay for short-term detection of mutagens (MutaGen assay). Mutat Res Genet Toxicol Environ Mutagen 535:55–72
130. Czyz A, Jasiecki J, Bogdan A, Szpilewska H, Wegrzyn G (2000) Genetically modified *Vibrio harveyi* strains as potential bioindicators of mutagenic pollution of marine environments. Appl Environ Microbiol 66:599–605
131. Czyz A, Szpilewska H, Dutkiewicz R, Kowalska W, Biniewska-Godlewska A, Wegrzyn G (2002) Comparison of the Ames test and a newly developed assay for detection of mutagenic pollution of marine environments. Mutat Res Genet Toxicol Environ Mutagen 519:67–74
132. Kuang Y, Biran I, Walt DR (2004) Living bacterial cell array for genotoxin monitoring. Anal Chem 76:2902–2909
133. Tani H, Maehana K, Kamidate T (2004) Chip-based bioassay using bacterial sensor strains immobilized in three-dimensional microfluidic network. Anal Chem 76:6693–6697

Hdb Env Chem (2009) 5J: 187–201
DOI:10.1007/698_2009_16

Adverse Effects in Aquatic Ecosystems: Genotoxicity as a Priority Measurement

Tamara Grummt, Petra Rettberg, Petra Waldmann, Jürgen Zipperle, and Peter-Diedrich Hansen

Abstract Some chemical substances are capable of modifying genetic material and thereby also damaging it. Aquatic genotoxicity testing represents an important contribution to environmental research today.

Sensitive testing methods for practical applications have been developed with reasonable certainty, which allow genotoxic (mutagenic) agents to be discovered in the environment. Risk levels for humans and the environment may be reliably assessed and the resulting consequences predicted.

Keywords Bacterial test systems, Eukaryotic indicator tests, Genotoxicity, Test strategies

Contents

T. Grummt (✉)
Federal Environment Agency, Bad Elster Branch, Heinrich-Heine-Str. 12, 08645, Bad Elster, Germany
e-mail: tamara.grummt@uba.de

P. Rettberg
Institute of Aerospace Medicine, Radiation Biology Department, Research "Group Astrobiology", Linder Höhe, 51147, Köln, Germany

P. Waldmann
AMMUG Johannes-Gutenberg- Universität Mainz

J. Zipperle
LUBW - State Institute for Environment, Measurements and Nature Conservation Baden-Württemberg

P.-D. Hansen
Technische Universität Berlin (BIT - Berlin Institute of Technology), Department of Ecotoxicology, Franklinstrasse 29 (OE4), 10587, Berlin, Germany

1 Introduction

The ever-increasing volume of regulations for water protection and drinking water standards in industrialized countries has led to a shift in focus from microbial problems to various kinds of chemical contamination of aquatic systems. Such chemical contamination is generalized under terms like "adverse or novel effects", and includes effects such as genotoxicity, immunotoxicity, neurotoxicity and reproductive toxicity.

Environmental contaminants that are suspected of being genotoxic take on special precedence on the basis of their biological consequences in a so-called reactionary chain of "exposure and effect". Genotoxic compounds (synthetic or naturally occurring) are substances that are capable of altering DNA and bringing about genetic changes, which usually result in adverse biological effects. Genotoxic substances can cause hereditary genetic mutations, cancer and embryonic malformations in humans.

Investigations on the influence of genotoxic threats to the population in an ecosystem have identified the following relevant effects (just to name a few): Decrease in vitality and a reduction of genetic diversity, which is associated with a reduced fitness of the population as a whole (Fig. 1).

The identified risks and consequential reaction in environmental and healthcare policy make the handling of the topic "genotoxicity" a priority research area, and demonstrate an immediate need for regulatory measures with the aim of minimizing exposure to genotoxic compounds or, when technically possible, avoiding exposure altogether.

The first significant step in this daunting task is the identification of genotoxic impact and the recognition of substance-specific mechanisms. Investigations of complex exposures in a low-dose-range, which are designed to reflect real conditions of exposure in the environment, entail quite a few methodological difficulties.

Nevertheless, several aquatic genotoxicity testing systems are available today that sufficiently provide systematic, needs-based genotoxicity testing, and under appropriate conditions allow for a reliable detection of potentially genotoxic substances.

Key questions of aquatic genotoxicity testing:

1. What substances are in the water, and in what concentrations do they occur?
2. What mechanisms of action do such substances have?

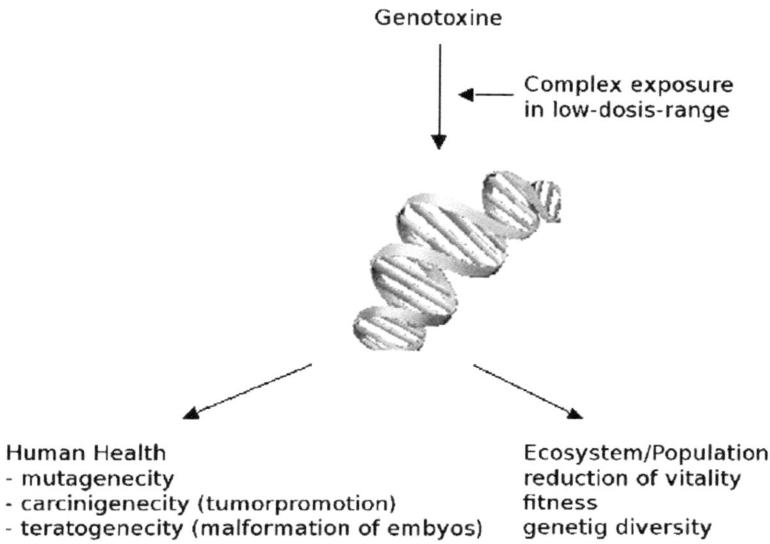

Fig. 1 Genotoxic effects and biological consequences: human health and environmental integrity

3. In order to keep health risks as low as possible, what are acceptable levels of potential exposure (threshold values)?

2 Methods of Aquatic Genotoxicity Testing

In genotoxicity testing a distinction is made between indicator and mutation testing (or assays). An indicator test demonstrates damage to the molecular structure of the DNA. The indicated DNA damage can result in a genetic mutation. A mutation assay, on the other hand, indicates already existing mutations (e.g. gene or chromosome mutations), which are presumably caused by exposure to genotoxic agents.

In the following we shall take a detailed look at such genotoxicity testing systems as they have been applied to "water" as a testing matrix, and in some cases been developed to address relevant issues effecting aquatic organisms, out of which primary cells or cell lines have been derived for research purposes (Table 1).

2.1 *Bacterial Assays*

Bacterial assays belong to the basic set of tools for genotoxicity testing of chemicals and environmental samples. There are several established methods that are easy to

Table 1 Genotoxicity and test organism

Test system	Test organism
Alkaline filter eluation	Mussels (*Corbicula fluminea*)
	Algae (*Chlamydomonas reinhardii*) [1]
DNA-unwinding assay	Fish cell line (RTG-2, PLHC-1)
	Fish larvae (*Danio rerio*, *Oncorhynchus mykiss*)
	Crustaceans (*Astacus astacus*, *Corophium volutator*, *Gammurus pulex*)
	Mussels (*Dreissena polymorpha*) [2]
Comet assay	Fish cell line (RTG-2, PLHC-1)
	Primary hepatocytes (*Danio rerio*, *Oncorhynchus mykiss*)
	Protozoa (*Acanthamöba castellani*)
	Algae (*Chlamydomonas reinhardii*)
	Mussels (*Corbicula fluminea*)
	Fish (*Danio rerio*) [3, 4]
UDS assay	Primary hepatocytes (*Oncorhynchus mykiss*) [5, 6]
In vitro micronuclei assay	Fish cell line (RTG-2, PLHC-1) [7]

administer and usually generate reproducible results. Of the higher eukaryotic cells, bacterial cells are distinguished by their genome organization and metabolic mechanisms. In higher organisms such as mammals, a sophisticated and highly variable xenobiotic metabolism relies on a system of enzymes to remove unwanted contaminants from the body. Through the addition of this enzyme system (such as the S9 extract from a rat or mouse liver) to the bacterial assay, the activation potential of indirect mutagens in the mammalian organism may be partially reconstructed in vitro. However, lessons learned from the bacterial assay may not always be directly transferred to eukaryotic systems, and require additional analysis.

Examples of bacterial indicator tests for screening mutagenicity are the umu-test, the SOS-Chromotest and the LUX-FLUORO-Test. These tests use the bacteria *Salmonella typhimurium* and *Escherichia coli* as test organisms. The tests are based on the measurement of a genotoxically determined induction of a group of regulated genes, or so-called SOS regulons. The SOS functions are complex biochemical processes, which counteract the degradation of bacterial genes and allow the survival of the cell. These include a somewhat fallible enzymatic DNA repair system, the formation of an excision repair complex and an inhibitor of cell division. The turning-on of an SOS regulon is in effect the cell's physiological response to a DNA-damaging impact. In all three tests the induction of an SOS-regulated promoter is measured with the help of a hidden (genetically engineered) reporter gene. The Umu and SOS Chromotests are mostly concerned with isolating the lacZ gene, which codes for the β-Galactosidase enzyme found in *E. coli*, and is easily identified by a colour reaction. In the LUX-FLOURO-test, the magnitude and kinetics of the SOS induction are determined by measuring the induced bioluminescence. In this case, the lux-Operon is the reporter that codes for the enzyme Luciferase and its substrate (Fig. 2).

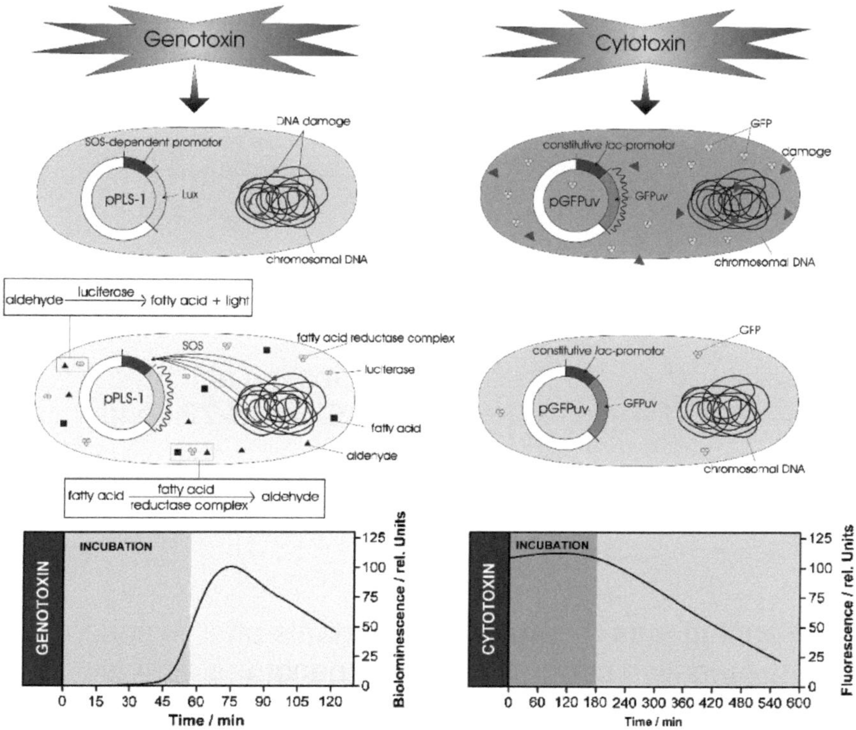

Fig. 2 Principle of the Lux-Fluoro Assay

Parallel to these tests, the cytotoxicity of the sample is determined by a decrease in fluorescence of the constitutively expressed GFP gene (green fluorescent protein). In addition, conventional turbidity measurements of the cell suspension are determined.

Bacterial indicator tests from a single strain of bacteria are thus able to determine a broad spectrum of differently functioning genotoxic agents. Moreover, the parallel use of different strains of bacteria, in which individual genes responsible for the repair of DNA damage are mutated, allows for specific analysis of any occurring DNA damage, while subsequent conclusions about the nature of the damaging agents can be made.

As mentioned above, indicator tests do not directly provide proof of mutations. However, they clearly demonstrate a high correlation between the induction of SOS regulon genes and the manifestation of mutations. The three indicator tests can be carried out using 96-well microplates, and are thus automated for routine applications. While the SOS Chromotest is a commercially available test-kit, the LUX-FLUORO-Test has predominantly been used in research projects until now, including high-throughput screening tests of potential new pharmaceuticals [8]. Meanwhile, the umu-test has already been standardized for the analysis of genotoxicity in water samples (according to DIN and ISO standards), and is used for

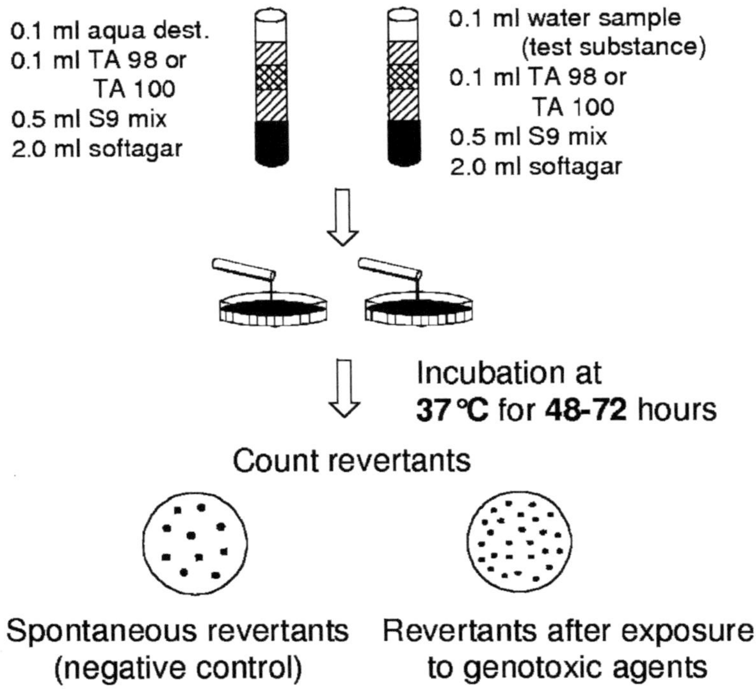

Fig. 3 Flow chart of the *Salmonella*/microsome test (AMES test)

research purposes as well as official governmental monitoring [9]. Through its mention in the Waste Water Ordinance (AbwV, 2002), the umu-test has also found its way into legal regulatory frameworks.

A true mutation test is the *Salmonella typhimurium* reverse mutation assay (Ames test, Fig. 3). Because of mutagenetic changes, the salmonella strains used in this test require the amino acid histidine for growth (his-). In addition, these strains contain mutations that increase the bacteria's sensitivity to some genotoxic agents. The rfa-mutation increases the cell membranes' permeability against larger, particularly hydrophobic molecules. The uvrB-mutation causes the failure of an enzyme system that resects certain DNA damages. In the presence of genotoxic agents, reverse mutations to the wild-type (his^+-phenotype) frequently occur. These reverse mutations grow on agar plates containing mere traces of histidine, and can grow up to colonies, which can then be counted. The number of reverse mutations serves as a measure of genotoxicity. Different salmonella strains were used to identify different types of mutagens or mutation types. In the standardized version of the *Salmonella typhimurium* reverse mutation test, strain TA 98 is used to detect frameshift mutations, while strain TA 100 is used to detect base substitution mutations. In addition to the umu-test, the Ames test has also been standardized according to DIN and ISO norms for aquatic screening [10].

2.2 Eukaryotic Assays

There are a number of available methods to practically identify genotoxicity in eukaryotic cells and aquatic organisms at various endpoints, such as DNA strand breaks, DNA repair and micronuclei. Methods for identifying DNA strand breaks include the Comet assay, the base filter elution test and the DNA alkaline unwinding assay. In these tests, single and double-strand breaks as well as alkaline or abasic points converted to strand breaks are summarily detected. Moreover, all three methods allow such endpoints to be identified in different cell types and tissues, regardless of the cell cycle. Strand breaks reduce the length of chromosomal DNA. Numerous genotoxic substances can cause DNA strands to break indirectly. The strand breaks occur (intermittently) with more frequency if the damaged DNA sites are repaired by base and nucleotide excision. The DNA damage is first visible when DNA repair is initiated in the test.

The Comet assay (or Single Cell Gel Electrophoresis assay) uses the different migration behaviours of intact and damaged DNA in an electric field [11]. In the Comet assay, single cells (either from permanent cell cultures, or isolated from the organism itself) are embedded in agarose on laboratory slides and lysed under alkaline conditions or in high salt concentrations. DNA is thereby released and then electrophoretically unravelled in a strong alkaline environment and tinted with specialized fluorescent dyes. The damaged DNA is then stretched into comet-shaped strands in the electric field, revealing a characteristic comet tail, after which the test is named (Fig. 4). The length of the comet tail and its genetic contents are dependent on the extent of the existing strand breaks or the frequency of alkali-labile sites in the DNA. In contrast, intact DNA does not migrate; rather it remains in the form of a circular core area. The extent of DNA damage is thus mostly determined by the DNA migration in the electric field, or, by the length of the comet's tail. In addition to tail length, fluorescence intensity, and therefore the amount of DNA in the head and tail areas, may be measured with the help of appropriate image evaluation technology.

The base filter elution test also confirms evidence of such DNA lesions. In this test reduced lengths of chromosomal DNA due to strand breaks can be detected [12]. In a strongly alkaline elution solution, damaged DNA fragments pass more rapidly through the filter (with defined pore size) than the longer, intact DNA strands. The eluate containing DNA is thus captured in fractions. While many short pieces of severely damaged DNA fall out in the first eluted fractions, only a few pieces are initially eluted in the case of marginally damaged DNA. The elution rate of the DNA is calculated according to a defined time. In addition, one can fluorometrically determine the DNA content in the eluate fractions as well as the amount of DNA remaining on the filters. Under the given experimental conditions, the alkaline elution of DNA almost appears as a first-order of kinetics. The decline of DNA on the filter, per unit of time, is proportional to the frequency of single-strand breaks. Damaged and undamaged DNA may therefore be distinguished by the different (negative) slopes of their elution curves. This procedure has been greatly simplified in recent years, so that measurements are also possible with microplates.

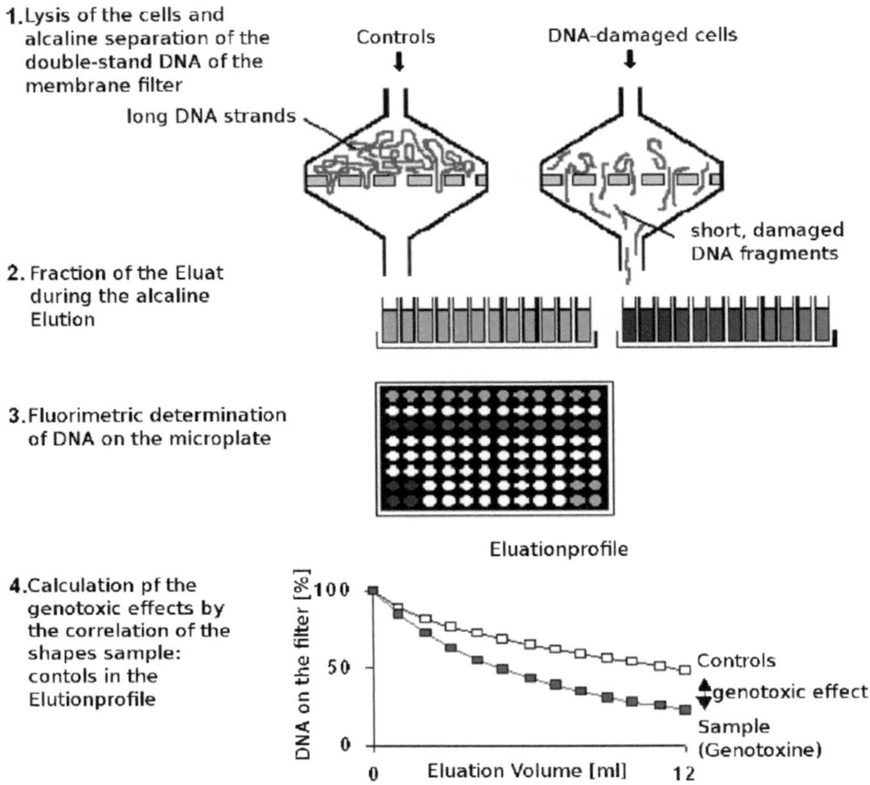

Fig. 4 Flow chart of the Alkaline Filter Eluation (AFE test)

The DNA alkaline unwinding assay exploits the fact that DNA strand breaks reveal additional unwinding sites in an alkaline environment, while the number of unwinding sites of intact DNA mainly reflects the number of DNA strand ends and physiologically determined topoisomerase interfaces along the DNA molecule. During an alkaline lysis, DNA with strand breaks tends to unravel faster than undamaged DNA because of the additional unwinding sites present. The default unwinding rate is measured over time, usually 30 min. After single and double-stranded DNA are separated from each other with the hydroxylapatite ion chromatography method and quantitatively calculated in the fluorometer, the relative fraction of double-stranded DNA (*F*-value) remaining after this given time period may then be determined [2].

The measurement of DNA repair synthesis (UDS) is a widely accepted indicator test for genotoxicity, and represents an important part of testing strategies for chemicals and pharmaceuticals [13]. The UDS test was developed with many different variations, using primary cells and cell lines such as lymphocytes, epithelial cells and liver cells. It covers a wide spectrum of applications, for example, using liver cells from rodents. Cells from aquatic organisms such as trout or zebra fish have also been used in water and wetland protection screening research.

Generally a cell is capable of detecting lesions in the DNA early on, excising the damaged sections and replacing these through neosynthesis. This process is called unscheduled DNA synthesis (UDS). The principle of UDS testing is based on the fact that physico-chemical contaminants can induce or activate DNA-Excision Repair. In so doing, the introduced cells, used together with the test substance or test mixture, will be provided with radioactive thymidine. This is formed during the repair synthesis in the newly synthesized section of DNA. The amount of built up radioactivity is for the most part autoradiographically determined. It is a measure to confirm that repairs have taken place. After the photochemical developing and staining of specimens, a semi- or fully automatic microscopic analysis of the cells follows. The number of silver grains on the nucleus region is calculated, as well as about one to three large plasma areas of similar size for comparison. The net silver rate is the average difference between the silver grain frequency (NNG) and the nuclei and the cytoplasm area. The greater the net silver rate of the sample, the more likely the sample should be evaluated as genotoxic.

The in vitro micronucleus test represents a relatively simple procedure for detecting chromosome damage [14, 15]. Micronuclei are structures containing chromatin in the cytoplasm. While formation in vivo and in vitro frequently occurs under exposure to physical or chemical contaminants, spontaneous formation occurs rather infrequently. Micronuclei arise during mitosis, when a chromosome or chromosome fragment fails to be incorporated into one of the daughter nuclei during cell division. When cells with chromosomal damage form in the anaphase of cell division, fragments lacking centromeres cannot reach the terminal ends of the cell and remain in the cell, outside the teleophase nucleus. Another cause of micronuclei formation is an abnormal function of the spindle apparatus, or centromere region of the chromosome, in which whole chromosomes in the spindle apparatus can no longer migrate to the prospective region of the nucleus. The chromosomes, or fragments thereof, are not incorporated in the daughter cell's nuclei during cell division and go on to develop their own nuclear membrane, resulting in what appears to be a tiny autonomous nucleus. Since it is presumed that the formation of micronuclei occurs during mitoses, a corresponding testing method for divisible eukaryotic cells is made available. On account of the possibility of automated image analysis, micronucleus testing, with its use of cell lines or cultured cells taken from (exposed) organisms, has become increasingly important in ecotoxicological screening for potential genotoxic subtances.

3 Examples of Practical Applications

Modern chemical analysis has allowed environmental contaminants to be identified in the nano-, pico- and even femtogramme ranges. To what extent analytically detectable contaminants are also toxicological threats, must be confirmed by data from biological testing systems. The evaluation of potential genotoxic risks in complex environmental media and their impact on humans and the environment cannot be limited to the consideration of individual substances. In addition to considerations about reasonable operation ease and expense, important criteria for the selection of relevant testing systems include the specificity and sensitivity to pollutant mixtures. In the following example, by using data from analytical measurements and biological tests (*Salmonella*/microsomal test), an attempt was made to establish justified principles for the application-specific assessment (concerning necessary remediation) of the potential risks of military waste deposits (typical explosive compounds, STV), as they occur as complex mixtures in water. STV causes the formation of methaemoglobin, reproductive disorders, mutations and cancer. Genotoxic effects of this substance class are represented by the nitro-functional group, which demonstrates an active structural principle of genotoxicity. Initial analytical measurements at the site of a former military arms manufacturer pointed to a diffuse and widespread contamination of adjacent surface waters, including nitro-aromatic STV from military waste deposits. In a 1-year measurement program, genotoxicity could be decidedly detected in monthly water screening tests (with reproducible results) only in samples with certain nitro-aromatic concentrations. Through this specific case study, a concentration-dependent correlation between complex chemical exposure and biological effects could be made. The combination of chemical analysis and biological testing led to a usage-based prioritization of pending remediation measures and, ultimately, to ecologically and economically acceptable proposals for an acute environmental problem.

The DNA unwinding test serves as a further example of an application-based genotoxicity testing procedure in the environment. In particular, its current range of applications includes monitoring of coastal and inland waters as an adjunct research interest of chemical residue testing, as well as the evaluation of damaged fish larvae as a consequence of UV-B exposure. The effects of biotoxins (such as the neurotoxins domoic acid and saxitoxin) can likewise be detected at the genetic level, both in the case of exposed individual organisms as well as populations and ecosystems. In addition to chemical residue testing data, monitoring studies on the Elbe River have delivered important data on genotoxic potential, as gathered in DNA unwinding tests on exposed zebra mussels (*Dreissena polymorpha*), which have led to important specifications for remediation of certain parts of the river. With the help of DNA unwinding tests on zebra mussels, a colour-coded genotoxicity table for water quality classification could be developed. This work will be included in the Federal Water Quality Atlas, conventional water quality maps and corresponding follow-up documentation for the European Water Framework Directive (WFD).

As an important part of sediment monitoring and the evaluation of impacts to estuaries and inland waters, DNA unwinding tests on long-term exposed mussels have helped provide important information about the environmentally safe deposition or disposal of contaminated sediments.

In situ exposure studies of UV-B radiation (from sunlight) on pelagic fish eggs have proven to be especially useful in the verification of genotoxic impacts on the ecosystem. DNA unwinding tests on fish eggs allow for quantitative registration of Thymindimere and DNA damage, in addition to deformity rates at the genetic level.

Similar to the DNA unwinding test and Comet assay, the base filter elution test can also be carried out on different test organisms from aquatic habitats. A few years ago, on commission from the State Rheinland-Pfalz, Ministry of the Environment, the base filter elution test was carried out to detect DNA strand breaks in golden orfe (*Leuciscus idus melanotus*) in connection with the fish-monitoring test (golden orfe test). After 14 days of exposure in the Rhine, golden orfe from the Rhine water station in Mainz were tested for genotoxic potential using the base filter elution test. The target tissue was taken from the liver of the fish. The golden orfe, a vertebrate, was later replaced by the freshwater clam, *Corbicula fluminea*, for ethical reasons. In this case the target tissue was taken from the gills. In order to compare test results gleaned from different test organisms, results were referenced to the genotoxic substance Benzo[*a*]pyrene (B[*a*]P), and specified in GSP (genotoxic impact potential, 1 GSP Ψ 0.2 μg B[*a*]P part per Litre).

Average genotoxicity levels in the Rhine near Mainz decreased almost continuously from 1987 to 1995. Over the course of the study, genotoxic potentials were measured with increasing infrequency, while the overall amount of the measured effects declined as well. Over the 8-year research period, the lowest genotoxic impact levels were recorded in 1990. During the same period, the percentage of adsorbable organic halides (Adsorbierbare organisch gebundene Halogene, AOX) also steadily declined. These parallel developments suggest a link between the two parameters, in that the reduction of AOX is consequentially accompanied by a reduction in genotoxic potential.

3.1 A Battery of Tests

Discussions in the field of chemical and pharmaceutical screening over the past 10–20 years have shown that there is no single ideal testing system for the detection of genotoxicity. However, a definitive testing strategy consisting of three consecutive steps has been generally accepted. The test assumes that two to three in vitro methods with several different biological endpoints are sufficient for achieving a satisfactory result in the first phase of testing contaminants for possible genotoxic effects. Environmental monitoring also seems to follow the uncontested dictum that a hierarchy of testing stages is necessary for the detection of genotoxicity. The aim of the first stage (the basic test) must be the positive detection of genotoxic potential in the water sample by using two to three short-term procedures. For qualitative screening, a crucial selection criterion in the choice of testing systems is a heightened sensitivity of the system. Special consideration is given to the fact that substance mixtures with unknown genotoxic potential will be tested.

The objective of the basic test is achieving qualitative conclusions on the genotoxicity of a sample. For this reason, in vitro methods are usually used in the first stage of genotoxicity testing. Literature indicates a high level of reliability in the

Table 2 Exposure scenarios, action plans and target in the aquatic genotoxicity assessment

Scenario	Action plans/Target values	Measures
Impact monitoring, waterways and lakes	Documentation of the status and trends of the waterways and lakes	Planning concepts and long-term decision support for water authorities and municipalities
Acute disasters	Real-time risk assessment, Risk communication	Risk management monitoring indicators and remediation strategies
Emissions and genotoxic effects in the recipient waters	Assessment and evaluation of genotoxic effects (dose-response relationship and characterisation of the effects and mechanism, Impact identification	Eco-toxicological and toxicological evaluation of the impact (risk description, risk estimation, risk characterisation)

informational value of in vitro testing [16]. The methods described above are carried out relatively quickly and can easily differentiate between genotoxic (yes) and non-genotoxic (no) results.

While standardized test protocols are already available for the microbial test systems (Umu and Ames tests), the current methodological validation and standardization of eukaryotic testing are in drastic need of improvement. Ultimately, these test versions are also to be defined and integrated in the battery of tests.

As long as such essential prerequisites (such as the procurement of quantitative evidence and the detection of specific mechanisms or impacts) have not been met, the decisive yes/no (genotoxic/non-genotoxic) result of the basic testing phase cannot be regarded as a definitive, scientifically validated risk assessment.

The basic test, with its precautionary character and qualitative predictive value, makes an important contribution towards protecting the environment from potentially genotoxic substances, and respectively minimizing their effects with reasonably cost-effective measures.

The basic test phase prescribes no specified testing strategy. The heterogeneity of environmental problems certainly does not allow this. Rather, the testing strategies are to be methodically adapted to the specific environmental question at hand, in accordance with existing exposure scenarios and related scientific knowledge (Table 2).

While the basic test may be considered adequate enough for emissions testing, it is clearly insufficient for timely risk assessments, the scientific presentation of evidence in the case of environmental damage, and in the evaluation of continuous emissions of substances with suspected genotoxic effects. It will be necessary to supplement the basic test with other testing systems (e.g. cytotoxicity or apoptosis assays) that can both quantitatively and qualitatively determine the respective impact potential of the substances in question.

In any case, the basic test will likely remain the cornerstone for initial assessments of potential genotoxic risks. The nature and extent of the necessary tests for risk assessment indicate the structure of the battery of tests thus far, and provide incentive for further testing strategies.

4 Future Prospects

At the current stage of development, two tasks must be considered with highest priority for the future:

- Implementation of existing methodological instruments into current environmental monitoring programs;
- An efficient integration of basic research (e.g. identification of biological target structures) and application-based developments in chemical analysis to identify key biological mechanisms associated with specific contaminants.

4.1 Implementation

The current practice in water monitoring shows that aquatic genotoxicity testing is still in a developmental stage that largely precludes a routine broad-based deployment. With knowledge of the uncertainties in the use of the battery of tests, methodological development and the collection of data should be otherwise undertaken.

The medium-term goal of current scientific work should therefore be to update and develop the available testing methods in order to process as much of the spectrum of specific issues in the aquatic field as possible and to be able to incorporate these into regulatory requirements. Despite existing objections to the integration of additional parameters into existing screening programmes, the current research concept aims at keeping testing programmes financially feasible, enabling innovative new developments to come about and thus creating opportunities for sustainable water protection in the context of potential health problems of the future.

4.2 Identification of Key Biological Mechanisms Associated with Specific Contaminants

Environmental toxicology greatly benefits from rapid advancements in molecular biology. The use of molecular biological methods will trigger a sea change in environmental risk analysis and assessment, at which point it is worth mentioning: contaminant-associated effects must be understood at the molecular level, if they are to be minimized or prevented.

A significant goal of current research activities is thus to adapt the instrumental equipment of analytical nuclear technologies to the latest achievements in biological and medical research (*the molecular approach*) as soon and as best as possible. The integral analytical procedures common to the fields of nucleic acid-, protein- and analytic function research undoubtedly possess enormous potential for development and application in the field of environmental toxicology.

The fascination with these technologies may not, however, obstruct the question of where the use of these new methodological approaches in environmental fields may lead.

For one thing, the accurate measurement of exposure proves to be a crucial criterion for the successful application of the procedures. In any case, it is a matter of one exposure in the context of complex mixtures, which can lead to the induction of complex toxicological effect profiles.

On the other hand, the allocation of macromolecules places enormous demands on the analytical method - from the collection of the sample (yield), to the sample preparation (stability of the effect complex), to the actual analytical procedure (sensitivity and specificity).

It is also to be made clear, that the substantiated interactions between biological target structures and environmental contaminants usually provide no proof of the biological relevance of such a finding. In most cases, causal chains of effects must be additionally determined in respective biotests.

5 Summary

The occurrence of genotoxic substances in aquatic systems is a serious problem because of the risk to both human and ecosystem health. The far-reaching consequences of possible genotoxic impacts (in situ) for environmental policy, trying to discourage exposure to genotoxic substances, require a refined experimental concept.

The bioassays, umu-test, Ames-test, Comet assay (in protozoans, green algae, mussels, primary fish hepatocytes, fish and mammalian cell lines), the DNA unwinding test (in fish larvae, crayfish, mussels and fish cell lines), the alkaline filter elution (in mussels and green algae), the UDS-test (in primary fish hepatocytes) and the micronucleus test (in fish and mammalian cell lines and primary hepatocytes) are sensitive enough to detect genotoxic effects in complex mixtures in a low dose range.

These tests seem to be suitable for routine testing and should therefore be used in the first stage of a graduated battery of tests for the detection of genotoxicity in aquatic systems in combination with effect-oriented chemical analysis.

References

1. Waldmann P (1993) Die alkalische Filterelution als Methode zur Erfassung von gentoxischen Potenzialen in Gewässern. Dissertation, University of Mainz
2. Wittekindt E, Matthess C, Gaumert T, Hansen P-D (2000) Die gentoxische Gewässergüte-Klassifizierung der Elbe - entwickelt mit Hilfe des DNA-Aufwindungstests mit der Dreikantmuschel (The genotoxic quality classification of the River Elbe - developed by means

of DNA unwinding assays with the zebra mussel *Dreissena polymorpha*). Hydrol Wasserbewirt 44:131–143

3. Weßler A, Erbes M, Krolla-Sidenstein P, Obst U (1998) Nachweis von DNA-Schäden in aquatischen Protozoen und Algen mit der Mikrogelelektrophorese (comet assay) für eine gentoxikologische Gewässergütebewertung von Oberflächenwasser. *Acta Hydrochim Hydrobiol* 26:82–89
4. Deventer K (1999) Untersuchungen zum Nachweis gentoxischer Wirkungen von Stoffen mittels Comet Assay und Mikrokerntest an der Regenbogenforelle (*Oncorhynchus mykiss*) und der Dreikantmuschel (*Dreissena polymorpha*). Dissertation, University of Karlsruhe
5. Walton DG, Acton AB, Stich HF (1984) DNA repair synthesis following exposure to chemical mutagens in primary liver, stomach, and intestinal cells isolated from rainbow trout. Cancer Res 44(3):1120–1121
6. Klaunig JE (1984) Establishment of fish hepatocyte cultures for use in in vitro carcinogenicity studies. Natl Cancer Inst Monogr 65:163–173
7. al-Sabti K, Metcalfe CD (1995) Fish micronuclei for assessing genotoxicity in water. *Mutat Res* 343(2–3):121–135
8. Ptitsyn LR, Horneck G, Komova O, Kozubek S, Krasavin EA, Bonev M, Rettberg P (1979) A biosensor for environmental genotoxin screening based on an SOS lux assay in recombinant *Escherichia coli* cells. *Appl Environ Microbiol* 63:4377
9. ISO 13829 (2000) Determination of the genotoxicty using the umu-test
10. ISO 16240 (2005) Determination of the genotoxicity of water - *Salmonella*/microsome test (AMES-test)
11. Tice RR, Agurell E, Anderson D, Burlinson B, Hartmann A, Kobayashi H, Miyamae Y, Rojas E, Ryu JC, Sasaki YF (2000) Single cell gel/comet assay: guidelines for in vitro and in vivo genetic toxicology testing. *Environ Mol Mutagen* 35:206–621
12. Waldmann P (1996) *Münchner Beiträge zur Abwasser- Fischerei- und Flussbiologie* 49:391
13. Madle S, Dean SW, Andrae U, Brambilla G, Burlinson B, Doolittle DJ, Furihata C, Hertner T, McQueen CA, Mori H (1994) Recommendations for the performance of UDS tests in vitro and in vivo. *Mutat Res* 312:263–285
14. Ma TH (2001) Tradescantia-micronucleus bioassay for detection of carcinogens. *Folia Histochem Cytobiol* 39(Suppl 2):54
15. Helma CS, Knasmüller S, Schulte-Hermann R (1994) Die Belastung von Wässern mit gentoxischen Substanzen I. Methoden zur Prüfung der Gentoxizität. UWSF-Z. Umweltchem. *Ökotox* 6:277–288
16. Grummt T (2002) Genetische toxikologie. In: Höll K, Wasser, 8. Aufl. (Hrsg. A. Grohmann) de Gruyter, Berlin, 529–549

Hdb Env Chem (2009) 5J: 203–226
DOI:10.1007/698_2009_17

Genotoxicity in the Environment (Eco-Genotoxicity)

P. D. Hansen, E. Wittekindt, J. Sherry, E. Unruh, H. Dizer, H. Tüg, H. Rosenthal, V. Dethlefsen, and H. von Westernhagen

Abstract Genotoxicity is a priority measurement for assessment and communication of genotoxic effects (DNA strand breaks) in the water matrix and the trophic levels of the ecosystem (Eco-Genotoxicity). Genotoxicity induction with single substances and surface water samples shows clearly the good sensitivity and reproducibility of the DNA-unwinding assay. The DNA alkaline unwinding assay detects levels of DNA fragmentation; the relative numbers of DNA strand breaks are described as the "–log *F*-value." The results of DNA fragmentation in genotoxin-exposed cells of a RTG-2 cell line, the mussel *Dreissena polymorpha*, and fish larvae of the fish species *Danio rerio* are discussed. The early life stages of the fish species *D. rerio* show the significant high sensitivity and reproducibility of the DNA alkaline unwinding assay. As a reference method for the validation of genotoxicity pulsed-field gel electrophoresis (PFGE) was investigated. The LOEC (Lowest Observed Effect Concentration) of the genotoxin 4-nitroquinoline-*N*-Oxid (4-NQO) was 78 $\mu g\ l^{-1}$ for both test organisms, cell lines and fish larvae. Native surface water samples from representative stations on the River Rhine and the River Elbe significantly induced genotoxicity in the RTG-2 fish cell line and in fish larvae from *D. rerio*. An increased sensitivity was measured

P.D. Hansen (✉), E. Wittekindt, E. Unruh, and H. Dizer
Technische Universität Berlin (BIT – Berlin Institute of Technology),
Department of Ecotoxicology, Franklinstrasse 29 (OE4), 10587, Berlin, Germany
e-mail: peter-diedrich.hansen@tu-berlin.de

J. Sherry
Environment Canada, Aquatic Ecosystem Protection Research Branch,
867 Lakeshore Road, Burlington, Ontario, Canada

H. Tüg, H.von Westernhagen
Alfred-Wegener-Institut für Polar- und Meeresforschung, Columbusstrasse,
27568, Bremerhaven, Germany

H. Rosenthal
World Sturgeon Conservation Society, Schifferstrasse 48, 21629, Neu Wulmstorf, Germany

V. Dethlefsen
Bundesforschungsanstalt für Fischerei, Deichstrasse 12, 27472, Cuxhaven, Germany

in *D. rerio* fish larvae 48 h after hatching during 3 h-exposure. The LOEC of the mussel *D. polymorpha* exposed in river water was less sensitive in the DNA alkaline unwinding assay: LOEC = 250 µg l^{-1} 4-NQO. In coastal waters the marine mussel *Mytilus edulis* is a good indicator organism for the detection of genotoxic effects and biotoxins. Beside immunotoxicity also genotoxic effects (DNA strand breaks) are found especially in UV-B (ultraviolet-B) exposed mussels (Luckas, B., J. Dahlmann, K.Erler, G. Gerdts, N. Wasmund, C. Hummert, P.-D. Hansen (2005) An Overview on Key Phytoplankton Toxins and their recent occurrence in North Sea and Baltic. Environmental Toxicology 20(1):1–17). To validate the influence of UV-B on exposed mussels (hepatopancreas and mussel hemocytes) UV-B exposure was investigated (Figs. 1 and 2) to induce and demonstrate eco-genotoxic effects in mussels and fish embryos. The fish eggs of the marine fish species *Limanda limanda* were experimentally exposed to UV-B radiation in a solar radiation simulator onboard the research vessel RV Walther Herwig III. The experimental design attempted to simulate present and future conditions in the context of eco-genotoxicity and increased UV-B exposure due to northern hemisphere ozone loss. The impact of low dosages of UV-B indicates an important genotoxic and ecological threat concerning ecosystem integrity status at the population level. In our experiments embryos of *L. limanda* taken from sites located along a gradient of chemical stress in the German Bight were exposed to solar ultraviolet-B light and allowed to recover for a defined period. The involved research groups from different disciplines and research organizations showed a significant time and dosage-dependent influence of UV-B radiation on mortality and sublethal eco-genotoxic effects (DNA damage in the environment). Our hypothesis was that there might be an interaction between exposure to chemical contaminants and the ability of embryos to recover from UV-induced genotoxic effects by repairment.

Keywords Alkaline unwinding assay, Biotoxins, *Danio rerio*, DNA, *Dreissena polymorpha*, Early Life Stages, Genotoxicity, *Limanda limanda*, RTG-2, Thymine dimers, UV-B radiation

Contents

1 Introduction

The estimation of environmental pollution in respect of genotoxicity effects on aquatic organisms is relevant for monitoring studies and risk assessment and risk management. More than 200 genotoxicity assays are available [1]. The main emphasis of genotoxicity assays is the detection of mutations as well as the induction of carcinogenic processes in order to estimate human risk [2, 3]. An environmental task of importance is the determination of ecotoxicological (eco-genotoxic) endpoints [5–11]. With regard to waste water and effluent assessment, bacterial test systems, i.e., the Ames-assay [4, 5, 12], the SOS-Chromotest [13, 14], and the umuC assay [9, 15], are the most applied assays for the detection of genotoxins.

Monitoring studies used in genotoxicity assessment of environmental matrices like surface water, sediments, and the organismic level [11, 16] require the development and implementation of standardized and internationally harmonized eukaryotic bioassays. Prokaryotic genotoxicity assays offer an excellent chance for real-time biosensor development (see Fig. 6). This will enhance the relevance at the population and ecosystem level. In eukaryotic cells, DNA strand breaks lead to the formation of DNA fragments that behave differently in electrophoresis than the entire chromosome. Differences in DNA migrations can be detected with the comet assay in single eukaryotic cells [7, 18, 19]; however, the comet assay is only an indicator assay. An alternative method for the assessment of primary DNA damage in tissue from exposed aquatic test organisms, which are of higher organized evolutionary order, is the DNA alkaline unwinding assay. The DNA of an exposed eukaryotic organism shows a higher quantity of DNA-unwinding points in comparison to untreated DNA. As a consequence, the DNA-unwinding process under alkaline conditions ($pH > 12$) is enhanced. A successive hydroxyapatite chromatography technique is used to separate single-strand (ss-DNA) and double-strand (ds-DNA) DNA fractions. The basics of this procedure were obtained from [20, 21]. This method has been commonly used for monitoring DNA fragmentation potential using various biological systems in ecotoxicological and toxicological research fields [19, 22].

For the assessment of eco-genotoxicity after environmental pollution, impacts of genotoxins as well as biotoxins, the DNA alkaline unwinding assay is a relevant tool for field and laboratory work. [7]. Fish species notably expressed high sensitivity for the detection of genotoxic damage [23], and enhanced effects can be detected when larval stages are investigated [24, 25].

Various studies were carried out with the cyprinidae *Danio rerio* Hamilton-Buchanan as a model organism. Effects of chemical-induced embryo–larval toxicity are measured by incubating newly fertilized eggs from *D. rerio* to test environmental samples [26]. Analogue applications have been developed for testing teratogenicity caused by sewage water. The test species *D. rerio* was used for monitoring genotoxicity and related effects, including tumor formation in adult stages of *D. rerio*, in comparison with mutagenicity effects [27]. The influence of Aflatoxin B1 metabolism on the formation of DNA adducts was studied in *D. rerio* [28] as well as TCDD modulation effects in this context [29]. The endpoints of the exposed marine fish embryos (*Limanda limanda*) were mortality, malformations, changes in buoyancy of embryos measured as changes in osmolarity of the perivitelline fluid [24], permeability and genotoxicity and repairment (DNA unwinding and Thymine dimers).

2 Genotoxicity and Organisms for Assessment

The DNA alkaline unwinding assay has been applied to the research field of toxicology testing of chemical products in cell cultures such as mouse lymphoma [30], rat and mouse liver tissue [31], human lymphocytes [22], and diploid fibroblasts [32]. In addition, there are many studies in the literature related to the use of the DNA alkaline unwinding assay for testing rainbow trout gonadal cell (RTG-2) permanent cultures. The first investigations were described by [33].

Filter-feeding organisms, such as the fresh water mussel *Dreissena polymorpha* [34] and the marine mussel *Mytilus edulis* [18, 35], are suitable for monitoring the biological impact of xenobiota and genotoxins. *D. polymorpha* [34, 36] is sessile, widely distributed in Europe and North America and expresses a relative resistance against environmental contaminants. The Zebra mussel is a commonly used test organism for passive and active biomonitoring for the detection and quantification of different environmental contaminants like heavy metals, chlorinated hydrocarbons, polycyclic aromatic compounds, and radionuclides as well as to assess their bioavailability in the ecosystem branches [34, 37]. This mussel was used in monitoring studies from the ARGE ELBE in the River Elbe and for analogue studies in the River Rhine [36]. The technique of the DNA alkaline unwinding assay is well adapted to exposure studies to describe environmental scenarios using the following test organisms and suborganismic cells of different trophic levels:

- Rainbow trout gonadal cell (RTG-2) permanent cultures
- The mussel *D. polymorpha* (soft tissue)
- Early life stages of *D. rerio*

In an extended study these three organism levels were investigated to measure the genotoxic potential of the relevant reference substances 4-NQO, 2-acetaminofluorene (2-AAF), benzo(a)pyrene (B(a)P), *N*-nitrosodimethylamine (NDMA), and nitrofurantoine (NF). The laboratory and field trial also investigated different representative surface water samples of the River Rhine and the River Elbe. For the measurement of different fractions of single-strand DNA and double-strand DNA from RTG-2 and *D. polymorpha* tissue homogenates, the method was modified with Bisbenzimide fluorescence dye, whereas the recorded low levels of DNA concentrations measured in *D. rerio* yolk sac larvae requires fluorometric measurement with the new Picogreen® fluorescence dye.

2.1 *Chemicals and Environmental Samples*

The surface water samples tested are from the German rivers Rhine and Elbe. The test program was continued over a period of 1 year (from February to June).

The samples were centrally collected and distributed. Transported at 4°C the samples reached the participants of this trial within 24 h. The reservoir "Wahnbachtalsperre" was studied as a reference station (negative control) and the River Wupper (Gas Works = positive control) for the water samples. The Lake Gatow was a reference station for the mussels *D. polymorpha*.

2.2 Fish Cell Line RTG2

RTG-2 is a permanent cell line derived from embryonic gonad tissue from rainbow trout (*Oncorhynchus mykiss*). This cell line was obtained from the "German Collection for Cell Cultures; DSMZ Braunschweig." The cultivation of the RTG-2 cell line has to be carried out in flasks at 20°C. Monolayer cultures with a confluence grade of 90% were collected after enzymatic dissociation. Therefore, the cells were washed with PBS (200 mg l^{-1} KH_2PO_4, 8,000 mg l^{-1} NaCl, 1,150 mg l^{-1} Na_2HPO_4; pH 7.2 ± 0.2), overlaid with trypsine solution (0.05% trypsine in PBS/0.02% EDTA) and incubated for 10 min. RTG-2 culture medium (MEM with Earle's salts, without sodiumbicarbonate, without L-glutamine) was enriched with 4,766 mg l^{-1} HEPES, 294 mg l^{-1}, 850 mg l^{-1} $NaHCO_3$, 50 mg l^{-1} neomycinsulfate, and the pH was adjusted to 7.2 + 0.2. After sterile filtration with 0.2-μm pore-filters, 10% FKS was added and the osmolarity was adjusted to 300 ± 20 m Osmol. The FKS was added to inhibit the trypsine activity. For the exposition experiments, a cell concentration of 2×10^5 cells ml^{-1} was used. Three parallels per test; two independent tests were carried out for each single substance dose, surface water and dilutions, respectively. Cell suspension (10 ml) was placed in Petri dishes and incubated overnight at 20°C. The culture medium was removed and the single substance solved in water or water samples mixed with 2× concentrated medium were placed onto the cells. Substances that were not well soluble in water were solved in DMSO. The maximum DMSO concentration on the cells was below 0.1%. For the single substance exposition for 1 h at 20°C was applied. During the water sample exposition experiments, the exposition time was increased to 2 h. Cells were incubated with 325 μg l^{-1} 4-NQO as the positive control. After the exposition, the test solutions were removed and the cells overlaid with 4-ml ice-cooled Dulbecco`s PBS (DPBS; 132 mg l^{-1} $CaCl_2 \cdot 2H_2O$, 200 mg l^{-1} KCl, 200 mg l^{-1} KH_2PO_4, 100 mg l^{-1} $MgCl_2 \cdot 6H_2O$, 8,000 mg l^{-1} NaCl, 1,150 mg l^{-1} Na_2HPO_4). The single layer cells were scratched from the Petri dishes using a wiper and the cell solution placed in a tube (on ice). Afterwards the cells were centrifuged for 5 min at 200× g at 4°C, the supernatant was removed and the cells were resuspended carefully into 200 μl DPBS.

2.3 Mussels: Dreissena polymorpha

The reference station for mussel species *D. polymorpha* is Lake Gatow in Lower Saxonia, Germany. The age of the organisms was 3–4 years. After transport the mussels were maintained at 5°C in a refrigerator under aeration conditions and fed with green algae, if the maintenance was longer than 4 weeks. In a preadaptation period before the exposure experiment starts, the mussels were maintained for 5 days in ISO water (ISO 7346/3) at 22°C with aeration. For each exposition experiment, 9–12 mussels were exposed for 72 h at 22°C in 1 l exposition medium or surface water samples. As positive control, the mussels were exposed in 500 μg l^{-1} 4-NQO (in 0.1% DMSO in ISO water). After the exposition, the mussels were anaesthetized for 10 min on ice and the whole tissue was pooled from three mussels. From each exposition pool, three parallels were tested. Two independent experiments were carried out for each single substance dose and surface water. Gills were analyzed after in vitro-exposition for 1 h pooling three gill pairs per test in 200 μl homogenization buffer without a homogenization step.

2.4 Fish Larvae (Yolk Sac), *Danio rerio*

The adult zebrafish. *D. rerio* were obtained from the Umweltbundesamt (Federal Environment Agency) Berlin. The adult fish (one female and two male) were maintained in ISO-water at 26–28°C with a 12-h light/dark-cycle. For the production of eggs, plant dummies on grade cover, settled on a glass container, were introduced into the aquarium. Eggs were incubated in ISO-water at 28°C for 50–54 h for embryonic development of fish larvae. Hatched fish larvae were maintained under the same conditions prior to exposition experiments.

For each exposition experiment, ten larvae (48 h after hatching) were used. Each experiment was carried out as two independent experiments in three parallels. The ISO-water was replaced by the exposition medium and the test organisms were exposed for 3 h at 28°C. As the positive control 625 μg l^{-1} 4-NQO was used.

3 DNA-Unwinding Assay for Monitoring

For testing cytotoxicity in the RTG-2 cell line, the neutral red test was applied. 100 μl of cell suspension (3 × 104 cells) was placed in the wells of a microtest plate. The plates were incubated overnight at 20°C. The medium was removed and 200-μl exposition medium (single substances and surface water samples, respectively) was placed onto the cells in three parallels. The concentration of fetal calf serum was 2% and 1× medium concentration was adjusted.

To avoid evaporation the plates were closed with a plate sealer and were incubated for 1 h at 20°C. The exposition time in the cytotoxicity test was the same as in the DNA alkaline unwinding assay. After the exposition time (1 h) the exposition medium was removed from the cells and overlaid with MEM including Earle`s salts containing neutral red solution 0.005%. The plates were incubated for 3 h at 20°C. The medium was removed and the cells were washed with 100-μl Dulbecco`s PBS

(DPBS). The plates were incubated for 15 min at room temperature after 150-μl acetic acid (1% in 50% ethanol) was placed in the wells. The optical density is measured at 540 nm against a reference wavelength of 690 nm. Exposition-induced inhibition of cell growth is calculated in percent.

The cytotoxic effect of the mussels was estimated visually during the exposure experiment. Mussels that have maximal opening of their shells and which express no reaction after mechanical stimulation were removed from the exposition test. Cytotoxicity effects in fish larvae were tested at the end of the exposition experiments. The mobility of the larvae was tested and toxic effects as well as lethality effects were documented.

All steps of the DNA-Unwinding assay were carried out on ice. After the preparation of tissue samples, the 3-fold volume of homogenization buffer (13.8 g l^{-1} Tris, 4.5 g l^{-1} NaCl, pH 8.5) was added immediately and the samples were homogenized. The cell debris sedimented for 15 s and 0.2 ml of the homogenized sample was filled in new tubes (control: 0.2 ml of homogenization buffer).

For the lysis 0.5-ml lysis buffer (0.15 mol l^{-1} NaOH, Titrisol® Merck; 7.1 g l^{-1} Na_2HPO_4; 58.44 g l^{-1} NaCl) was added slowly. The samples were incubated for 30 min in the dark (cell lines, mussels, fish larvae).

Afterwards the 0.5-ml neutralization buffer (0.18 mol l^{-1} HCl; 5.0 mg l^{-1} phenol red) and 0.5-ml 0.5% SDS (RTG-2) or 0.5-ml 0.5% SLS with 7.46 g l^{-1} Na-EDTA (aquatic organisms), respectively were added. This was followed by an ultrasound treatment with a tip sonicator for 10 s (40 W).

For the DNA separation process, hydroxyapatite (BioRad DNA-Grade Bio Gel HTP; HAP; 1 Vol. HAP + 2 Vol. 10-mM potassium phosphate buffer) was heated for 30 min at 100°C shortly before use to eliminate CO_2. The stock solution of potassium phosphate buffer included 68.05 g l^{-1} KH_2PO_4 and 87.09 g l^{-1} K_2HPO_4. Afterwards 400 μl of cooled hydroxyapatite solution was added to the samples. With an interval of 3 min the samples were resuspended four-times on a vortex-shaker. The samples were centrifuged quickly at 3,000×*g* for one second. The pellet was washed twice by resuspending in 1.25-ml washing buffer (10 mM potassium phosphate buffer; 20% formamide) under the shaking and centrifuging conditions mentioned above.

Purification of formamide before use: 10 g charcoal plus 50 g mixed-bed ion exchanger were added to 1 l formamide. The solution was stirred for 2 h at 4°C and filtered through a double paper filter. The pH value of a 1:2 dilution with water should be below 7.5 and the conductivity below 40 μS. Otherwise the procedure has to be repeated.

For the elution of DNA from the hydroxyapatite, 1.0-ml elution buffer I (125-mM potassium phosphate, 20% formamide) was added to the HAP-DNA pellet. With an interval of 3 min the samples were shaken with a vortex-shaker and centrifuged quickly at 3,000×*g* for one second. Afterwards 0.1 ml (cell lines, mussels), 50 μl of the supernatant (ss-DNA from fish larvae) respectively, were filled in four wells (four parallels) of a microtest plate. The remaining HAP-DNA pellets were eluted with 1-ml elution buffer II (500-mM potassium phosphate, 20% formamide) under the same elution and centrifuge conditions as described for the elution of ss-DNA. Afterwards 100 μl of the supernatant (ds-DNA from cell lines and mussels), 50 μl (ds-DNA from fish larvae) respectively, were filled in four microplate wells (four parallels).

The DNA concentration is estimated by Bisbenzimide fluorescence staining (cell lines and mussels). The microplates are closed by films and incubated 5 min at 80°C in a water bath. The plates are cooled immediately in ice water. Then 200-μl Bisbenzimide solution (250 μM Bisbenzimide, dissolved in 0.05 M potassium phosphate buffer enriched with 2.2 M NaCl, pH 7.4) is added. After 45 min incubation at room temperature the fluorescence is measured with a fluorimeter (excitation: 360 nm, emission: 450 nm).

The DNA concentration for the tests with fish larvae is estimated by Picogreen® fluorescence staining. Therefore, the DNA is pretreated with RNAse (DNA-free, Boehringer Mannheim). 25 μl of firstly prepared RNAse solution (2 μl ml^{-1} RNAse in TE-buffer (10 mM Tris, 1 mM EDTA, pH 7.5) is added to the DNA and the plates are incubated for 15 min at room temperature. The microplates are closed by films and incubated for 5 min at 80°C in a water bath. Afterwards 25-μl Picogreen solution (1% Picogreen® ds-DNA Quantitation Reagent, Molecular Probes, INC, Göttingen in TE-buffer) is added to the microplates. The plates are incubated for 5 min at room temperature and the fluorescence is measured with a DYNATEC Laboratories, Type Fluorolite 1000 fluorescence photometer (excitation: 480 nm, emission: 520 nm).

Calculation of the – log F-value

After subtraction of the blank-fluorescence signal, the relative fluorescence values of the elute fractions were calculated. This F-value represents a measure of the remaining ds-DNA compared with the total DNA amount (ds + ss-DNA) in a sample:

$$F = \frac{ds - DNA}{(ss - DNA + ds - DNA)}$$

ds = double stranded
ss = single stranded.

The negative logarithm of the value F ($-\log F$) is proportional to the relative number of the DNA-unwinding points. By calculation of the factor p value from the proportion of the –log-values of exposed and control organisms, it is possible to quantify the DNA-fragmentation potential:

$$Factor\ p = \frac{-\log F(\text{exposed organisms})}{-\log F(\text{control organisms})}$$

For measurements using Picogreen® fluorescence staining, the fluorescence signal of the ss-DNA is not proportional to that of the ds-DNA. For the calculation of the $-\log F$ values the blank-corrected values of the ds-DNA have to be corrected with the factor 1.13.

3.1 Statistical Analyses

Both the experiments with single substances as well as those with surface water samples were carried out in two independent experiments.

The results in genotoxicity from single-substance experiments were evaluated by application of the following factor of significance (*SF*), calculated by the following analysis of variance:

$$\text{Factor of significance: } SF = 1 + t\frac{\sqrt{\dfrac{s^2_{sample}(n-1)+s^2_{control}(m-1)}{n+m-2}}}{\sqrt{\dfrac{n\cdot m}{n+m}}}$$

The value *t* represents the 97.5%-percentile of the *t* distribution (degrees of freedom $n + m-2$). *SF* represents the upper limit of the 97.5% for the difference of the arithmetic mean of the induction factors in the DNA-unwinding assay (factor *p* values), if sample and control variations are normally distributed. The probability that $\overline{IF}_{\text{control}} \leq SF$ is 95% was calculated to indicate a significant genotoxic result.

Statistical significance of genotoxicity effects in surface water samples:
Statistical significance of surface water samples, calculated by comparing the results for –log *F*-values of the exposed and nonexposed organisms (controls) were analyzed separately for each experimental run by the Dunnett-Test. The $P < 0.05$ level was considered significant in a single experiment. Taking into account the results of the two independent experiments, samples are defined as genotoxic, if at least one experiment expresses a significant positive result and/or a twofold-enhanced –log *F* value above the standard deviation of the control –log *F* values is detected in combination with at least one significant result.

3.1.1 Results of the Study

To examine the inducibility of genotoxic effects caused by single substances, the DNA alkaline unwinding assay was performed with the cell line RTG-2, *in vitro* and *in vivo* assays with *D. polymorpha* (gills and soft tissues), and with *D. rerio* fish larvae (Table 1).

3.1.2 Exposure Experiments with the Fish Cell Line (RTG-2)

With exception of 4-NQO (with a detection limit of 78 μg l^{-1}), genotoxicity effects induced by single substances in the RTG-2 cell line were detected in higher concentration ranges (>7.8 mg l^{-1} for 2-AA; 5 mg l^{-1} for B(a)P). The highest detection limit was measured with the single substance DMNA (>125 mg l^{-1}; Table 1).

Table 1 Genotoxins and LOEC data for RTG-2, *D. polymorpha*, and fish larvae (yolk sac) of *D. rerio* by the DNA-unwinding assay

	Single substance				
Test organism	2-AAF	B(a)P	DMNA	NF	4-NQO
RTG-2 (1 h exposition)	7,812.5	5,000	125,000	7,800	78
Dreissena polymorpha- Gills (1 h in vitro exposition)	n.m.	25*	n.m.	n.m.	n.d.
Dreissena polymorpha-soft tissue (72 h in vivo exposition)	(25,000)*	n.m.	(10,000)*	n.m.	250
Danio rerio-yolk egg stages, static experiments (3 h exposition)	250	100	n.d.	1,250	78

Concentrations are given in [μg l^{-1}]

Abbreviations: *2-AAF* 2-Acetaminofluorene; *B(a)P* Benzo(a)pyrene; *DMNA* *N*-nitrosodimethylamine; *NF* Nitrofurantoine; *4-NQO* 4-Nitroquinoline-*N*-oxide

n.d.: not detectable, n.m.: not measured, *: $P < 0.05$ (Dunnett-Test)

The detection limits on the basis of the significance factors (95%-Percentile) for the tests with the cell line RTG-2; *Dreissena polymorpha* and *Danio rerio* (yolk sac larvae stages; 48 h after hatching) with $N = 2$ independent experiments with $N > 3$ parallels per test were calculated. The *in vitro* experiments with gills from *Dreissena polymorpha* analyzed with the DNA alkaline unwinding assay were statistically evaluated with $P < 0.05$ in the Dunnett-test for both experiments

*Not evaluated by the significance factor SF, statistically significant with $P < 0.05$ in the Dunnett-test

3.1.3 Exposure Experiments with the Mussel: *D. polymorpha*

In the *in vitro* exposition experiments with gills of the mussel *D. polymorpha* the highest sensitivity was measured with the single substance benzo(a)pyrene with a detection limit of 25 μg l^{-1}. In the *in vivo* exposition experiments, the single substance 4-NQO induced genotoxicity with 250 μg l^{-1} in a good dose-response relationship. With benzo(a)pyrene and nitrofurantoine nonenhanced DNA fragmentation was observed in soft tissues after in vivo exposition. In addition, significant enhanced DNA fragmentation was detected with 2-acetaminofluorene in high concentration ranges ≥20 mg l^{-1} (resp. *N*-nitrosodimethylamine ≥10 mg l^{-1}). Enhanced sensitivity was detected in the *in vivo* exposition experiments in soft tissue in comparison with the in vitro experiments with gills (Table 1).

3.1.4 Exposure Experiments with Fish Larvae (Yolk Sac Larvae): D. rerio

The single substance 4-NQO induced genotoxicity in *D. rerio* yolk sac fish larvae with 78 μg l^{-1} whereas 2-acetaminofluorene was genotoxic in concentration ranges ≥250 μg l^{-1} (Table 1). The substance benzo(a)pyrene induces enhanced DNA fragmentation from 100 μg l^{-1} and the antibiotic nitrofurantoine expressed genotoxicity with ≥1,250 μg l^{-1}.

In all chosen test organisms, 4-NQO expressed the highest genotoxic effect in a good dose-response relationship (Table 1).

3.1.5 Genotoxicity Induced by Surface Water Samples

To test genotoxicity in different surface water samples, the results of the DNA alkaline unwinding assay using the test systems with the RTG-2 cell line, the mussel *D. polymorpha* (soft tissues), and yolk sac fish larvae (*D. rerio*) were compared. To measure genotoxicity, surface water samples were analyzed from February to November. Two genotoxic samples were evaluated with the RTG-2 cell line (River Elbe, Schnackenburg in April and the positive control at the Gas Works, a gasoline influenced station). The cell line RTG-2 expressed high sensitivity to genotoxic load in surface water.

Two samples from the River Rhine (Karlsruhe, July) and the supplementary sample from the River Elbe (Schmilka, April) expressed significant enhanced DNA fragmentation in *D. polymorpha* soft tissue.

In summary, eight water samples from stations express enhanced DNA fragmentation. Low variation coefficients compared with high sensitivity of test organisms in the presence of genotoxic load (surface water) were measured and confirmed by the calculated factors of significance ($SF = 1.38$; $N = 65$). The best conditions regarding the exposition time of *D. rerio* larvae was estimated to be 3 h when 48 h old larvae were used in the exposure.

The standard deviation was taken low. No false-positive results were seen.

3.2 Genotoxicity and Repairment: Influence of Solar Ultraviolet-B on Pelagic Fish Embryos: DNA Fragmentation and DNA Thymine Dimers

To investigate genotoxicity and repairment in environmental samples we collected embryos of the fish species *L. limanda* from sites in the zone of influence of the Rivers Rhine and Elbe, and from relatively unpolluted coastal and offshore sites. There are two main fluvial discharges into the German Bight: the Rivers Rhine and Elbe. Both rivers pass through heavily populated and industrialized watersheds and consequently, have long associations with contaminant loadings (references). Secondary fluvial discharges into the German Bight include the Rivers Eider and Weser. Chemical stressors delivered to the sea via those fluvial sources have been associated with impacts in pelagic and bottom-feeding fish [19, 24]. Our hypothesis was that there might be an interaction between exposure to chemical contaminants in the zones of influence of the Rivers Rhine and Elbe and the ability of embryos to repair UV-induced DNA damage. In which case, embryos from the chemically stressed sites would be expected to have more UV-induced DNA damage as reflected by the formation of thymine dimers and DNA strand breaks. The natural unwinding of DNA's double helix starts at either strand ends or at strand nicks, which are generated by topoisomerases. Additional unwinding points occur at apurinic and apyrimidinic (AP) sites, strand breaks (SBs), and repair patches. Unwinding can also start at reactive oxygen species (ROS) induced lesions in the

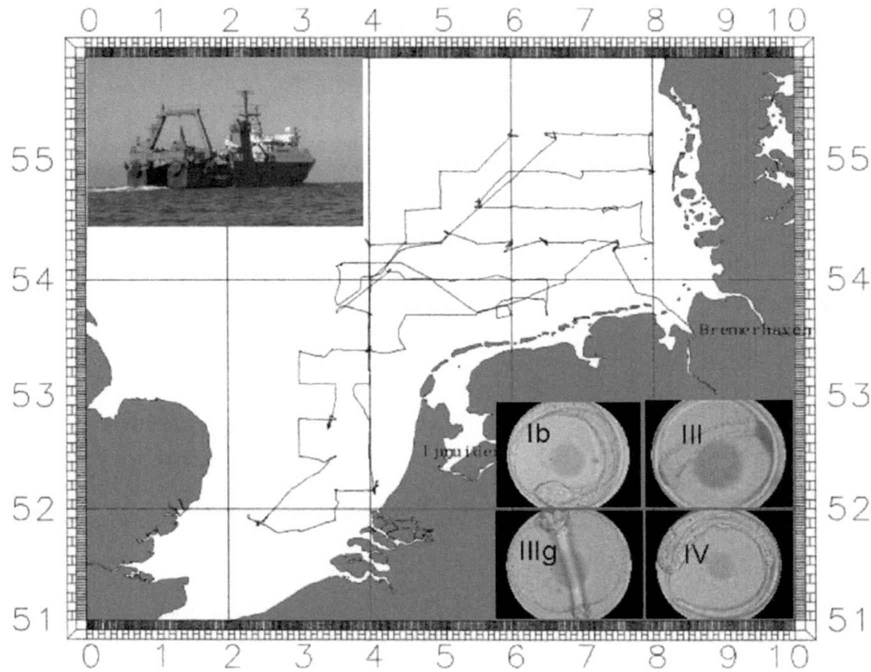

Fig. 1 Cruises of the Research Vessel RV Walther Herwig III and sampling stations. *L. limanda:* ontogenetic stages Ib-IV, 9 days, UV-B exposure und incubation for genotoxin monitoring in the Southern North Sea

DNA-double helix or in the genetic code. Those additional unwinding points combine to enhance DNA's normal unwinding rate under alkaline conditions. DNA damage in the form of thymine dimers differs distinctly from DNA-strand breaks. Thymine dimers, which tend to be stable, form when the H-bonding bridge between thymine and adenine is replaced by a cyclobutane covalent bridge between two adjacent thymine molecules.

Fish embryos (*L. limanda*) were experimentally exposed to different amounts of UV-B irradiation in a sunshine simulator and in exposure experiments onboard of a research vessel (Figs. 1, 2). The experiments were carried out with defined and realistic irradiation qualities and quantities. The following endpoints of the exposed fish embryos were studied: mortality, malformations, and changes in buoyancy of embryos measured as changes in osmolarity of the perivitelline fluid. The permeability was measured and the genotoxicity by DNA unwinding and the thymine dimers by an ELISA.

3.3 *Fish Embryos: Limanda limanda and UV-Exposure*

To obtain information concerning relevant sublethal effects for the fish-egg embryos beside the DNA damage (Fig. 3) the uptake of fluorescein, the accumula-

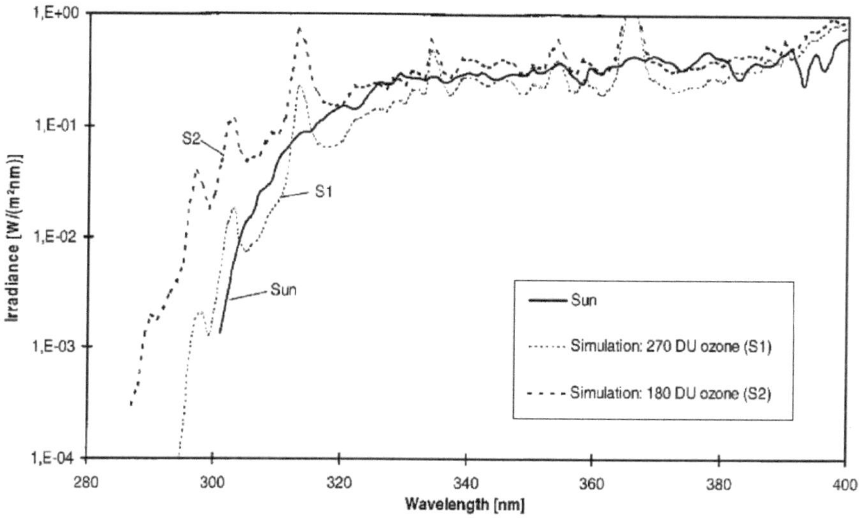

Fig. 2 Sunshine simulators (SONSI) S1 and S2 and adjustment over three cruises with comparison to the solar reference spectrum measured onboard RV Walther Herwig III at 54.5°N, 7.0°E

tion of Rhodamin B (Rh B) combined with the specific ionic blocker, Verapamil (Ver) were used to estimate the export activity of MXRtr in fish eggs. The developmental stages of fish-egg embryos and the uptake of fluorescein were investigated as a parameter to quantify the permeability by the load of fluorescein in the fish-egg embryos (green fluorescence, 450–490 nm). The penetration of fluorescein (Fig. 4) was measured by fluorescent microscopy and microfluorometry. The red fluorescence of accumulated Rh B in the eggs was measured as well by contact microfluorometry (excitation 510–560 nm, dichroic mirror 575 nm, barrier filter >590 nm) with a 16× objective and mirror (25 μm in diameter). The fish-egg embryos were incubated with 0.5 μM Rh B and 10 μM of Ver for 10 min. The mean difference between Rh B fluorescence with and without inhibitor Ver is proportional to the amount of Rh B eliminated by the MXRtr system per hour.

In the present study, an optimized version of the method of [19] was investigated to measure unwinding sites in the DNA of UV-exposed embryos. The embryos were homogenized and then lysed under alkaline conditions, after which the suspension of partially unwound DNA was rapidly neutralized, dispersed, and then ultrasonicated to prevent renaturation. SS-DNA and ds-DNA were then separated by ion exchange chromatography on hydroxyapatite. The DNA contents of the ss- and ds-fractions were quantified by DNA-specific fluorescence. The degree of unwinding was calculated as F (the proportion of double-stranded DNA), and most commonly expressed as $-\log F$, which is proportional to the number of unwinding points. Increases in the -log F value of embryos from treatment sites compared to control site values are indicative of genotoxic impact. The genotoxic agent

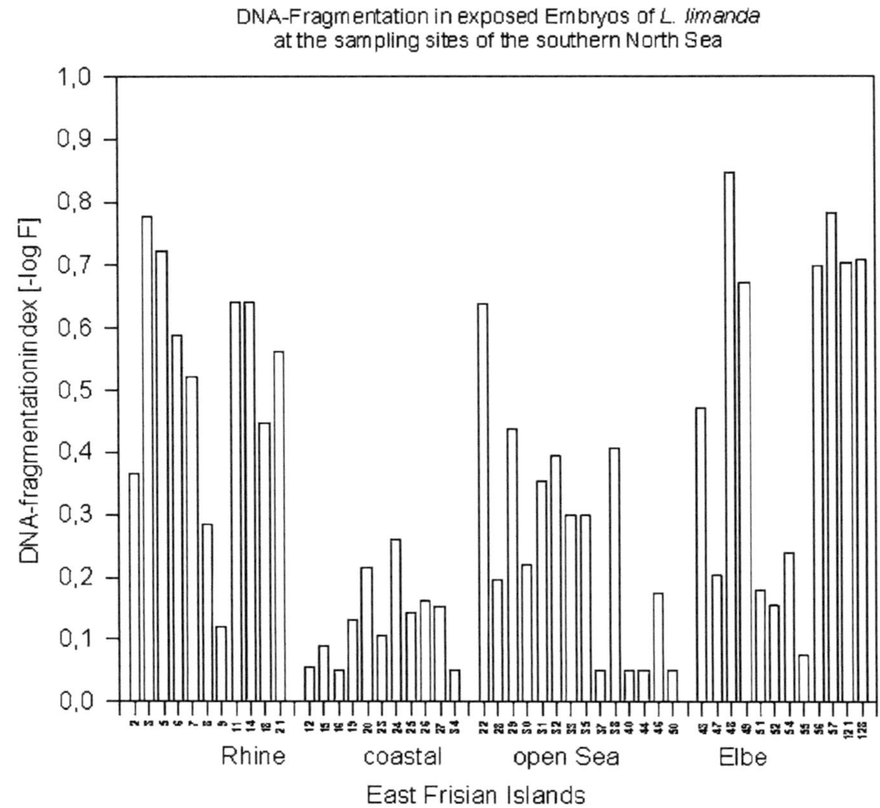

Fig. 3 DNA-fragmentation in exposed embryos of *L. limanda* at the North Sea sampling sites (see Fig. 1) in the Rhine and Elbe Estuary, coastal areas, and open sea

4-nitroquinoline-*N*-oxide or 2-aminoanthracene was the positive control, and a reagent blank was processed with each sample set.

The eco-genotoxic impact on stressed fish-egg embryos exposed to UV light was assessed by DNA-Unwinding. The unwinding of the DNA double helix starts at natural unwinding points which are strand ends and strand nicks caused by enzymes, i.e., topoisomerases. Additional unwinding points can be found in exogenously damaged DNA. These can be apurinic/apyrimidinic sites, strand breaks, and repair patches. They enhance the average unwinding rate under controlled conditions. After a defined time, alkaline lysates of tissue homogenizates or cells are neutralized, and the single-stranded and double-stranded fractions are separated by ion exchange chromatography using hydroxyapatite.

The DNA concentration of each fraction is quantified by DNA-specific fluorescence stain in a microplate fluorometer.

The enzymes DNA-ligases and DNA-polymerases activated by UV-light are responsible for repair of DNA strand breaks. These enzymes are monitoring the

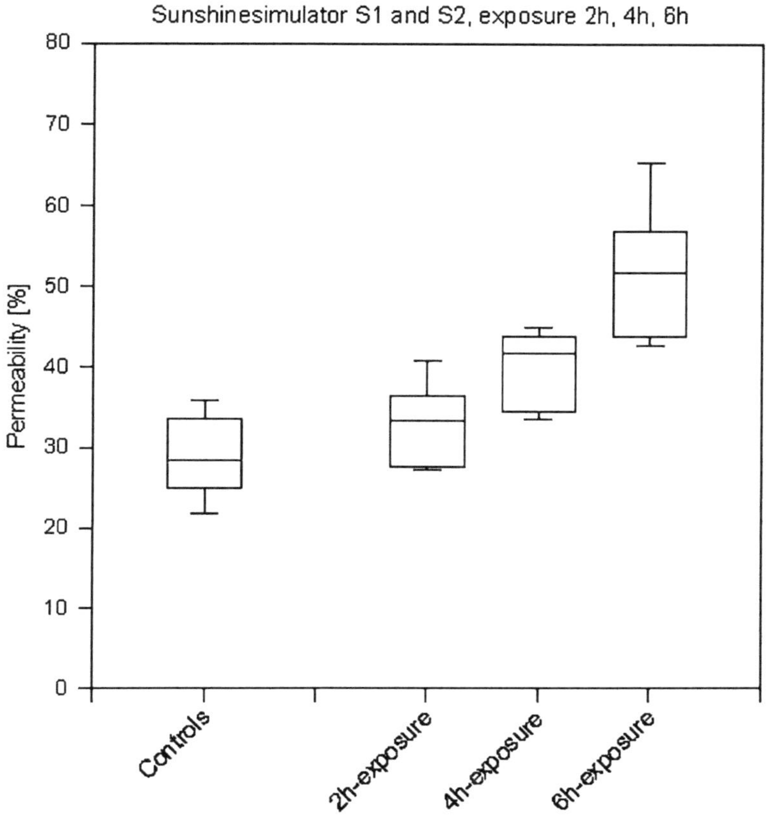

Fig. 4 Permeability [%] of exposed fish eggs (*L. limanda*) after 2 h, 4 h, 6 h

DNA-molecule to check the H-bridging covalent binding sites and the nucleoside gaps to replace these gaps by the corresponding nucleotide. In general there are lesions on the DNA-double helix or in the genetic code because of the ability of the reactive oxygen species including some oxygen-centered free radicals to modulate genes and protein expression and alter DNA polymerases.

Thymine dimers cause no lesions in the DNA-molecule like DNA-strand breaks. The formation of thymine dimers is a result of the breaking of H-bridging covalent binding sites between thymine and adenine and replacement by a covalent binding cyclobutane thymine dimer between two thymin molecules, which are located side-by-side at the DNA strand. The cyclobutane-thymine dimers are very stable. The repairment is improved by special enzymes like DNA-polymerases and topoisomerases that perform the genetic code repair.

To measure possibly UV-mediated DNA damage in fish-egg embryos both procedures, the DNA-unwinding assay and the assay for the determination of thymine dimers, were applied. The DNA-Unwinding Assay (DNA alkaline denaturation assay) monitors the genotoxicity in the eukaryotic tissues of the fish-egg embryos. In this study the assay was optimized and applied to homogenized embryos of *L. limanda* (for a detailed method see [19]).

The protocol for the alkaline unwinding assay was originally derived from procedures published in [20]. The alkaline DNA unwinding assay has three analytical steps (see Sect. 3); the actual unwinding of the DNA in the fish-egg embryo under controlled alkaline conditions, the separation of single-stranded DNA from double-stranded DNA by ion exchange chromatography, and the quantitative measurements of the DNA fractions by fluorometry. After the lysis and the time-related defined end of the unwinding process by neutralization of the lysate the samples are sonicated to prevent rejoining of complementary DNA strands. For positive controls and internal standard a mixture of homogenized tissues of fish eggs and a dissolved genotoxic substance, e.g., 4-nitroquinoline-*N*-oxide or 2-aminoanthracene were used. Natural unwinding points are the DNA strand ends, topoisomerase nicks, and spontaneous DNA lesions.

The biological significance of UV-mediated DNA lesions are their premutagenic and carcinogenic potential. The DNA unwinding is measuring genotoxicity, i.e., DNA-damage and DNA repair. Examples of DNA-damage are single- and double-strand breaks, DNA-crosslinks, oxidative lesions (ROS), loss of bases, and adduct-formation. UV-B and UV-C light generate lesions in DNA like cyclobutyl pyrimine dimers and 6–4 pyrimidione dimers which block DNA replication and transcription. Because of the importance of these premutagenic lesions and their lethal consequences and carcinogenic potential if unrepaired an immunochemical method (ELISA) was developed to measure the thymine dimers.

After lyses of the fish-embryos the samples were stored at −80°C. The DNA was purified by RNAses degradation, phenol-extraction, extraction by chloroform and isoamylalcohol and finally by precipitation in ethanol. The analytical steps are the denaturation of the DNA and a competitive ELISA with thymine dimer antibodies [25]. Beside damage to the DNA by increasing strand breaks and photoproduction of dimers, increase of the transcriptional processes in the fish-egg embryo may also lead to an increase in DNA repair.

3.4 Embryonal Development and UV-Exposure

Extended experiments were carried out with different developmental stages of the fish-egg embryos to determine UV-light stress-related change of the permeability (fluorescein uptake). There was a high deviation of 5–40% in the early stages especially in stage Ia, a lower deviation of the embryonal stages Ib, IIα, IIβ, IIγ, and IIIα (around 30%). The mean permeability of stage IIIβ was approximately 20%. The permeability of the later stages IIIγ, IVα, and IVβ increased to 35%.

The DNA unwinding results were calculated as *P*-values (P = -log *F*x/-log *F*0) and only results >1 are discussed as effects (see Sect. 3). Most of the investigated samples for genotoxicity from 1997, 1998 and 1999 are individual samples so there is no knowledge about the range of -log F values and the deviation of thymine dimers in the controls and the exposed samples. Analytical studies in 1999 completed the measurements and determined the analytical range of the controls and the dosage data [25].

In the final phase of the experiments the results demonstrated the effect of the UV light in increasing DNA repair (Fig. 5b) through the increase in transcriptional processes in the fish-egg embryos. The developmental stage of the fish-egg

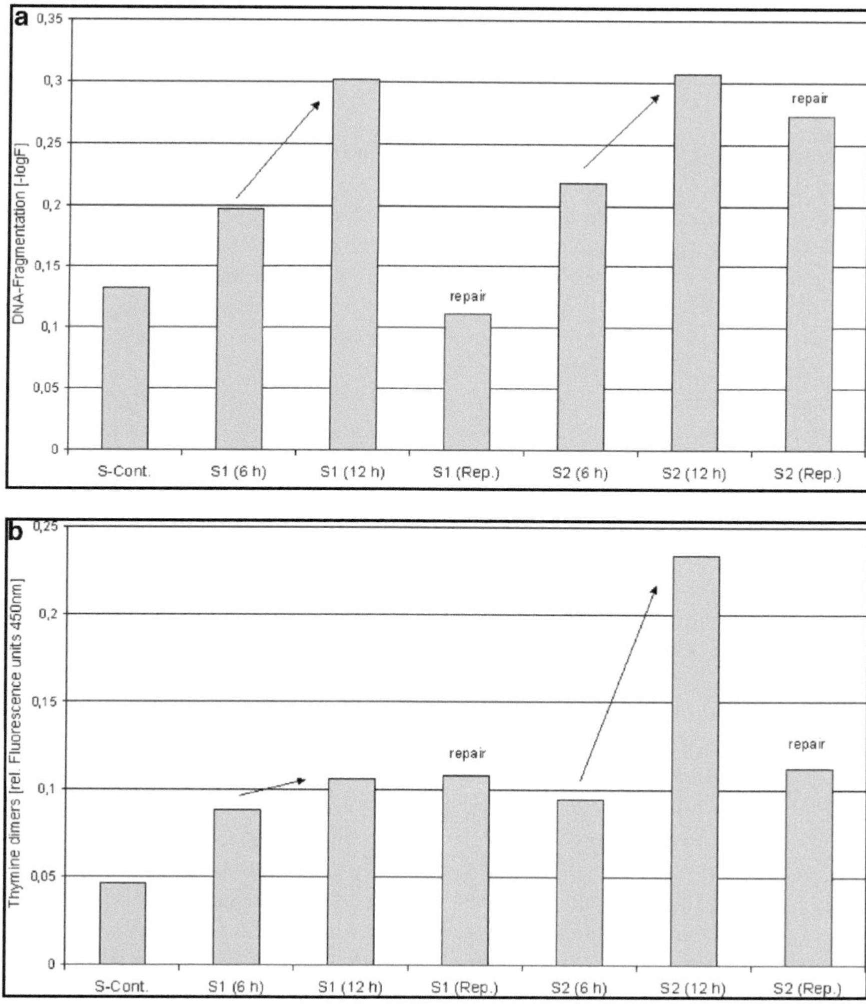

Fig. 5 (**a**) Sunshine simulator (SONSI) S1 and S2: DNA-fragmentation of sequential UV-B-exposed embryos (*L. limanda*) and repairment. (**b**) Sunshine simulator (SONSI) S1 and S2: DNA thymine dimers of sequential UV-B-exposed embryos (*L. limanda*), early ontogenetic stages Ib-II, two-times 6 h in total 12 h exposure and 18 h repairment

embryos at the beginning of the experiment was Ibβ and at the end of the experiment IIIβ-IIIγ in the controls.

The data show permeability (Fig. 4) of the embryos and DNA damage (Figs. 3 and 5a) measured by DNA unwinding [19] and thymine dimers (Fig. 5b) measured by ELISA. The results showed a time and dosage-dependent influence of UV-B radiation on the sublethal parameter genotoxicity (DNA damage). The repairment of the DNA strand breaks and the thymine dimers were investigated in [38] though the results are questionable (postreplication repair in embryo cells).

3.5 *Mussels: Mytilus edulis*

Some algal toxins such as saxitoxins (STX) and microcystin LR may strongly induce the repair mechanisms of DNA strand breaks at low concentrations. Conversely, another algal toxin, domoic acid (DA), led to clear increases in DNA strand breaks at similar concentrations [35]. Biotoxins (Red Tide) represent an ecological and economic problem concerning aquaculture and tourism in coastal areas world wide [39]. The genotoxic, neurotoxic, and immunotoxic effects by domoic acid are produced by the diatom *Pseudonitzschia multiseries* and the biotoxin saxitoxin causes paralytic shellfish poisoning (PSP) and is produced by *Alexandrum catenella* and *Alexandrum ostenfeldii*. In addition, the flagellate *Dinophysis acuta* causes diarrhetic shellfish poisoning (DSP). The need for increasing quality control at different critical steps of the food production chain has to be addressed by developing a set of rapid biomarker and sensor systems. The subacute effects of the toxins on the immune defense mechanisms of hemocytes and DNA fragmentation in cells of gills and hepatopancreas of several commercially relevant mussel species have been investigated. The PSP toxins were quantitatively analyzed by the Institute for Nutrition in Jena (Germany). Little is known about the metabolism and accumulation pathways of saxitoxins in mussels. The acute toxic effect of saxitoxins on bivalves has also rarely been investigated. In this study the neurotoxic, genotoxic, and immunotoxic effects of saxitoxins on different tissue of the blue mussels *M. edulis* were investigated under laboratory conditions. The neurotoxic, immunotoxic, and genotoxic effects of domoic acid on *M. edulis* were also investigated by laboratory exposure experiments. Mussels were intramuscularly injected with DA at concentrations 1, 10, 100, and 500 ng g^{-1} body weight (bw). After incubation eco-toxic parameters including cholinesterase (ChE) activity in the gills of the mussels, phagocytosis activity of mussel hemocytes, and DNA strand breaks (DNA unwinding) in the hepatopancreas of the mussels were investigated. PSP toxins were extracted from the digestive glands of the mussel *M. edulis* to check the immune defense mechanisms of the hemocytes and DNA fragmentation in cells of gills and hepatopancreas. The toxins caused no change in the number of hemocytes, but there was an alteration of the membrane integrity. Phagocytosis activity of the hemocytes was strongly suppressed by PSP toxins. PSP toxins diminished physiological DNA fragmentation in gills and hepatopancreas of mussels due to induction of the repair pathways for primary DNA damage. The causes and effects of biotic and abiotic parameters, i.e., UV-B exposure, were discussed [39]. The final aim was to develop a bivalve toxicity assay as a commercial immunotoxicity test kit for algal toxins.

Primary DNA damage in the blue mussel *M. galloprovincialis* can be reduced through antioxidants like alpha-tocopherol. The heterogeneity of DNA strand breaks in animals analyzed by field experiments are obviously based on the dose, synergetic, or antagonistic impacts of chemical contaminants as well as individual alteration in the activity of repair pathways of DNA damage.

Surprisingly, fish and mussel samples collected in relatively clean areas, e.g., the Islands of Helgoland and Spiekeroog (Germany), indicated also relatively high genotoxic responses. Although the greatest level of most chemical parameters was observed in soft tissue of *M. edulis* from the estuary of the Elbe River, mussels from the "reference site" the Island of Helgoland have been moderately contaminated with a number of chemical compounds. In particular, residues with relatively high genotoxic potency such as polychlorinated biphenyl (PCB) congeners were determined in mussel and sediment samples at the Island of Helgoland, which correlates with genotoxic stress of mussels at this site.

4 Concluding Discussion and Future Trends for Real-Time Assessment

There are a number of different assays that measure DNA strand breakage in alkalis including alkaline elution, DNA alkaline unwinding [18], and the microgel electrophoresis "comet assay," for application in chemical testing as well as environmental research fields. In the present study, different optimization steps led to sufficient sensitivity regarding the detection of genotoxicity and decreased detection limits (LOEC values) in tests of single substances. For the single substance 4-NQO, Duis et al. [33] described a detection limit of 156 $\mu g\ l^{-1}$ in comparison with 78 $\mu g\ l^{-1}$ in the present study (both exposition of mono layer).

Different test organisms react with different sensitivity when exposed to genotoxic substances (the detection limit for 4-NQO was 78 $\mu g\ l^{-1}$ for RTG-2 and *D. rerio* fish larvae compared with 250 $\mu g\ l^{-1}$ when *D. polymorpha* was chosen). Different physiological and molecular causative relationships have an influence on the sensitivity of test organisms including detoxification, accumulation of contaminants, excretion, and biotransformation as well as the activation of DNA repair mechanisms.

The analyzed surface water samples had been judged for genotoxicity in *D. polymorpha* with the pulsed-field gel electrophoresis method in parallel experiments. With this method, false-positive genotoxic effects could be excluded. On the other hand, additional water samples were judged as positive in the PFGE compared with the DNA alkaline unwinding assay. These observations suggest the possibility of "borderline effects" especially in the DNA alkaline unwinding test procedure with the mussel *D. polymorpha*. The use of the mussel *D. polymorpha* as a test organism for long-term monitoring experiments is indicated because of high correlation between the parameter "DNA fragmentation potential" measured by the DNA alkaline unwinding assay when compared with chemical residues of Hg(II) and the sum of HCH isomers in soft tissue during an eight-week exposition interval, estimated by Spearman rank-correlation analysis.

Contrary to these observations, no correlation relationship was found when the DNA fragmentation potential (–log *F* values) from surface water exposition was compared with chemical parameters of chlorophenols, trihalogenemethanes, and HCH isomers in the surface water. Residue contents of contaminants in *D. polymorpha* tissue seem to be more suitable parameters for correlation studies of genotoxicity due to accumulation events. In addition, correlation studies with the parameter "DNA alkaline unwinding" using identical protocols with *D. polymorpha* and the parameter AOX did not express a significant relationship between the data. This test organism has a number of defense mechanisms to prevent exposure to pollutants, such as closing the shell, metabolization rate, and excretion; kinetics modifies the amount of genotoxicity. The mussels have an important role in their trophic level as part of the aquatic food chain for surface water systems. Laboratory exposure data for *D. polymorpha* had shown a relatively higher variability compared with cell-culture tests (RTG-2) and *D. rerio* early life stages. The variability of test replicates under laboratory conditions was higher compared with cell cultures (RTG-2) and *D. rerio*. For exposure monitoring experiments a control station with comparable environmental conditions (climate, nutrient content) is indicated. Recent test results from single substance and surface water samples of different pollution grade can be demonstrated by the application of the DNA alkaline unwinding assay to real-world samples to assess quality criteria in genotoxicity for surface and tap water in regulatory processes. To monitor genotoxicity in aquatic systems, especially the early life stages of the fish larvae *D. rerio* show a relatively high sensitivity and reproducibility with the DNA alkaline unwinding assay and should be adapted for application to long-term exposures. A combined test battery using cell lines, mussels, and fish stages allows a more complete characterization of amount of genotoxic pollution in surface waters.

The use of test organisms of different trophic levels introduced in the DNA alkaline unwinding assay allows a differential exposition procedure to assess acute and chronic effects in relation to physiologic adaptation and accumulation in aquatic environments. The detection of DNA strand breaks by alkaline unwinding followed by hydroxyapatite batch elution is a rapid and inexpensive technique especially for analysis of a large batch elution of organisms in field experiments. The lack of a nonpolluted reference site as a control and great variance of DNA strand breaks in analyzed tissues of test animals from selected sampling sites makes it necessary to analyze the physiological range of variance in control test organisms from nonpolluted areas, i.e., in the North Sea and Baltic Sea.

Immune defense reactions in the hemolymph of mussels include the proliferation of hemocytes and their mobilization from certain tissues to the contaminated periphery corresponding with modulation of phagocytosis activity. The immune defense response of *M. edulis* to domoic acid contamination at a concentration of 1 μg g^{-1} bw included the rapid proliferation of hemocytes within 48 h of incubation. The immune defense activity of the Pacific oyster (*Crassostrea gigas*) was also induced by 4 h exposure of the mussels to domoic acid and the diatom *P. pungens f. multiseries*. With high doses of cadmium and fluoranthene the number of hemocytes increased significantly. Conversely, the injection of PSP toxins led to increasing heterogeneity of hemocyte number and significant suppression of their phagocytosis activity [35]. Significant suppression of phagocytosis activity in the hemolymph of mussels was also noted following an exposure to phenol or high doses of selected heavy metals.

Saxitoxins have been seen to induce the DNA repair system at concentrations of 100 and 500 ng g^{-1} bw. The main results are the apparently significant induction of the DNA repair mechanism following short-term exposure of mussels to saxitoxins, domoic acid, and microcystin-RL. An isolate of saxitoxins from the gills, digestive organs, and adductor muscle of mussels (*M. chilensis*) received from the Medical Institute of the Salud Aysen in the XI. region of Chile was used for investigations and reference. The chemical differentiation of the isolates carried out in the Institute of Food Chemistry of the Friedrich-Schiller-University in Jena suggested the presence of different saxitoxin and gonyautoxins compounds. After chemical analysis and quantification of the saxitoxins, a stock solution of the isolate was prepared containing 1 mg ml^{-1} saxitoxins. The chemical instrumental analysis shows clearly an extremely high level of saxitoxin equivalents.

Exposure experiments with fish embryos were carried out to demonstrate the effect of UV light in increasing DNA repair through an increase in transcriptional processes in the fish-egg embryos. We employed two main scenarios, a reduction to 270 (S1) and to 180 (S2) Dobson units (DU) in single or repetitive exposure of 2, 4, or 6 h. The developmental stage of the fish-egg embryos at the beginning of the experiment was Ib and at the end of the experiment IIIβ-IIIγ in the controls. In SONSI 2 the development of the fish-egg embryos was slightly delayed to IIβ-IIIα. In summary there are sublethal effects at the molecular level that can be demonstrated by DNA unwinding and thymine dimers. However, in the experiments with only one exposure in the sunshine simulator and in field experiments where radiation was low there was not always enough information for a clear interpretation. In future bio-analytical work the deviation of the data in the matrix fish-egg embryo will be investigated.

Monitoring the effects of environmental genotoxic substances has gained increasing importance in recent times. Assessment of environmental samples for the presence of genotoxins (mutagens and carcinogens) has a high priority in protecting inland and coastal waters. Many tools have been meanwhile standardized and harmonized by [17] – determination of genotoxicity using the umu-test, [40] – determination of the genotoxicity of water – *Salmonella*/microsome test (AMES-test), [41]– Evaluation of genotoxicity by measurement of the induction of micronuclei – Part 2: "Mixed population" method using the cell line V 79 and many other indicator assays like the comet assay but the missing tools are the real-time assays in genotoxicity. In the field of bio-analytical approaches [42] promising on

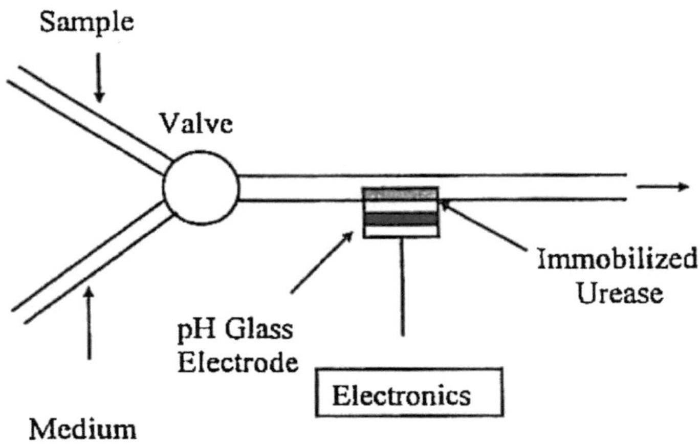

Fig. 6 Whole-cell biosensor for the assessment of a genotoxic potential modified after [2, 17]

line formats are discussed and in progress. There are already some robot systems according to [17] on the market and flow-through concepts (Fig. 6) are on the way. New testing strategies and concepts will provide information on genotoxins in the environment. Extended exposure of organisms to environmental genotoxins results in physiological disorder such as reproduction impairment and other related abnormalities.

Acknowledgments The authors wish to thank the Federal Ministry for Science and Technology (BMBF) in Germany and the European Commission for funding the case studies 1, 2, and 3 via the projects ALGAETOX (ENV4-CT98–0784), SANDRINE ENV4-CT98–0801), BIOMAR EV5V-CT94–0550), BEEP (EVK3-CT2000–00025), and CITY FISH (EVK1-CT1999–00009). We thank Birgit Fischer for technical assistance and Jürgen Büsing from the European Commission for his competent contributions in stimulating many frontier approaches und ideas.

References

1. Houk VS, Waters MD (1996) Genetic toxicology and risk assessment of complex environmental mixtures. Drug Chem Toxicol 19:187–219
2. Hansen P-D (2008) Biosensors and ecotoxicology. Eng Life Sci 8:1–7
3. Hansen P-D (2008) Biosensors for environmental and human health. In: Young J Kim, U. Platt (eds) Advanced environmental monitoring, Sect. 4, Chap. 23. Springer, Heidelberg, New York, pp 297–311
4. Rao SS, Burnison BK, Efler S, Wittekindt E, Hansen P-D, Rokosh DA (1995) Assessment of genotoxic potential of pulp mill effluent and an effluent fraction using AMES- mutagenicity and umuC-genotoxicity assays. Environ Toxicol Water Qual 10:301–305
5. Rao SS, Neheli TA, Carey JH, Herbert A, Hansen P-D (1996) DNA alkaline unwinding assay for monitoring the impact of environmental genotoxins. Environ Toxicol Water Qual 11:351–354
6. De Maagd PGJ, Vethaak AD (1998) Biotransformation of PAHs and their carcinogenic effects in fish. In: Neilson AH (eds) The Handbook of environmental chemistry, 3J PAHs and related compounds. Springer, Berlin, pp 265–310

7. Hansen P-D, Herbert A (1998) Small-scale in vitro genotoxicity tests for bacteria and invertebrates. In: Wells PG, Lee K, Blaise C (eds) Microscale aquatic toxicology – advances, techniques and practice. CRC, Florida, pp 237–252
8. De Maagd PGJ, Tonkes M (2000) Selection of genotoxicity tests for risk assessment of effluents. Environ Toxicol 15(2):81–90
9. Dizer H, Wittekindt E, Fischer B, Hansen P-D (2002) The cytotoxic and genotoxic potential of surface water and wastewater effluents as determined by bioluminescence, umu-assays and selected biomarkers. Chemosphere 46:225–233
10. Hansen P-D (2003) Biomarkers. In: Markert BA, Breure AM, Zechmeister HG (eds) Bioindicators and biomonitors, principles, concepts and applications. Elsevier, Amsterdam, pp 203–220
11. Hansen P-D, Blasco J, De Valls A, Poulsen V, van den Heuvel-Greve M (2007) Biological analysis (Bioassays, Biomarkers, Biosensors) In: Damia Barceló and Mira Petrovic (eds) Sustainable management of sediment resources, Vol 2, Sediment quality and impact assessment of pollutants. Elsevier, Amsterdam, pp 131–157
12. Maron DM, Ames BN (1983) Revised methods for the salmonella mutagenicity test. Mutat Res 113:173–215
13. Quilardet P, de Bellecombe C, Hofnung M (1985) The SOS chromotest, a colorimetric bacterial assay for genotoxins: validation study with 83 compounds. Mutat Res 147:79–95
14. Côté C, Blaise C, Delisle C, Meighen EA, Hansen P-D (1995) A miniaturized Ames test employing bioluminescent strains of *Salmonella typhimurium*. Mutat Res 345:137–146
15. Oda Y, Nakamura S, Oki I, Kato T, Shinagawa H (1985) Evaluation of the new system (Umu-test) for the detection of environmental mutagens and carcinogens. Mutat Res 147:219–229
16. Hansen P-D (1996) Bioassays on sediment toxicity. In: Calmano and Förster (eds) Sediments and toxic substances - environmental effects and ecotoxicity. Springer, Berlin, Heidelberg, pp 179–196
17. ISO 13829 (2000) Determination of the genotoxicity using the umu-test
18. Nacci D, Nelson S, Nelson W, Jackim E (1992) Application of the DNA alkaline unwinding assay to detect DNA strand breaks in marine bivalves. Mar Environ Res 33:83–100
19. Herbert A, Hansen P-D (1998) Genotoxicity in fish eggs/embryos. In: Wells PG, Lee K, Blaise C (eds) Microscale aquatic toxicology – advances, techniques and practice. CRC, Florida, pp 491–505
20. Kanter P-M, Schwarz HS (1978) A hydroxylapatite batch assay for quantitation of cellular DNA damage. Anal Biochem 97:77–84
21. Ahnström G, Erixon K (1981) Measurement of strand breaks by alkaline denaturation and hydroxyapatite chromatography. In: Friedberg EC, Hanawalt PC (eds) DNA repair: a laboratory manual of research procedures. Marcel Dekker, New York, pp 403–418
22. Holmberg M, Lageberg M, Niejahr B, Rödin L (1988) Simultaneous detection of DNA strand breaks and unscheduled DNA synthesis in mutagen-treated human lymphocytes in the absence of hydroxyurea. Mutat Res 202:111–118
23. Shugart LR (1988) Quantitation of chemically induced damage to DNA of aquatic organisms by alkaline unwinding assay. Aquat Toxikol 13:43–52
24. Dethlefsen V, von Westernhagen H, Tüg H, Hansen P-D, Dizer H (2001) Influence of solar ultraviolet-B on pelagic fish embryos: osmolality, mortality and viable hatch, Helgol. Mar Res 55:45–55
25. Hansen P-D, Sherry J, Dizer H, Tüg H, Dethlefsen V, von Westernhagen H (2009) Influence of solar ultraviolet-B on pelagic fish embryos: DNA Fragmentation and DNA Thymindimers, Journal of Applied Idthyology, 25, 5 in press
26. Nagel R (1993) Fish and environmental chemicals – a critical evaluation of tests. In: Braunbeck T, Hanke W, Segner H (eds) Fish. Ecotoxicology and ecophysiology, vol 1. VCH Verlagsgesellschaft, Weinheim, pp 47–156
27. Belitzky GA, Lytcheva TA, Khitrovo IA, Safaev RD, Zhurkov VS, Vyskubenko IF, Sytshova LP, Salamatova OG, Feldt EG, Khudoley VV, Mizgirev IV, Khovanova EM, Lugnivenko EG, Tanirbergenov TB, Malinovska KI, Revazova Yu A, Ingel FI, Bratslavsky VA, Terentyev AB,

Shapiro AA, Williams GM (1994) Genotoxicity and carcinogenicity testing of 1,2-dibromopropane and 1,1,3-tribromopropane in comparison to 1,2-dibromo-3-chloropropane. Cell Biol Toxicol 10(4):265–279

28. Troxel CM, Reddy AP, O'Neal PE, Hendricks JD, Bailey GS (1997) In vivo aflatoxin B1 metabolism and hepatic DNA adduction in Zebrafish (Danio rerio). Toxicol Appl Pharmacol 143:213–220
29. Troxel CM, Buhler DR, Hendricks JD, Bailey GS (1997) CYP1A Induction by ß-Naphthoflavone, Aroclor 1254, and 2,3,7,8-Tetrachlorodibenzo-p-dioxin and it`s influence on Aflatoxin B1 metabolism and DNA adduction in Zebrafish (Danio rerio). Toxicol Appl Pharmacol 146:69–78
30. Garberg P, Akerblom EL, Bolcsfoldi G (1988) Evaluation of a genotoxicity test measuring DNA-strand breaks in mouse lymphoma cells by alkaline unwinding and hydroxyapatite elution. Mutat Res 203:155–176
31. Dragani AT, Barale R, Parodi S, Taningher M, Zucconi D, Della Porte G (1990) Negative results of short-term genotoxicity tests with 1,4-bis82-(3,5-dichloropyridyloxy)]benzene. Carcinogenesis 11:1153–1157
32. Nygren J, Cedervall BS, Eriksson M, Dusinska M, Kolman A (1994) Induction of DNA strand breaks by ethylene oxide in human diploid fibroblasts. Environ Mol Mutat 24:161–167
33. Duis K, Unruh E, Hansen P-D (1996) Validation of a cytotoxicity assay with the Fish cell line RTG2 as an replacement of the fish assay (Validierung eines Zytotoxizitätstests als Ersatzmethode zum Fischtest nach DIN 38412 Teil 31: Phase II; Final Report, BMBF AZ. 0310314B
34. Reincke H (1992) Biological effect monitoring in the river Elbe using the zebra mussel Dreissena polymorpha. In: Neumann D, Jenner HA (eds) The zebra mussel Dreissena polymorpha. Gustav Fischer Verlag, Germany
35. Dizer H, Unruh E, Bissinger V, Hansen P-D (2001) Investigation of genotoxicity and immunotoxicity for monitoring marine pollution in the Baltic sea and Mediterranean Sea. In: Garrigues Ph, Barth H, Walker CH, Narbonne J-F (eds) Biomarkers in marine organisms: a practical approach. Elsevier, Amsterdam, pp 237–257
36. Borcherding J (1992) Morphometric changes in relation to the annual reproductive cycle in Dreissena polymorpha – a prerequisite for biomonitoring studies with Zebra Mussels. In: Neumann D, Jenner HA (eds) The Zebra Mussel Dreissena polymorpha. Limnologie aktuell, Bd. 4, Gustav Meyer Verlag, Stuttgart, pp 87–99
37. Neumann D, Jenner HA (eds) (1992) The zebra mussel Dreissena polymorpha. Gustav Fischer Verlag, Germany
38. Hochegger H, Sonoda E, Takeda S (2004) Post-replication repair in DT 40 cells: translesion polymerases versus recombinases. BioEssays 26:151–158
39. Luckas B, Dahlmann J, Erler K, Gerdts G, Wasmund N, Hummert C, Hansen P-D (2005) An overview on key phytoplankton toxins and their recent occurrence in North Sea and Baltic. Environ Toxicol 20(1):1–17
40. ISO 16240 (2005) Determination of the genotoxicity of water - Salmonella/microsome test (AMES-test)
41. ISO 21427 - 2 (2006) Evaluation of genotoxicity by measurement of the induction of micronuclei - Part 2: "Mixed population" method using the cell line V 79
42. Bilitewski U, Brenner-Weiss G, Hansen P-D, Hock B, Meulenberg E, Müller G, Obst U, Sauerwein H, Scheller FW, Schmid R, Schnabl G, Spener F (2000) Bioresponse-linked instrumental analysis. Trends Anal Chem 19(7):428–433

Hdb Env Chem (2009) 5J: 227–272
DOI:10.1007/698_2009_18

Liquid Chromatography–Mass Spectrometry Methods for Analysis of Endocrine-Disrupting Chemicals in Wastewaters

Cristina Postigo, Marina Kuster, Marta Villagrasa, Sara Rodríguez-Mozaz, Rikke Brix, Marinel la Farré, Miren López de Alda, and Damià Barceló

Abstract Endocrine disrupters (EDs) continue to be the focus of extensive research. Their effects and mode of action as well as their occurrence and fate in the aquatic environment are still largely unknown and need investigation. Wastewater treatment plant effluents are for most of them the main source of contamination in water bodies. A large number of EDs are relatively polar compounds that pass easily wastewater treatments and reach the aquatic environment, where they can exert undesirable effects on both wildlife and humans. Their removal during wastewater treatment, very often inefficient, is usually estimated from the comparison of the concentrations detected at the influent and the effluent and the technique of choice for their analysis is in many cases liquid chromatography–(tandem)mass spectrometry (LC–(tandem)MS).

Herein we review the analytical methods, including sample preparation and analysis, based on LC–(tandem)MS described in the literature for the determination of different classes of EDs (pesticides, phthalates, bisphenol A, alkylphenolic compounds, perfluorinated compounds, pharmaceuticals, and estrogens) in wastewater and their occurrence in this kind of matrix.

Keywords Analysis, Emerging contaminants, Endocrine disrupters, LC-MS, Wastewater

C. Postigo, M. Kuster, R. Brix, M. Farré, M. López de Alda (✉), and D. Barceló
Department of Environmental Chemistry, IDAEA-CSIC, C/Jordi Girona
18–26, 08034, Barcelona, Spain
e-mail: mlaqam@cid.csic.es

M. Villagrasa, S. Rodríguez-Mozaz, and D. Barceló
Catalan Institute for Water Research (ICRA), Parc Científic i Tecnològic de la Universitat de Girona, Edifici Jaume Casademont, 15 17003 Girona, Spain

Contents

Abbreviations

ACN	Acetonitrile
AcOH	Acetone
AMPA	Aminomethylphosphonic acid
APCI	Atmospheric pressure chemical ionization
APs	Alkylphenols
APECs	Alkylphenoxy carboxylates
APEOs	Alkylphenol ethoxylates
API	Atmospheric pressure ionization
BBP	Butyl benzyl phthalate
BEHP	Bis(2-ethylhexyl) phthalate
BIO	Biochanin-A
BPA	Bisphenol A
CNS	Central nervous system
COU	Coumestrol
DAI	Daidzein
DBP	Di-*n*-butyl phthalate
DCM	Dichloromethane
DEHP	Di-(2-ethylhexyl) phthalate
DEP	Diethyl phthalate
DES	Diethylstilbestrol
DMP	Dimethyl phthalate
DOP	Di-*n*-octyl phthalate

DPP	Dipentyl phthalate
E1	Estrone
E1S	Estrone-sulfate
E2	Estradiol
E3	Estriol
ED	Endocrine disrupter
EE	Ethynyl estradiol
EI	Electron ionization
EO	Ethoxylate
EPI	Enhanced product ion
ESI	Electrospray
ETAc	Ethyl acetate
Et_2O	Diethylether
FA	Formic acid
FASs	Fluorinated alkyl substances
FOR	Formononetin
6:2 FtS	6:2-Fluorotelomer-sulfonate
GC	Gas chromatography
GEN	Genistein
GLY	Glycitein
GPC	Gel permeation chromatography
HE	Hydroxyl estrone
HOAc	Acetic acid
IT	Ion trap
LC	Liquid chromatography
LOD	Limit of detection
LOQ	Limit of quantification
MeOH	Methanol
MES	Mestranol
MIPs	Molecularly imprinted polymers
MM	Mixed mode
MS	Mass spectrometry
MS/MS	Tandem mass spectrometry
MTBE	Methyl-*tert*-butylether
NH_4Ac	Ammonium acetate
NH_4HCO_2	Ammonium formate
NI	Negative ionization
NP	Nonylphenol
NPECs	Nonylphenoxy carboxylates
NPE1C	Nonylphenoxy monocarboxylate
NPE2C	Nonylphenoxy dicarboxylate
NPEOs	Nonylphenol ethoxylates
NP1EO	Nonylphenol monoethoxylate
NP2EO	Nonylphenol diethoxylate
NSAID	Nonsteroidal antiinflamatory drugs
OP	Octylphenol

OPE2C	Octylphenoxy dicarboxylate
OPEOs	Octylphenol ethoxylates
PAEs	Phthalic acid esters (phthalates)
PCPs	Personal care products
PFASs	Perfluorinated alkyl substances
PFBS	Perfluorobutane sulfonate
PFCs	Perfluoroalkyl carboxylates
PFCAs	Perfluorocarboxylic acids
PFO	Perfluorooctane
PFOA	Perfluorooctanoic acid
PFOS	Perfluorooctane sulfonate
PFOSA	Perfluorosulfonamide
PFOSAA	*N*-ethyl perfluoro-octanesulfonamidoacetate
PFSs	Perfluoroalkyl sulfonates
PFSAs	Perfluorosulfonic acids
PI	Positive ionization
Q	Single quadrupole
QqQ	Triple quadrupole
Q-TOF	Quadrupole-time of flight
Q-trap	Quadrupole-ion trap
RP	Reverse phase
SEC	Size exclusion chromatography
SIM	Single ion monitoring
SPE	Solid-phase extraction
SRM	Selected reaction monitoring
STP	Sewage treatment plant
TCAA	Trichloroacetic acid
TEA	Triethylamine
TET	Triethyl tin
TFA	Trifuoroacetic acid
TOF	Time of flight
TRI	4′,6,7-Trihydroxyisoflavone
TrBA	Tri-*n*-butylamine
UPLC	Ultra-performance liquid chromatography
WWTP	Wastewater treatment plant
XAPECs	Halogenated alkylphenoxy carboxylates
XAPs	Halogenated alkylphenols

1 Introduction

Endocrine disrupters (EDs) are exogenous agents that interfere with the synthesis, secretion, transport, binding, action, or elimination of natural hormones in the body that are responsible for the maintenance of homeostasis, reproduction, development, and/or behavior.

Under the category of EDs there are a wide variety of chemical substances that include both natural but mainly anthropogenic compounds. Over the years various national and international governmental and nongovernmental organizations have compiled lists of EDs. These lists together with information extracted from the scientific literature have formed the basis of the inventory created by the Institute for Environment and Health [1], which includes a total of 966 compounds or elements purported to be EDs.

Herein we focus on seven compounds or compound classes, which are amenable to LC-MS analysis: (1) pesticides, (2) phthalates, (3) bisphenol A (BPA), (4) alkylphenolic compounds, (5) perfluorinated compounds, (6) pharmaceuticals, and (7) estrogens. These compounds or compound classes have been associated to a variety of effects (persistent estrus, mammary, testis and thyroid tumors, reduced fertility, alteration of the luteinizing hormone, cortisol, insulin, and/or testosterone, etc.) but most of them are suspected to have estrogenic and/or androgenic activity (phthalates, bisphenol A, alkylphenols, estrogens) or to disrupt thyroid function (some pesticides, pharmaceuticals, and perfluorinated surfactants). Nevertheless, their effects and mode of action are still largely unknown, as is the occurrence and fate in the aquatic environment of those compounds, such as pharmaceuticals, that have been only recently discovered or recognized as environmental contaminants (known as emerging contaminants).

Some of the studied compounds are included in the list of priority substances in the field of water policy and are thus subject to environmental quality standards in surface waters [2]. This is the case of the pesticides alachlor, atrazine, chlorfenvinphos, chlorpyrifos, diuron, endosulfan, simazine, and trifluralin, the alkylphenolic surfactants nonylphenol (NP) and octylphenol (OP), and the phthalate di(2-ethylhexyl)-phthalate (DEHP). Others, such as the pesticides bentazon, dicofol, mecoprop, glyphosate and AMPA (glyphosate degradation product), bisphenol A, and the perfluorinated compound perfluorooctane sulfonate (PFOS) are included in the list of substances subject to review for possible identification as priority substances or priority hazardous substances [2].

For most of the reviewed compounds wastewater treatment plant (WWTP) effluents are the most important source of contamination in water bodies [3]. Most of them are relatively polar substances that pass easily wastewater treatments and natural filtration systems. Their removal during wastewater treatment, very often inefficient, is usually estimated from the comparison of the concentrations detected at the influent and the effluent. However, wastewater samples are very complex matrices and the determination of microcontaminants at very low concentrations in them is a difficult task. In addition, since some EDs have been recently discovered as environmental contaminants, analytical methods for their determination in the various water bodies are scarce.

The following sections discuss the main properties, uses, and effects of the compound classes under review, the procedures described in the literature for their analysis in wastewaters, distinguishing between sample preparation and analysis, and the levels reported in both influent and effluent wastewater samples.

2 Specific Groups of EDs

2.1 *Pesticides*

Many pesticides and metabolites are suspected or claimed to be EDs. Residues of these chemicals and their transformation products are likely to occur in the environment as some of them are persistent and the ones that are relatively labile are extensively used and may end up in the aquatic media through direct run-off, leaching, careless disposal of empty containers, equipment washing, and effluents [4, 5].

According to the EDs compilation report made by the Institute for Environment and Health [1] a total of 228 biocidal products, such as insecticides, herbicides, and fungicides are established or potential EDs. These compounds belong to very different chemical classes; however, the majority of them correspond to the chemical group of the persistent organochlorine pesticides and metabolites, such as DDT and its metabolites, endosulfan, aldrin, dieldrin, heptachlor, chlordane, metoxychlor, mirex, lindane, hexachlorobenzene, and chlordecone (kepone), among others.

The next most important group of pesticides in terms of endocrine-disrupting activity comprise the organophosphorous compounds. Inside this group common use pesticides and insecticides such as chlorfenvinphos, chlorpyrifos, diazinon, dichlorvos, dimetoate, fenitrothion, malathion, and parathion are included.

Different types of herbicides belonging to various chemical classes have also shown endocrine-disrupting effects. These substances are phenoxyacid compounds (2,4,5-T and 2,4-D), amides (acetochlor and alachlor), triazines (atrazine, cyanazine, and simazine), dinitroanilines (trifluraline), triazoles (amitrole), phenylureas (diuron), and ureas (linuron), among others.

Pesticides in general have been subjected to regulation and control for many years [6]. In the EU, for instance, the maximum admissible concentration of any individual pesticide (including the active substances and their relevant metabolites, degradation and reaction products) in drinking water and ground water is 0.1 $\mu g\ L^{-1}$. For surface waters, the European Parliament has recently adopted environmental quality standards (EQS), which include both annual average (AA) and maximum allowable concentrations (MAC) [2]/105/EC). These EQS are very low for some compounds such as endosulfan (AA concentration of 5 $ng\ L^{-1}$ in inland surface water and 500 $pg\ L^{-1}$ in other surface waters) and less restrictive for other compounds such as alachlor, atrazine, diuron or simazine, with AA concentrations of 0.3, 0.6, 0.2, and 1 $\mu g\ L^{-1}$, respectively, in both inland and other surface waters.

2.2 *Phthalates*

Phthalic acid esters (phthalates, PAEs) are widely used as polymer additives in the manufacture of plastics, rubber, and styrene to make them flexible and, with minor applications, in cosmetics, medical products, and insecticide carriers. PAEs can enter the environment through losses during manufacturing processes or by leaching from final products since they are not chemically bonded to the polymeric matrix [7]. Phthalates have low acute and chronic toxicity, which generally decreases with increasing alkyl chain length [8]. The reproductive system, and in general the endocrine system, has been described to be affected by these kinds of contaminants [9, 10]; the endocrine-disruptive potency (relative to 17β-estradiol) for a large number of phthalates ranges from 10^{-4} to 10^{-7} in in vitro and in vivo assays [11, 12]. The massive use, environmental ubiquity, and estrogenic character of phthalates have inspired disposal criteria, occupational exposure limits, and air and water concentration regulations. The US Environmental Protection Agency (EPA) has established a maximum admissible concentration in water of 6 μg L^{-1} for di-(2-ethylhexyl) phthalate (DEHP) [13]. In surface water, the EU has established an annual average concentration of 1.3 μg L^{-1} for the same compound [2].

2.3 Bisphenol A

Bisphenol A (4,4′-isopropylidenediphenol) is used for the production of polycarbonate plastic and epoxy resins, both utilized in a wide variety of applications in food and drink packaging, as powder paints, as additives in thermal paper, in dental fillings and as antioxidants in plastics [14]. It has also been used as an inert ingredient in pesticides, as a fungicide, antioxidant, as a flame retardant, rubber chemical, and polyvinyl chloride stabilizer. Human exposure to bisphenol A eventually released from these materials poses a health risk of increasing concern due to its endocrine-disrupting activity. The estrogenic as well as antiandrogenic activity of BPA has been shown both in in vitro and in vivo experiments [15–22]. BPA is approximately 10,000-fold less potent than 17β-estradiol [23, 24]. In addition, some studies indicate that humans might be more sensitive than lab rats to bisphenol A: Human liver microsomes do not glucuronidate BPA as extensively as the rat liver microsomes do, and therefore, estrogen target tissues in humans may be subject to greater exposure to BPA [25].

2.4 Alkylphenolic Compounds

Alkylphenol ethoxylates (APEOs) are a class of low-cost nonionic surfactants, which have previously been used extensively in both households and industry. But, because of environmental considerations, they have voluntarily been fazed out of the majority of household products during the last few decades.

However, as they are low-cost and are multifunctioning (detergents, deemulsifiers, solubility-, defoaming- and dispersion agents) they are still used in large quantities in the industry. Their primary use is in aqueous solutions and they enter easily into the environment through wastewater discharges, often through WWTPs. Alkylphenolic compounds with longer ethoxylate chains have not demonstrated estrogenicity, but are readily transformed into their shorter chain metabolites or carboxylates in the environment or in the WWTPs [26, 27].

Nonylphenol and octylphenol, along with their carboxylates and ethoxylates (with 1–2 ethoxylate units) have all been shown to be weak estrogenic compounds with an estrogenic effect 10^4–10^6-times lower than that of 17β-estradiol [28–30].

However, as concentrations generally are 10^4–10^5-times higher than endogenic estrogens, alkylphenolic compounds often have significant effect on the estrogenicity of natural water samples [31, 32, 33] and are as such of major environmental concern.

2.5 Perfluorinated Compounds

Fluorinated alkyl substances (FASs) are chemicals that contain atoms of both carbon and fluorine. Because of their unique properties, FASs compose a family of manmade chemicals that have been used for decades to make products that resist heat, oil, stains, grease, and water. FASs include perfluoroalkyl sulfonates (PFSs), nonionic PFSs, and perfluoroalkyl carboxylates (PFCs) (Table 1).

Table 1 Perfluorooctane sulfonate and related substances

Designation	Name	Formula
PFSs	Perfluoroalkyl sulfonates	
PFBuS	Perfluorobutane sulfonate	$C_4F_9SO_3^-$
PFHxS	Perfluorohexane sulfonate	$C_6F_{13}SO_3^-$
PFOS	Perfluorooctane sulfonate	$C_8F_{17}SO_3^-$
PFDS	Perfluorodecane sulfonate	$C_{10}F_{13}SO_3^-$
Nonionic PFSs	Nonionic perfluoroalkyl sulfonates	
PFOSA	Perfluorooctanesulfonamide	$C_8F_{17}SO_2NH_2$
N-EtFOSA	*N*-ethyl perfluorooctanesulfonamide	$C_8F_{17}SO_2NHC_2H_5$
N-MeFOSA	*N*-methyl perfluorooctanesulfonamide	$C_8F_{17}SO_2NHCH_3$
t-Bu-PFOS	*N*-tert-butyl perfluorooctanesulfonamide	$C_8F_{17}SO_2NHC_4H_9$
N-EtFOSE	2-(*N*-ethylperfluoro-1-octanesulfonamido)-ethanol	$C_8F_{17}SO_2N(CH_2CH_3)CH_2CH_2OOH$
7:1 FTOH	Pentadecafluoro-1-octanol	$CF_3(CF_2)_6CH_2OH$
PFCs	Perfluoroalkyl carboxylates	
PFBA	Perfluorobutyric acid	$C_3F_7CO_2H$
PFPeA	Perfluoropentanoic acid	$C_4F_9CO_2H$
PFHxA	Perfluorohexanoic acid	$C_5F_{11}CO_2H$
PFHpA	Perfluoroheptanoic acid	$C_6F_{13}CO_2H$
PFOA	Perfluorooctanoic acid	$C_7F_{15}CO_2H$
PFNA	Perfluorononanoic acid	$C_8F_{17}CO_2H$
PFDA	Perfluorodecanoic acid	$C_9F_{19}CO_2H$
PFUnA	Perfluoroundecanoic acid	$C_{10}F_{21}CO_2H$
PFDoA	Perfluorododecanoic acid	$C_{11}F_{23}CO_2H$
PFTDA	Perfluorotetradecanoic acid	$C_{13}F_{27}CO_2H$
PFHxDA	Perfluorohexadecanoic acid	$C_{15}F_{31}CO_2H$
PFOcDA	Perfluorooctadecanoic acid	$C_{17}F_{35}CO_2H$
8:2 FTCA	2H,2H-perfluorooctanoic acid	$C_6F_{13}CH_2CO_2H$
10:2 FTCA	2H,2H-perfluorododeocanoic acid	$C_{10}F_{21}CH_2CO_2H$
8:2 FTUCA	2H-perfluoro-2-decenoic acid	$C_8F_{16}CHCO_2H$
10:2 FTUCA	2H-perfluoro-2-dodecenoic acid	$C_{10}F_{20}CHCO_2H$

Because of their high production worldwide, in October 2000 the US EPA proposed a significant new rule (SNUR) for 88 perfluorooctane (PFO)-related substances. Perfluorooctane sulfonate (PFOS) and related substances have also been on the agenda of the Organization for Economic Cooperation and Development (OECD) since the year 2000 [34]. Some legislation that generally applies to the release of substances to the environment may be relevant for PFOS. Thus, the IPPC Directive 96/61/EC includes fluorine and its compounds in the "indicative list of the main polluting substances to be taken into account if they are relevant for fixing emission limit values." Very recently, PFOS has been included in the list of substances subject to review for possible identification as priority substances or priority hazardous substances [2]. PFOS is suspected to be a thyroid and reproductive disrupter.

Many of the degradation products of FASs have been found in the environment throughout the world, but PFOS and perfluorooctanoic acid (PFOA) are the compounds more widely detected and these compounds are environmentally persistent substances that have been detected worldwide in human blood, surface water, wastewater, soils, sediments, air, and biota samples. The origin of FASs in WWTPs is unclear, but it is possible that these chemicals are introduced into WWTPs through the cleaning and care of surface treated products, and their use in industrial processes.

2.6 *Pharmaceuticals*

During the last decade pharmaceutically active compounds (PhACs) have been considered to constitute a health risk for humans and aquatic ecosystems. Drugs are continuously being introduced in the environment, first of all, because variable amounts, up to 95% of the administered dose [35, 36], can be excreted unmetabolized and discharged into domestic wastewater. Unwanted or expired medications can be improperly disposed of directly in wastewater.

Several pharmaceuticals can therefore reach WWTPs in substantial amounts and a portion of them can be released into surface water. Moreover, several pharmaceuticals are excreted as conjugates and can release the active moiety by cleavage during treatment processes [37]. Pharmaceuticals are designed to have a specific action on humans and life organisms. Thus, their presence in natural waters can be expected to induce an undesirable effect on nontarget organisms exposed to them. In most cases, a clear cause-effect relationship has not been established yet, but subtle effects have been attributed, for instance, to tricyclic antidepressants (effects on development, spawning, and a wide array of other behaviors in shellfish, ciliates, and other aquatic organisms) and calcium-channel blockers (dramatic inhibition of sperm activity in certain aquatic organisms). Examples of pharmaceutical EDs are alprazolam (benzodiazepine, inhibits corticotrophin secretion), carbamazepine (anticonvulsant, affects circulating thyroid hormones), diazepam (benzodiazepine, affects male reproductive organs), and sulfamethoxazole (antibacterial and antipneumocytic, alters thyroid function).

2.7 *Estrogens*

Estrogens are female steroid sex hormones based on a cholesterol structure and are mostly naturally excreted as their biologically inactive glucuronide and sulfate derivatives. Besides the main endogenous hormones estradiol, estrone, and estriol, other compounds have been synthesized to mimic their method of action, usually for medical reasons (e.g., ethynilestradiol and diethylstilbestrol). In addition several compounds naturally occurring in some plants have been found to cause estrogenic response, the so-called phytoestrogens.

The fate of estrogens during wastewater treatment has been investigated in several studies as WWTP effluents are the major source of contamination by estrogens in urban surface waters. In order to correctly estimate this burden a variety of analytical methods have been developed to assess the levels at which those compounds are present in the WWTP influents and effluents. Currently LC–(tandem)MS preceded by SPE (solid-phase extraction) is the most frequently applied technique [38].

3 Analysis of EDs in Wastewater

3.1 *Sample Preparation*

3.1.1 Pesticides

Occurrence of endocrine-disrupting pesticides in sewage water has been rarely studied as compared to other environmental matrices such as ground and surface waters. The analysis of these compounds in water is usually done with multiresidue methodologies that determine simultaneously different classes of organic pollutants. An overview of the methodologies reported to analyze endocrine-disrupting pesticides in sewage water based on liquid chromatography–mass spectrometry (LC–MS) is shown in Table 2.

Residues of persistent organochlorine pesticides have been analyzed so far in sewage water samples by gas chromatography (GC) coupled to mass spectrometry or electron-capture detectors (ECD) [40, 44–46]. Because of their physico-chemical properties these chemicals are not amenable to LC–MS analysis, therefore, they are not further discussed.

Table 2 Overview of LC-MS analytical methodologies developed for determination of endocrine-disrupting pesticides in sewage water

	Sample preparation			LC		MS		
Reference	Volume	Method	Cartridge/loading column	Chromatographic column	Mobile phase	Detector (interface)	Acquisition mode	Observations
[39][a]	1 mL	On-line SPE	Hypersil GOLD (20 × 2.1 mm, 12 μm)	Hypersil GOLD™ (50 × 2.1 mm, 3 μm)	Water (0.1% FA)–MeOH	QqQ (ESI)	SRM	Matrix effects study Occurrence in real wastewater samples Quantitation by internal standard addition
[40][b]	300 mL	SPE	Strata C-18	Uptispher C18 3HDO (100 × 2 mm, 3 μm)	Water (10 mM NH_4Ac) – MeOH	QqQ (ESI)	SRM	5 OC pesticides detected by GC–MS Matrix effects study Occurrence in real wastewater samples Quantitation by internal standard addition
[41][c]	200 mL	Automated SPE	Oasis HLB (200 mg)	Zorbax, SB-C18 (250 × 3 mm, 5 μm)	PI mode: Water (0.1% FA) - ACN, NI mode: Water - ACN	QTRAP (ESI)	SRM/EPI	Matrix effects study
					PI mode: Water (0.1% FA) - ACN, (5% water + 0.1% FA) NI mode: Water (0.05% NH_4HCO_2) - ACN	TOF (ESI)	SCAN	Occurrence in real wastewater samples TOF analysis used for extra confirmation and identification of unknowns

(continued)

Table 2 (continued)

		Sample preparation		LC		MS		
Reference	Volume	Method	Cartridge/ loading column	Chromatographic column	Mobile phase	Detector (interface)	Acquisition mode	Observations
[42][d]	1 L	SPE	Oasis HLB (60 mg)	Neutral pesticides: Nucleodur C18 Gravity (125 × 2 mm, 3 μm)	Neutral pesticides: Water (0.1% FA) – MeOH(0.1% FA)	QqQ (ESI)	SRM	Matrix effects study
				Acidic pesticides: Polymer coated C18 (150 × 2 mm, 3 μm)	Acidic pesticides: Water (0.6% FA) – MeOH(0.6% FA)			Quantitation by internal standard addition
[43][e]	400 mL	Automated sequential SPE	Isolute C-18 ec (500 mg	Superspher 100 RP-18 (250 × 2 mm, 4 μm)	Water (0.1% Acetic acid) – ACN	Q (ESI)	SIM	Occurrence in real waste-water samples
	300 mL		Oasis HLB (200 mg)					

ED pesticides included in the different multiresidue methods:
[a]Atrazine, Simazine, Cyanazine, 2,4-D
[b]Dimethoate, 2-Imidazolidinethione, Tebuconazole, Hexaconazole
[c]Atrazine, Simazine, Chlorfenvinphos, Diuron
[d]Atrazine, Diazinon, Diuron, Metolachlor, Simazine
[e]Atrazine, Simazine, 2,4-D, 2,4,5-T, Carbaryl, Diuron, Linuron, Trifluralin, Chlorfenvinphos, Chlorpyrifos, Alachlor, Propanil
FA formic acid; *ACN* acetonitrile; *MeOH* methanol; *NH4HCO2* ammonium formate; *NH4Ac* ammonium acetate; *QqQ* triple quadrupole; *QTRAP* quadrupole-ion trap; *TOF* time-of-flight; *Q* single quadrupole; *SRM* selected reaction monitoring; *SIM* single ion monitoring

The technique most routinely used to preconcentrate this type of pesticides is solid-phase extraction. Prior to the SPE procedure, sewage water samples are decanted [43] or filtrated through 0.45–0.7 μm glass fiber filters [39–42] in order to avoid clogging of the SPE cartridges by suspending matter. In some published methodologies the concentration and clean-up procedures are performed automatically using devices such as the ASPEC XL [41] and the AutoTrace [43]; however, a fully automated on-line method based on column-switching has been recently developed by Viglino et al. [39].

C18 or polymeric sorbents are usually the materials used in the different SPE procedures. Optimum sample volumes loaded into SPE cartridges range between 1 mL (in the on-line approach developed by Viglino et al. [39]) and 1 L [42, 47].

3.1.2 Phthalates

The very low levels of PAEs and the complex matrix of many environmental samples, including sewage water, make sample preconcentration necessary for a reliable determination of these compounds [48]. Several preconcentration techniques such as liquid–liquid extraction (LLE) and, especially, solid-phase extraction, have been developed. Using a hemicelles-based solid-phase extraction before liquid chromatography–atmospheric pressure chemical ionization–ion trap–mass spectrometry (LC–APCI–IT–MS) separation and quantitation, high concentration factors (up to 2,500) and thus low detection limits were achieved by López Jiménez et al. [13] for butyl benzyl phthalate (BBP), DBP, and DEHP in sewage water (see Table 3). A priority goal when analyzing phthalates is to reduce sample treatment to the minimal steps in order to avoid or minimize PAE contamination through laboratory materials like tubing, pipette tips, septa, solvents, etc. [13, 52, 53]. In recent years, solid-phase microextraction (SPME) has been extensively investigated to simplify sample treatment prior to GC analysis

Table 3 LC–MS methods for quantitative determination of phthalate esters in wastewaters

Matrix	Compound	Extraction	Separation and detection	LOD (ng L^{-1})	Reference
Industrial effluent	DBP, DMP, BEHP	SPE, Lichrolut EN	LC–APCI–MS	60–100	[49]
Urban wastewaters	DEP, DBP, DMP, DEHP	SPE, Lichrolut RP-18	LC–ESI–MS	10–110	[50]
Wastewaters	DEHP, BBP, DBP	SPE, Hemimicelles	LC–IT–MS	0.035–0.080	[13]
Urban wastewaters	DEP, BEHP, DOP, DMP, DBP, BBP	SPE, Lichrolut RP-18	LC–APCI–MS	12–57[a]	[51]

[a]*LOQ* Limit of quantification

LOD limit of detection; *DEHP* di-(2-ethylhexyl) phthalate; *DEP* diethylphthalate; *BBP* butyl benzyl phthalate; *DBP* di-*n*-butyl phthalate; *DPP* dipentylphthalate; *BEHP* bis(2-ethylhexyl) phthalate; *DOP* di-*n*-octylphthalate; *DMP* dimethyl phthalate

Table 4 LC-MS methods for quantitative determination of BPA in wastewaters

Matrix	Extraction	Separation and detection	LOD (ng L^{-1})	Reference
Urban wastewaters	SPE LiChrolut C18	LC–MS	52[a]	[51]
Sewage treatment plant and receiving waters	SPE LiChrolut RP C18	LC–MS	50	[50]
Wastewater	SPE Oasis HLB	LC–MS/MS	2	[56]
Textile industry and corresponding WWTP effluents and receiving rivers	SPE Oasis HLB	LC–MS/MS	2	[57]

[a] *LOQ* limit of quantification; *LOD* limit of detection

of PAEs [54, 55]. The use of SPME reduces the risk of secondary contamination during sample handling, however, some drawbacks such as low extraction yields, long extraction times, and/or poor precision still remain [13].

3.1.3 Bisphenol A

There are several analytical methods based on LC-MS for the determination of BPA among other organic contaminants and/or EDCs in water. Some of them, applied specifically in wastewater matrices, are gathered together in Table 4. Sample pretreatment previous to LC-MS analysis usually comprises sample preconcentration and clean-up SPE. As is shown in Table 4, silica-based C18 and polymeric HLB are the sorbents commonly selected for SPE.

3.1.4 Alkylphenolic Compounds

Alkylphenolic compounds have been analyzed by a wide range of chromatographic analysis techniques, ranging from GC-MS over normal-phase liquid chromatography with fluorescence detection to reverse-phase liquid chromatography with MS or MS/MS detection. An overview of the sample pretreatment in a number of selected methods dealing with the analysis of wastewater is presented in Table 5. The majority of methods use solid phase extraction as preconcentration. However, in two methods liquid-liquid extraction (LLE) with dichloromethane was chosen, because of higher observed recoveries.

For the SPE methods, wastewater was filtered through 0.45 μm membrane filters, before loading onto cartridges. A variety of solid phases and eluents have been used, but the majority employ C_{18} (Lichrolut or Isolut) with Oasis HLB as a close second.

3.1.5 Perfluorinated Compounds

One of the major problems associated with FAS analysis is background contamination in the analytical blanks. The use of glass bottles for the water sample collection

Table 5 LC–MS and LC–MS/MS methods used to determine alkylphenolic compounds in wastewater samples

Compounds	Extraction method	Separation and detection method	LC column	Mobile phase	MS system	LOD (ng L^{-1})	Reference
NPEOs, NPECs	SPE: ENVI-Carb: acidified	RP-LC–ESI–MS/MS $[M+NH_4]^+$, $[M-H]^-$	Zorbax C_8 (150 × 4.6 mm, 5 μm)	MeOH + 0.1% formic acid/H_2O + 10 nM NH_4Ac	Quattro Ultima (Micromass)	10–50	[58]
APs, APEOs, APECs	SPE: poly(styrene–divenylbenzene) LLE: dichloromethane (APECs)	MM–LC–ESI–MS/MS	Mixed mode (size exclusion and RP)				[27]
APs, APECs, XAPECs, XAPs, APEOs, XAPEOs	SPE: LiChrolut C_{18}: MeOH	RP-LC–ESI–MS $[M+Na]^+$, $[M-H]^-$	LiChrospher 100 C_{18} (250 × 4 mm, 5 μm)	ACN–H_2O (NI) MeOH–H_2O (PI)	HP 1100 (Agilent)	20–100	[59]
APs	SPE: Isolute C_{18}: propan-2-ol	RP-LC–ESI–MS/MS $[M-H]^-$	Nucleosil C_8 (250 × 2 mm, 5 μm)	ACN–H_2O	Quattro LC (Micromass)	20–80	[60]
APEOs	LLE: dichloromethane	RP-LC–ESI–MS	Phenomex Luna (250 × 2 mm, 5 μm)	MeOH–H_2O (both with 5mM NH_4Ac + 0.05 nM TCAA)	Esquire	1,000	[61]
APs, APEOs, APECs	SPE: Isolute C_{18}: MeOH	RP-LC–ESI–MS $[M+Na]^+$, $[M-H]^-$	Purospher STAR RP-18 (125 × 2 mm, 5 μm)	MeOH–H_2O	HP 1100 (Agilent)	5–100	[62]
APs, APECs, APEOs	SPE: Oasis HLB: MeOH/AcOH/ETAc + 0.1% formic acid	RP-LC–ESI–MS/MS $[M+NH_4]^+$, $[M-H]^-$	Synergi Polar-RP (150 × 2 mm, 4 μm)	ACN–H_2O (NI) ACN-H_2O + 20 mM NH_4Ac	Quattro Micro (Micromass)	1–100	[57]
APEOs	SPE: Oasis HLB: MeOH	RP-LC–ESI–MS/MS $[M+Na]^+$, $[M-H]^-$	Purospher STAR RP-18 (125 × 2 mm, 5 μm)	MeOH–H_2O	Quattro LC (Micromass)	2–50	[56]
APs, NPEOS	SPE: LiChrolut C_{18}: MeOH/DCM	RP-LC–ESI–MS $[M+Na]^+$, $[M-H]^-$	LiChrospher 100 C_{18} (250 × 4 mm, 5 μm)	ACN–H_2O (NI) MeOH–H_2O (PI)	HP 1100 (Agilent)	60–120	[50]

(continued)

Table 5 (continued)

Compounds	Extraction method	Separation and detection method	LC column	Mobile phase	MS system	LOD (ng L^{-1})	Reference
APs, APEOs, APECs	SPE: Oasis HLB: MeOH + DCM	RP–LC–ESI–MS/MS $[M+NH_4]^+$, $[M-H]^-$	Zorbax Eclipse XDB–C_8 (150 × 2.1 mm, 5 μm)	MeOH/H_2O (both with 1.5 mM NH_4Ac)	API 3000 (Applied Biosystem)	0.04–12	[157]
APEOs, APECs, CAPECs	SPE: black carbon: MeOH + DCM/MeOH	RP-LC–ESI–MS	Altima(150 × 4.6 mm, 5 μm)	ACN–H_2O (both with 0.1 mM TFA)	–	–	[63]
APs, APEOs	SPE: Oasis HLB: MeOH-Et_2O	RP-LC–ESI–MS $[M+Na]^+$, $[M-H]^-$	Hypersil Green ENV (150 × 4.6 mm, 5 μm)	ACN/MeOH–H_2O (both with 0.5% HOAc + 0.1% TEA)	SSQ-7000 (Finnigan)	100–500	[64]
APs, APEOs	SPE: C_{18}: AcOH + MeOH/MTBE	RP-LC–ESI–MS/MS	–	–	–	10–20	[65]
APs	SPE: LiChrolut EN: ACN	RP-LC–ESI–MS $[M-H]^-$	Lichrospher 100 RP-18 (250 × 4 mm, 5 μm)	ACN–H_2O + 0.5% HOAc	HP1100 (Agilent)	8–300	[66]
APs	SPE: Oasis HLB: 2-propanol-MTBE	RP-LC–ESI–MS/MS $[M-H]^-$	Phenomex Luna (100 × 2 mm, 3 μm)	ACN–H_2O (both with 1 mM NH_4–formate)	1100 LC–MSD Trap	1–20	[67]
APs, APECs	SPE: SDS hemicelles-based: MeOH	RP-LC–ESI–MS $[M-H]^-$	Hypersil AV-1634 C_{18} (150 × 4.6 mm, 5 μm)	MeOH–H_2O –	1100 LC–MSD Trap	99–165	[68]
APs	SPE: LiChrolut C_{18}: Et_2O:MeOH	RP-LC–APCI–MS $[M-CH_2]^+$	Gemini C_{18} (150 × 4.6 mm, 5 μm)	ACN–H_2O + 1% HOAc	ZQ 2000 (Micromass)	26–28	[51]

ACN acetonitrile; *AcOH* acetona; *APs* alkylphenols; *APECs* alkylphenoxy carboxylates; *APEOs* alkylphenol ethoxylates; *DCM* dichloromethane; *ESI* electrospray; *ETAc* ethyl acetate; *Et_2O* diethylether; *HOAc* acetic acid; *LC* liquid chromatography; *MeOH* methanol; *MM* mixed mode; *MS* mass spectrometry; *MTBE* methyl-*tert*-butylether; *NH_4Ac* ammonium acetate; *NI* negative ionization; *PI* positive ionization; *RP* reverse phase; *SPE* solid-phase extraction; *TCAA* trichloroacetic acid; *TEA* triethylamine must be avoided because the target analytes in aqueous solutions may bind to the glass. The samples are usually collected in polyethylene or propylene bottles rinsed with different solvents prior to use [69–71]. Filtration of water is not considered a viable sample preparation step because, as reported in several studies [71, 72], some filters remove the analytes from water and some filters themselves present levels of perfluorinated alkyl substances (PFASs). In order to solve the drawbacks presented by filtration, centrifugation has been selected as an alternative pretreatment step. SPE is the technique most extensively used for the extraction of PFASs from environmental water samples. SPE extraction has usually been carried out with cartridges in the off-line mode. Because of the different polarities of the analytes, their extraction requires different strategies. The ionic short chain (C4–C6) compounds [perfluorocarboxylic acids (PFCAs) and perfluorosulfonic acids (PFSAs)] require moderately polar media (Oasis wax SPE) for efficient extraction. For longer chains, less polar or nonpolar SPE phases (C18 and Oasis HLB) may be applied. The nonionic compounds have usually been extracted from the matrix by nonpolar media (C18), and also by moderately polar media (Oasis HLB and Oasis WAX SPE) [69, 70]. LLE has also been used for extraction of FASs from wastewater samples. By using LLE, it is possible to extract samples with or without prior filtration and hence it is possible to determine concentrations in the aqueous fraction or the whole sample [73]. The extraction procedures described above are often time consuming and imprecise, as additional clean-up steps are usually required. To overcome these drawbacks and simplify the sample pretreatment procedures, direct LC–MS/MS analysis of municipal wastewater has been tested by Schultz et al. [71]. These authors discarded the use of SPE as a sample concentration step because critical SPE experiments with C18 cartridges yielded low recoveries for some target analytes. This technique allowed an increase in the sensitivity that permitted the detection of FASs at environmentally relevant levels in influent and effluent wastewaters.

3.1.6 Pharmaceuticals

Several methods have been developed for the determination of drugs and their metabolites in the lower ng L^{-1} range. Most of them use SPE followed by analysis by GC-(tandem)MS or LC–(tandem)MS. Several review articles have been published on this topic: [74–76].

The vast majority of the methods developed in the past focus on specific therapeutic classes [77–82], with antibiotics being the most studied ones due to the concern about their potential for antibiotic resistance [83–85]. Because of the elevated polarity of some groups of pharmaceuticals, LC–MS and LC–(tandem)MS is often the technique of choice for analysis of these compounds in wastewater since it avoids the derivatization step required in GC–MS methods, specially for acidic drugs.

A survey of the most recent and representative multiresidue LC–MS/MS methods developed for the determination of regularly used pharmaceuticals in wastewater is given in Table 6.

Table 6 Summary of the most representative multiresidue LC-MS/MS methods for quantitative determination of pharmaceuticals in wastewater

Compounds	Matrix	Pretreatment	Extraction	Elution solvents	Column	Mobile phase	MS system	LOD (ng L^{-1})	Reference
30 Compounds: Antibiotics, anticancer, antiinflammatories, bronchodilator; cardiovascular, CNS, gastrointestinal drugs; diuretics, estrogens, and lipid regulators	Urban wastewaters	Two methods: (1) neutral drugs at pH = 7, (2) another group at pH = 2	(A) Lichrolut EN (B) Oasis MCX	(1) MeOH/ethyl acetate (2) MeOH/NH_4OH and NaOH in MeOH	C8	ESI(+) Aq. formic acid/ACN ESI(−)Aq. TEA/ACN	Triple quadrupole	30–70	[86]
16 Compounds: Antibiotics, β-blockers, psychiatric drugs, antiinflammatories	Hospital wastewater	pH adjustment (pH 7)	Oasis HLB	MeOH	C18	ESI(+) Aq. formic acid/ACN ESI(−) Water/CAN	Triple quadrupole ESI	0.3–5.6 (LOQ)	[13]
28 Compounds: Antiinflammatories, lipid regulators, anti-ulcer agents, antihistaminics, antibiotics, and β-blockers	Urban and industrial wastewaters	Natural water pH	Oasis HLB	MeOH	Acquity UPLC™ BEH C18	ESI(+) ACN/MeOH (2:1)/NH4Ac/HOAc ESI(−) MeOH/H2O	Triple quadrupole ESI-TOF	100–250	[87]
28 Compounds: Antiinflammatories, lipid regulators, anti-ulcer agents, antihistaminics, antibiotics, and β-blockers	River and wastewaters	Natural water pH	Oasis HLB	MeOH	C18	ESI(+) ACN/MeOH (2:1)/NH4Ac/HOAc ESI(−) MeOH/H2O	Triple quadrupole ESI	1	[88]

12 Compounds: Hormones, hormone conjugates and macrolide antibiotics	Wastewaters	No pretreatment	DVB-phobic Speedisk + clean-up with SEC	SPE: Tert-butyl methyl ether/MeOH	RP-MAX RP-18	MeOH/H2O (hormones) NH4Ac aq./ACN (antibiotics)	Triple quadrupole ESI	25–50 (LOQ in WW)	[89]
11 Compounds: Antibiotics – tetracyclines and sulfonamides	Wastewaters	Na2EDTA and citric acid pH < 3	Oasis HLB	MeOH	C18	ESI(+) 0.1% aq. HCOOH/AcN	Triple quadrupole and ion trap ESI	20–50	[90]
12 Compounds: Antiinflammatories, lipid regulators and triclosan	Surface and wastewaters	Acidified pH = 2–2.5	Oasis HLB	MeOH	Phenyl– Hexyl	ESI(−) MeOH/H2O // solvent A with TrBA and acetic acid	Triple quadrupole and ion trap ESI	0.1–5.2 (LOQ)	[82]
14 Compounds: Lipid regulators, analgesics and antiinflammatories, antineoplastics, vasodilators, and antiepileptic	Wastewaters	Acidic comp. (anal., antiinfl., lipid regulators) pH = 2 Basic comp. pH = 7.5	Lichrolut RP-18 (acidic) Supelco LC-18 (basic)	MeOH (acidic and basic)	RP-18 (acidic and basic)	ESI(−) (acidic) ACN/MeOH (40:60)// 20 mM NH4Ac ESI (+) (basic) ACN in aq.	GC-MS (acidic) and triple quadrupole ESI (acidic and basic)	4–47	[91]
27 Compounds: Neutral and acidic pharmaceuticals, EDCs and PCPs	Surface and wastewaters	Acidified pH = 2	Oasis HLB	MeOH/MTBE (10:90, v/v)/MeOH	C12	ESI(+) and ESI(−) Aq. formic acid/MeOH	Triple quadrupole ESI/APCI	15–500	[92]
Ten compounds: Antiphlogistics, antiinflammatories, antidiabetics, antiulcer, and psychiatric drugs	River and wastewaters	pH adjustment (pH = 7–7.5)	Isolute C18	MeOH	RP-18	ESI(+) 20 mM NH4Ac in H_2O/ACN	Triple quadrupole ESI	1–60	[78]

(continued)

Table 6 (continued)

Compounds	Matrix	Pretreatment	Extraction	Elution solvents	Column	Mobile phase	MS system	LOD (ng L^{-1})	Reference
19 Compounds: Antibiotics - penicillins, chloramphenicol, sulfonamides, and macrolides	Surface and wastewaters	Acidified (pH = 3)	Lichrolut EN and C18	MeOH	RP-18	TET 20 Mm aq. oxalic acid/CAN The others 10 mM aq. NH_4Ac/ ACN	Triple quadrupole ESI	0.6–15 (LOQ)	[83]
Four Beta blockers, an antiepileptic drug and three fluoroquinolone antibiotics	Surface and wastewaters	Natural water pH	Oasis HLB	MeOH	XDB-C18	ACN and 1% acetic acid	Triple quadrupole ESI	3.5 ng L^{-1} in sewage influent LOQ	[93]
29 Multiclass pharmaceuticals: NSAIDs, lipid regulators, psychiatric drugs, antihistaminics, antiulcer agent, antibiotics, and β-blockers	Surface and wastewaters	Natural water pH	Oasis HLB	MeOH	RP-18	NI mode, (A) MeOH (B) water PI (A) a mixture of ACN-MeOH (2:1) (B) NH4Ac 5 mM/HAc at pH 4.7	Triple quadrupole ESI	1–160 ng L^{-1}	[87]
38 Pharmaceuticals and 10 of their most frequent metabolites, six pesticides, and two disinfectants	Wastewater	pH 8	Oasis HLB	MeOH	RP-18	PI (A) CAN (B) water with 0.1% formic acid NI (A) ACN (B) Water	LC-QTRAP–MS	0.04 and 50	[41]

Nine Basic pharmaceuticals	Wastewater	pH adjustment (pH = 7–7.5)	Speedisk phenyl and a NH2 solid-phase extraction tube for sample clean-up	MeOH	Pentafluorophenyl column (PFP-column)	Eluents were H2O/ACN [95:5, by vol. (A) and 5:95, by vol. (B)]	Triple quadrupole ESI	LOQ 0.05–10	[94]
12 Antidepressant compounds	Wastewater	Acidification	Oasis HLB	70% methanol in 2% acetic acid	XDB-Phenyl column	Water with 0.1% (v/v) formic acid * ACN	Triple quadrupole ESI	0.19–0.45	[95]
20 Pharmaceutical compounds	Wastewater	pH 4	Strata-X cartridges	MeOH	RP-18	(A) 20:80 ACN:water with 0.1% ammonium acetate (v/v) at pH 6.2 (B) 80:20 ACN:water with 0.1% ammoniumacetate (v/v)	Triple quadrupole ESI	0.002–0.855 in influent and 0.001–2.478 in effluent	[96]

The most recent analytical methodologies address the simultaneous extraction of all target compounds in a single SPE step using Oasis HLB cartridges working in general at neutral pH. The main advantages of the extraction with polymeric sorbents at neutral pH are: (1) simplified sample handling and no clean-up for the removal of humic and fulvic acids, (2) less risk of acidic hydrolysis of other compounds in multianalyte analysis procedures, (3) and possibility of on-line extraction using large sample volumes.

C18 cartridges have been also widely used. However, when using this material, depending on the nature of the compounds investigated, sample pH adjustment prior to extraction is usually required.

Less common cartridges used are Lichrolut ENV+, Oasis MCX, and StrataX. The first one is generally recommended for the extraction of polar organic compounds at low pH values, but it can also retain neutral drugs at pH 7, such as carbamazepine (antiepileptic) and macrolides, through hydrophobic interactions. Oasis MCX has been used to extract acidic, basic, and neutral compounds at low pHs. In this material, basic compounds are retained through cation exchange mechanisms and acidic and neutral ones through reversed-phase interactions. Finally, StrataX, as Oasis HLB, can retain a wide spectrum of analytes through hydrophilic and hydrophobic interactions.

However, the simultaneous extraction of several groups of compounds often requires a compromise which does not obtain the best conditions for each compound. An alternative is to use two or more SPE materials operating in series to classify target analytes according to their physico-chemical properties. The advantage of these procedures is that better conditions are achieved for the various target analytes, but they are time consuming.

Other relevant improvements in the extraction of pharmaceuticals from wastewater come from new materials such as the molecularly imprinted polymers (MIPs). An example is the method recently developed by Gros et al. [97] to determine eight β-blockers in wastewaters. The advantages offered by MIPs in terms of selectivity and specificity were compared with the polymeric material most commonly used Oasis® HLB. Even though recoveries achieved with both sorbents were similar, MIPs provided lower LODs than Oasis® HLB, due to their specificity for the target analytes and closely related analogues.

Another attractive process is on-line SPE. In on-line SPE the whole sample is transferred to the chromatographic system thus increasing the method sensitivity and reducing the volume of sample required. Other advantages are high reproducibility and less risk of contamination. An example is that presented by Quintana et al. [98] for analysis of nonsteroidal antiinflamatory drugs (NSAID) in surface water, wastewater, and urine.

3.1.7 Estrogens

Wastewater samples are complex matrices composed of organic matter, bacteria, and all kinds of chemical pollutants which are present at high concentrations and may interfere during all steps of the analytical procedure. The pretreatment of the sample is thus very important, especially when very low concentrations have to be detected, as is the case for estrogens.

Table 7 shows examples of complete analytical methods based on LC-MS used for the analysis of estrogens in wastewater samples. As can be seen, the most studied compounds in wastewater samples are the natural hormone estradiol (E2) and its metabolites, estrone (E1) and estriol (E3). A variety of glucuronide and sulfate derivatives (conjugates) of the previously mentioned hormones, hydroxyl estrone (HE) and the synthetic compounds ethynyl estradiol (EE), diethylstilbestrol (DES), and mestranol (MES) have also been analyzed. Concerning phytoestrogens, daidzein (DAI), genistein (GEN), and biochanin-A (BIO) have been reported in a few studies, whereas only one work has addressed the analysis of 4',6,7-trihydroxyisoflavone (TRI), formononetin (FOR), coumestrol (COU), glycitein (GLY), and the two conjugates daidzin and genistin.

Extraction of these compounds from environmental samples can be performed by sorptive or adsorptive techniques. Adsorptive SPE is very versatile and allows good extraction recoveries with cartridges containing varying adsorbents. Polymeric (e.g., Oasis HLB), modified silica-based (e.g., C18) and graphitized carbon-based (e.g., Carbograph) adsorbents are commonly used for wastewater samples (see Table 7). An important advantage of this technique is the possibility of automation, yielding highly robust and sensitive methods [119]. Adsorptive disks have also found application [89, 104, 105]. Some authors include an additional clean-up procedure after SPE (e.g., GPC, SPE, LLE) to obtain more pure extracts. One example of the application of a sorptive extraction technique is that reported by Mitani et al. [122] based on fully automated in-tube solid-phase microextraction coupled with LC–MS/MS with which limits of detection (LOD) in the low ng/L range for the main estrogens have been achieved.

3.2 *Instrumental Analysis*

3.2.1 Pesticides

The procedures for chromatographic separation are very similar and most of them rely on the use of reversed-phase C18 columns and the application of a gradient binary mobile phase (see Table 2). Ionization of the compounds is performed with an electrospray ionization source (ESI) working in the negative ionization (NI) mode when analyzing acidic pesticides such as phenylureas and phenoxyacid compounds, and in the positive ionization (PI) mode when analyzing the rest.

The main drawback of using this interface is the enhancement or suppression of the analyte MS signal due to matrix effects. Each pesticide may be affected to a different extent by this phenomenon in complex matrices such as sewage water, even from sample to sample; therefore quantitation of the compounds by the internal standard method [39, 42] or by the standard addition method [40] is indispensable to achieve reliable results [42]. Most of the peer-reviewed methodologies evaluate matrix effects in wastewater [39–42]. Ion suppression occurs for practically all analytes investigated, reaching up to 65% in the case of diazinon [42]; signal enhancement has been reported by Baugros et al. [40] for a series of endocrine-disrupting fungicides and OP compounds.

Table 7 LC–MS and LC–MS/MS methods for analysis of estrogens in wastewater influent (I) and effluent (E) samples

Compounds	Matrix	Extraction SPE material	Clean-up	LC separation Column	Mobile phase	Interface	Detection	Reference
E1, E2, E3, EE	E	Oasis HLB	–	Zorbax Extended-C18 (50 × 2.1 mm, 5 μm)	ACN/Water (10 mM NH_4OH)	ESI(NI)	MS/MS	[99]
E1, E2, E3, EE, E1S	I, E	tC18	GPC-SEC + SPE (NH2)	Gemini C18 (100 × 2 mm, 3 μm)	MeOH/Water (0.1% NH_4OH)	ESI(NI)	MS/MS	[100]
E2, EE, E1	E	Oasis HLB	–	Acquity UPLC BEH C18 (100 × 1 mm, 1.7 μm)	ACN/Water (0.1% formic acid)	ESI(PI)	MS/MS	[101]
E1, E2, E3, EE	E	Oasis HLB	–	Synergi RP-C12 (250 × 4.6 mm, 4 μm)	MeOH/Water (0.1% formic acid)	APCI(PI)	MS/MS	[102]
E1, E2, E3, EE, DAI, GEN, BIO	I, E	IST C18 + Sep-Pak®Plus NH_2 Propyl	–	Purospher STAR-RP-18e (125 × 2 mm, 5 μm), and Acquity C18 (50 × 2.1 mm, 1.7 μm)	ACN/Water	ESI(NI)	MS/MS	[103]
E1, E2, E3, EE, HE, MES, +3 conj	E	DVB-phobic (Speedisk)	GPC–SEC	Synergi RP-MAX (150 × 2 mm, 4 μm)	MeOH/Water	APCI/ ESI(NI)	MS/MS	[89, 104]
E1, E2, E3, EE	E	PolarPlus C18 (Speedisks)	–	BetaBasic C18 (150 × 2.1 mm, 3 μm)	ACN/Water (10 mM *N*-methylmorpholine)	ESI(NI)	MS/MS	[105]
E1, E2, E3, EE	I, E	Carbograph-4	–	Alltima C-18 (250 × 4.6 mm, 5 μm)	ACN/water (+MeOH/NH_3 post column)	ESI(NI)	MS/MS	[106, 107]
E1, E2, EE	I, E	Bakerbond C18	SPE (silica gel)	RP-C8 Hypersil MO5 (100 × 2.1 mm, 5 μm)	ACN/(MeOH)/water	ESI(NI)	MS/MS	[108]
DAI, GEN, BIO, FOR, GLY, TRI, COU + 2 conj	I, E	Oasis HLB	–	Alltima C-18 (25 × 4.6 mm, 5 μm)	ACN/water (10 mM TFA)	ESI(PI)	MS/MS	[109]
E1, E2, E3, EE + 5 conj	I, E	Sep-Pak Vac tC18	–	Synergi 4 μ Hydro-RP (75 × 2 mm, 4 μm)	MeOH/water	ESI(NI)	MS	[110]

6 conj	I, E	Oasis HLB	Anion exchange (Nucleogen 60–7, 125 × 4.6 mm)	Zorbax Extend-C18 (100 × 2.1 mm, 3.5 μm)	ACN/water (13.5 mM NH_4OH)	ESI(NI)	MS/MS	[111]
E1, E2	I, E	LiChrolut EN + C18	Immunoaffinity extraction	Betasil C18 (150 × 2.1mm, 3 μm)	ACN/water	ESI(NI)	MS	[111, 112]
E1, E2, E3, EE + 9 conj	I, E	Oasis HLB	SPE (Sep-Pak Plus Florisil and NH_2)	Zorbax Extend-C18 (150 × 2.1 mm)	ACN/water (1 mM NH_4OH)	ESI(NI)	MS/MS	[113]
DAI, GEN, BIO	I, E	Oasis HLB	–	Alltima C-18 (25 × 4.6 mm, 5 μm)	ACN/water (10 mM TFA)	ESI(PI)	MS/MS	[114]
E1, E2, E3, EE	I, E	Oasis HLB	–	Alltima C-18 (25 × 4.6 mm, 5 μm)	ACN/water (+NH_3 post column)	ESI(NI)	MS/MS	[114]
E1, E2 (α + β), E3, EE + 10 conj	E	Autoprep EDS-1	SPE (Florisil)	Zorbax Extend-C_{18} (150 × 1 mm, 3.5 μm)	ACN/water (100 mM TEA)	ESI(NI)	MS/MS	[115]
E1, E2, EE	E	C18	LLE + SPE (Florisil)	Hypersil BDS C_{18} (250 × 2.1 mm, 3 μm)	ACN/water	ESI(NI)	MS/MS	[116]
E1, E2, E3 + 8 conj	I, E, U	Carbograph 4	–	Alltima C-18 (25 × 4.6 mm, 5 μm)	ACN/water (10 mM formic acid)	ESI(NI)	MS/MS	[117]
E1, E2, E3, EE + 8 conj	I, E	Carbograph-4	–	Alltima C18 (250 × 4.6 mm, 5 μm)	ACN/water (+NH_3 post column)	ESI(NI)	MS/MS	[118]
E1, E2, E3, EE, DES, MES	E	on-line PLRP-s	–	LiChrospher 100 RP-18 (250 × 4 mm, 5 μm)	ACN/water	ESI(NI/PI)	MS	[119]
E1, E2, E3, EE, DES	I, E	LiChrolut RP-18	–	LiChrospher 100 RP-18 and 60 RP-Select B (250 × 4 mm, 5 μm)	ACN/water	ESI(NI/PI)	MS	[120]
E1, E2, E3, EE	I, E	ENVI-CARB	–	Alltima C-18 (250 × 4.6 mm, 5 μm)	ACN/water	APCI(PI)	MS/MS	[121]

GPC Gel permeation chromatography; *SEC* Size exclusion chromatography

Mass spectrometric detection of these substances in sewage water is mainly done with triple quadrupole analyzers (QqQ) operating in the selected reaction monitoring (SRM) mode. Some authors have also used single quadrupole instruments to identify the target analytes [43, 123]; however, these instruments do not offer enough selectivity to determine trace levels in such complex matrices.

Martinez-Bueno et al. [41] developed an analytical methodology based on the use of a hybrid quadrupole-ion trap instrument (QTRAP) operating under SRM and enhanced product ion (EPI) mode for target analyte identification and unequivocal confirmation of simazine, atrazine, diuron, and chlorfenvinphos, and a time-of-flight (TOF) spectrometer for extra confirmation and identification of nontarget or unknown metabolites.

3.2.2 Phthalates

The most common technique for PAE determination in environmental samples has been gas chromatography with detection through electron capture (ECD) and/or mass spectrometry. Reversed phase LC with UV or MS detection, combined with SPE, has been also extensively used for PAE analysis [13, 124, 125] (see Table 3). LC–ESI–MS based on the formation of sodium adducts can provide molecular ion information and has been found to be a reliable tool for quantitative analysis of phthalate esters in various matrices [53]. Phthalate esters show a characteristic peak at *m/z* 149 (protonated phthalic anhydride), except for the dimethyl phthalate (DMP) that shows the most abundant fragment at *m/z* 163 ($[M+H-2CH_3]^+$) [51–53]. An LC–ESI–MS/MS study showed that the two major pathways of phthalate fragmentation in ESI(+)–MS/MS are similar to those occurring in electron ionization (EI)–GC–MS [53]. One reaction is dominated by the loss of one of the substituents leading to the formation of monoester sodium adducts and another by the formation of sodiated phthalic anhydride ions $[(C_8H_4O_3Na]^+$ with *m/z* 171. Study with H^+ as mobile phase modifiers showed that protonated molecular ions are more reliable for the identification of phthalates by the MS-MS technique [53].

3.2.3 Bisphenol A

The methods most frequently used for the determination of BPA in water samples are GC and LC, both coupled to MS. In recent years, different bioassays have also been developed as a cheap and quick alternative to traditional chromatographic methods for high-throughput analysis or real-time monitoring. Some of these bioassays have been applied to the determination of the potential biological impact of bisphenol A, based on the measurement of toxicity [126] or estrogenicity [127], whereas other studies are based on immunoassays [128, 129], and immunosensors [130, 131].

LC–MS analytical methods described for the determination of BPA and other organic contaminants in wastewater matrices are listed in Table 4. Particularly interesting is the analytical method developed by Zafra-Gomez et al. [51] for the simultaneous determination of different EDCs including bisphenol A and its chlorinated derivatives (monochloro, dichloro, trichloro and tetrachlorobisphenol A). BPA shows a base peak at *m/z* 227 which corresponds to the molecular ion $[M–H]^-$ measured in negative ion mode. Using an ESI interface, the main fragment ion, in addition to the base peak, is observed at *m/z* 212, which results from the cleavage of one of the CH_3 groups. Using APCI [132], additional fragments are observed at *m/z* 211 $[(C_6H_4)C_3H_6(C_6H_4OH)]^-$ and *m/z* 133, which is assigned to $[(C_6H_4)C(CH_3)CH_2]^-$. Quantitative LC–MS/MS analysis of BPA is carried out in the SRM mode with two characteristic precursor–product ion transitions: 227 → 133 and 227 → 211 [56, 57]. Method detection limits achieved with LC–MS and LC–MS/MS-based methods have been around 2 and 50 ng L^{-1}, respectively.

3.2.4 Alkylphenolic Compounds

All but one of the methods reviewed employ reverse-phase (RP) chromatographic separation for analysis of alkylphenolic compounds in wastewater (see Table 5). The remaining method uses a mixed mode column utilizing both size exclusion and reverse-phase principles.

When working with reverse-phase separation it is not possible to separate APEOs with different numbers of ethoxylate (EO) units, but separation between OP and nonylphenol NP as well as their ethoxylates and carboxylates is obtained. Normal-phase chromatography, on the other hand, can separate APEOs as a function of their number of ethoxylate units, but can not separate the different alkylphenol species.

Many methods report the sum of long-chained APEOs instead of the single species, obtaining higher intensities and thus sensitivity. But even though the APEOs coelute, it is still possible to extract individual peaks when using MS or MS/MS detection.

ESI has been used almost exclusively because it has demonstrated better sensitivity than APCI towards the short-chain APEOs and the alkylphenols (APs) themselves. When using ESI in positive mode the APEOs almost solely form sodium adducts and these are often chosen as the parent ion. Unfortunately there is still rather poor sensitivity towards alkylphenol mono- and di-ethoxylates ($APEO_{1+2}$) which is often improved by measuring $[M+NH_4]^+$ instead of $[M+Na]^+$.

Limits of detection are reported from 1 to 1,000 ng L^{-1} with the lowest obtained in the SRM mode, using MS/MS.

3.2.5 Perfluorinated Compounds

As mentioned before, sample contamination can occur during sample pretreatment and also depends on the instrumental technique used. It is recommended to replace the PTFE HPLC tubing with stainless steel and PEEK [71]. According to the literature of wastewater reviewed, LC separation of fluorinated compounds has been mainly carried out with C18 columns; only one study reported the use of C8 columns.

Regarding mobile phases, mixtures of acetonitrile-water and methanol-water, often modified with ammonium acetate (2 mM) to improve LC separation and MS sensitivity have usually been used. LC in combination with MS or tandem MS has been the technique used for analysis of FASs in wastewater samples. MS detection is performed with an ESI interface operating in the negative ion mode. In the fragmentation pattern of FASs, the deprotonated molecules $[M-H]^-$ are the predominant ions. Typical ions and fragmentations monitored for PFOS and related substances correspond to $[SO_3]^-$, $[FSO_3]^-$, and $[M-SO_3]^-$ ions. For perfluorosulfonamide (PFOSA) and PFOA, $[SO_2N]^-$ and $[MCOOH]^-$ ions are the most abundant ones, respectively [69, 70, 73]. Today it is widely accepted that ESI ionization is susceptible to matrix effects. Matrix effects result from coeluting residual matrix components affecting the ionization efficiency of the target analytes and can lead to erroneous results. Wastewater matrixes are complex and their analysis typically requires sample enrichment and reduction of particle load. In the quantitative analysis by LC–(tandem)-MS, several strategies will be appropriate to diminish the analytes signal suppression of the analyte. The use of internal standards compensates well the matrix effects [70, 71, 73]. Other methodologies such us dilution of the extracts and adding a filter step before LC–MS analysis have been tested in the analysis of FASs in wastewater samples, but these methodologies did not compensate the matrix effects in influent wastewater samples [69].

3.2.6 Pharmaceuticals

LC–(tandem)MS via atmospheric pressure ionization (API) interfaces, especially ESI, has become the analytical technique of choice for analysis of pharmaceuticals in wastewater, since most of them are polar compounds, not amenable to GC analysis.

Chromatographic separation is in general achieved by means of C18 columns using acetonitrile, methanol, or mixtures of both solvents, and water as the mobile phase. To increase retention efficiency and MS sensitivity, mobile phase modifiers, buffers and acids, are recommended and widely used in concentrations from 2 to 20 mM (see Table 6), since it has been observed that higher concentrations can lead to a reduction of the signal intensities [133].

Generally, for acidic pharmaceuticals, which are determined in the NI mode, organic solvents and ultrapure water or aqueous ammonium acetate are preferred; whereas for neutral and basic compounds, analyzed in the PI mode, the use of buffers or modifiers is more common at neutral and acidic pH. When mobile phases are the same for both ionization modes, which are specific cases, pH of the mobile phases are usually adjusted to acidic or neutral conditions.

A critical aspect in quantitative analysis with LC–(tandem)MS is the occurrence of matrix effects producing ion suppression due to the presence of organic and inorganic components in the sample that coelute with target analytes. Since matrix components can be rather variable between samples—in nature and in amount—their effects are difficult to be predicted. Matrix effects may be reduced with proper pretreatments and efficient chromatographic separation, but the best manner to compensate them is using internal standards or standard addition. These methods, however, increase significantly the cost and time of analysis.

With respect to the type of MS detector, QqQ instruments have been by far the most widely used analyzers for quantitation of pharmaceuticals in wastewater. Nevertheless, in recent years, hybrid quadrupole-linear ion trap (Q-trap) and quadrupole-time of flight (Q-TOF) detectors have also been tested for this purpose (see Table 6).

Q-trap instruments outperform the conventional ion trap systems and deliver the functionality of QqQ [41, 95].

The main advantage of the TOF analyzers is the availability of full-scan mass spectra throughout each chromatogram and the accurate mass measurements that provide qualitative information for unequivocal identification of analytes present in the samples [134]. The most important drawback for quantitative measurements is the effective linear dynamic range which is significantly lower than the quadrupole instruments dynamic range (typically four orders of magnitude). Marchese et al. [80] compared the performance of a QqQ and a hybrid Q-TOF instrument for the determination of analgesics in water. The LOQs obtained with the TOF experiment were approximately three to five times higher than those obtained using the QqQ system operating in SRM mode.

More recently, Petrovic et al. [87] have developed a new multiresidue method based on ultraperformance liquid chromatography (UPLC–Q–TOF–MS/MS) for screening and confirmation of 29 pharmaceutical compounds. In another work by the same group, 32 compounds were included and the LOQs ranged from 0.1 to 15 ng L^{-1} in tap water, and from 2 to 300 ng L^{-1} in wastewater [135].

3.2.7 Estrogens

The analysis of estrogens by LC–(tandem)MS allows their selective and sensitive determination without previous derivatization (as in GC–MS). Separation of the analytes is mostly performed using C18-based reversed-phase columns. The mobile phase usually consists of acetonitrile or methanol with water. Some authors prefer methanol over acetonitrile affirming that a significant increase of signal intensity occurs when using methanol [100, 111, 118], although in most cases acetonitrile is reported to give better results. As the interface to MS detection ESI in NI has found larger application than APCI operated in PI (see Table 7). However, several authors have reported problems with matrix effects and prefer to employ APCI [89, 102, 121]; unfortunately APCI does not ionize the more polar estriol and sulfate derivatives efficiently [89].

The addition of basic or acidic compounds to the mobile phase is performed in about half of the methodologies reviewed. Ionization of estrogens using ESI in the NI has shown an improvement by the postcolumn addition of a weak base or weak basic conditions during chromatography [99, 100, 105–107, 111, 113–115, 118].

The LC of phytoestrogens can be performed with trifuoroacetic acid (TFA) [109, 114] or without acid addition [103].

Regarding MS detection, most works rely on the use of triple quadrupole tandem mass spectrometry due to its high selectivity and sensitivity when working in SRM mode. Detection limits are in the low ng/L range (commonly below 20 ng L^{-1}) even for wastewater influent samples. In a few cases TOF and Q-TOF instruments have been employed successfully to identify the compounds using their exact mass [103, 111].

4 Levels in Wastewater

4.1 *Pesticides*

There are very few data on the presence of endocrine-disrupting pesticides in sewage water and the compounds targeted vary considerably from one study to another.

In a study conducted in Spain, where atrazine, simazine, chlorfenvinphos, and diuron were investigated in the effluents of five sewage treatment plants (STPs), diuron was found to be the most ubiquitous and one of the most abundant compounds (average concentration 163 ng L^{-1}) and atrazine the least frequent and least abundant compound (average concentration 9 ng L^{-1}). Chlorfenvinphos and simazine were positively identified in 20% of the collected samples with average concentrations of 163 and 16 ng L^{-1}, respectively [41]. Similar trace levels of atrazine to those found in Spain have been determined in influent and effluent STP samples in the area of Montreal in Canada [39].

In the north of Italy, atrazine (41 ng L^{-1}) and diuron (165 ng L^{-1}) were the only disrupting pesticides detected in one of the effluent samples investigated in the course of a study conducted in three STPs [43].

In France the endocrine-disrupting fungicide tebuconazole and the insecticide dimethoate were frequently found in STP influents with levels ranging from 2 to 43 ng L^{-1} and from 2 to 20 ng L^{-1}, respectively. Another fungicide, hexaconazole, was found in treated sewage water at levels between 26 and 32 ng L^{-1} but was below the method limit of quantification in influent water. In the same study, a distinctly high concentration (2,287 ng L^{-1}), but in only one of the influent samples investigated, was detected for the carbamate pesticide degradation product 2-imidazolidinethione [40].

4.2 *Phthalates*

Phthalates have low water solubility and when released into the aquatic environment, they tend to absorb strongly to suspended particles and sediments [7, 11].

Biodegradation is expected to be the dominant degradation removal mechanism for phthalates entering STPs or released to surface waters, since a wide range of bacteria and actinomycetes can degrade phthalates under both aerobic and anaerobic conditions. Ultimate degradation of phthalates with sewage sludge inocula ranged from >50 to 99% in a study published by Staples et al. [8], while primary degradation half-lives in surface waters were estimated to range from less than 1 day to 2 weeks. Their degradation rate appears to decrease with increasing alkyl chain length because of the corresponding increase in log K_{ow} (which ranges from 1.46 to 13.1) [7]. DEHP and diethylphthalate (DEP) constitute the most ubiquitous phthalates found in wastewater, which correlates with their high production, widespread use and physico-chemical properties (low solubility and relatively high K_{ow}). There are several studies about the occurrence of phthalates in the aquatic environment. The most systematic of them reported levels of DEHP and DBP for 116 surface water samples, 35 sediments from rivers, lakes and channels, 39 sewage effluents and 38 sewage sludges collected in Germany [7]. The phthalate burden was mainly from DEHP, whilst di-*n*-butyl phthalate (DBP) was found in minor concentrations and BBP at concentrations near the detection limit. The concentrations found ranged from 0.3 to 98 g L^{-1} (surface water), 1.7–182 g L^{-1} (sewage effluent), 28–154 mg kg^{-1} (sewage sludge), and 0.2–8.4 mg kg^{-1} (sediment). Table 8 brings together environmental phthalate levels found for the case of different wastewaters, which is one of the major sources of these contaminants in the aquatic environment.

Table 8 Levels of phthalates in WWTPs

Compounds	WWTP influent (μg/L)	WWTP effluent (μg/L)	Geographical location	Reference
DEHP	Na	1.74–182	Germany	[7]
DMP	Na	0.6–3	Portugal	[49]
DEP	9.08–16.2	13.3	Spain	[50]
DBP	<LOD- 22	<LOD- 1.9	Spain	[13]
BBP	<LOD- 1.2	<LOD- 0.76		
DEHP	1.1–2.9	<LOD		
DEP	0.15–0.98	na	Spain	[51]
DEHP	0–0.15			
DOP	0–0.1			
DEP	0.059–0.389	na	Australia	[136]
DBP	0.025–0.138			
BBP	0.006–0.062			
DEHP	<LOD- 20.3			
DPP	0.07	0.008	Denmark	[137]
DBP	1.03	0.91		
BBP	0.39	0.13		
DEHP	35.4	0.96		

na Not analyzed

4.3 Bisphenol A

BPA can enter the environment through leaching from final products and during manufacturing processes [7]. WWTPs appear to be one of the major sources of pollution of BPA, which have led to elevated levels in both surface waters and sediments [138]. Table 9 shows the concentration values of BPA detected in different WWTPs. Despite the high removal rate in WWTPs, BPA is nevertheless released into the environment and commonly found in river waters [7, 138, 144–147] and occasionally in ground waters and drinking waters [145, 146, 148]. Some studies show that BPA may decompose microbially in water, with a half-life between 2.5 and 4 days [14], although persistence and bioaccumulation of BPA has also been reported [149]. Other studies pay attention to the chlorination derivatives formed during chlorination of drinking waters and wastewaters [51].

4.4 Alkylphenolic Compounds

During the last decade a large number of monitoring studies have been carried out in order to determine the fate of alkylphenols in wastewaters. Table 10 gives an overview of the results for the nonylphenol species found in such studies.

During the wastewater processes long-chain APEOs are degraded into short-chain APEOs (mainly di and mono ethoxylates) and the corresponding carboxylates (APECs) and then finally into NP or OP, which are not readily degradable.

In correspondence with this theory it is observed that influent concentrations of nonylphenol ethoxylates (NPEOs) are significantly higher than effluent concentrations. It can also be seen that the concentrations of nonylphenoxy carboxylates (NPECs) are not reduced from influent to effluent, which is attributed to the circumstance that NPECs are formed as rapidly as they are removed. The concentrations of nonylphenol are also reduced during the wastewater processes; however, this is believed to be due to adsorption to sludge more than elimination.

Table 9 Levels of BPA in WWTPs

WWTP influent (μg L^{-1})	WWTP effluent (μg L^{-1})	Geographical location	Reference
na	0.702	Germany	[7]
na	2.01–24–32	Spain	[139]
149.20	1.080	Canada	[140]
1.55	<LOD	Portugal	[56]
na	0–0.006	Belgium and Italy	[57]
0.542–3.01	0.162–0.258	Germany	[141]
0.378–0.890	0.012–0.028	UK	[142]
<LOD- 2.14	<LOD-1.10	Greece	[143]
0.06–1.51	<LOD -0.27	Spain	[50]

LOD limit of detection; *na* non analyzed

Table 10 Overview of environmental levels of alkylphenols in the last decade

Compounds		Country	Conc. (μg L^{-1})	Reference
NPEO	Influent	Austria (NP1EO)	2.6–35	[65]
		Austria (NP2EO)	1.2–5.8	[65]
		Italy	29–221	[150, 151]
		Spain	27–2,120	[152–154]
		Spain (NP1EO)	0.54–125	[62]
		Spain (NP2EO)	0.71–33	[62]
		Switzerland	96–430	[155]
		USA Sum(0–3EO)	47.6–262	[27]
		USA Sum(4–16EO)	265–745	[27]
		Denmark	6.5–344	[61]
		Croatia	5–392	[156]
	Effluent	Germany (NP1EO)	0.072–0.175	[157]
		Germany (NP2EO)	0.09–0.21	[157]
		Spain (NP1EO)	0.73–156	[62]
		Spain (NP2EO)	0.12–4.4	[62]
		USA Sum(0–3EO)	1.58–32.3	[27]
		USA Sum(4–16EO)	0.042–8.42	[27]
		Denmark	4.67–444	[61]
NPEC	Influent	Spain	0.16–450	[62]
		USA	0.28–3.23	[27]
		Croatia (NPE2C)	0.001–4.37	[156]
		Croatia (NPE1C)	0.001–3.2	[156]
	Effluent	Spain	1.3–307	[62]
		USA	7.58–99.3	[27]
		Germany (NPE1C)	0.8–1.12	[157]
NP	Influent	Belgium	0.4–219	[158]
		Italy	0.352–40	[60]
		Spain	0.5–251	[62, 153, 159]
		Austria	1.05–8.6	[65]
		The Netherlands	0.1–40	[160]
		Croatia	0.46–4.4	[156]
		Norway	0.002–5.2	[161]
	Effluent	Germany	0.14–00242	[157]
		Italy	0.081–0.88	[60]
		Spain	0.22–5	[62]

OP and OP with two carboxylate units (OPE2C) are found at concentrations 10-times lower than the corresponding nonalkylphenols. This reflects the fact that octylphenol ethoxylates (OPEOs) only make up for 10% of the overall use of alkylphenols.

4.5 Perfluorinated Compounds

Discharge of municipal wastewater effluents is one of the principal routes for introducing organic chemicals that are used in domestic and industrial settings into aquatic environments. The literature on FASs levels in wastewater samples is scarce. The main FASs reported in wastewater are the PFCs perfluorohexanoic acid (PFHxA), perfluoroheptanoic acid (PFHpA), PFOA, perfluorononanoic acid (PFNA) and perfluorodecanoic acid (PFDA), the PFSs perfluorobutane sulfonate (PFBS), perfluorohexane sulfonate (PFHxS), PFOS, perfluorodecane sulfonate (PFDS) and 6:2 fluorotelomer-sulfonate (6:2 FtS), and the nonionic PFSs perfluorooctanesulfonamide (PFOSA), *N*-ethyl perfluorooctanesulfonamide (N-Et-FOSA), 2-(*N*-ethylperfluoro-1-octanesulfonamido)-ethanol (N-Et-FOSE), and *N*-ethyl perfluoro-octanesulfonamidoacetate (PFOSAA) (see Table 1 for acronyms). Figure 1 shows ranges of concentration found in wastewater samples reported in the literature reviewed. Most of the analytes determined correspond to short-chain compounds because of the possible biodegradation of potential precursors entering the wastewater treatment plants. The highest concentrations were exhibited by PFOS, 6:2 FtS, and PFOA in both influent and effluent samples. The decrease of PFOS observed in WWTPs can be attributed to sorption into sludge [162]. The opposite situation occurs with PFOA, which is found at higher concentrations in effluents than influents [69, 71]. Biodegradation of precursors (fluorotelomer alcohols) could be responsible for the observed increases [163]. The distribution and fate of FASs during wastewater treatment are not well-known. More studies are needed in order to understand the behavior of FASs in WWTPs.

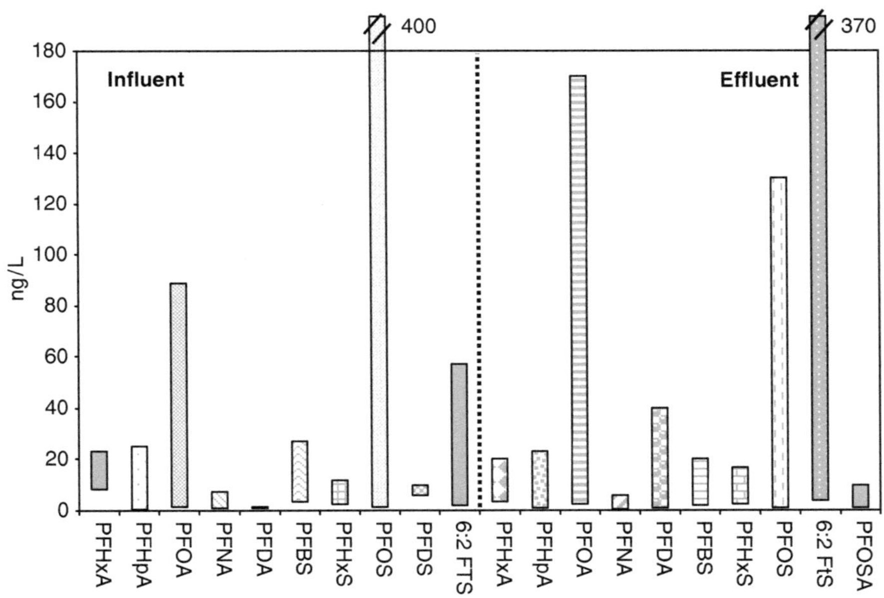

Fig. 1 Levels of fluorinated compounds in wastewater samples

4.6 Pharmaceuticals

Human and veterinary drugs are continuously being released into the environment mainly as a result of manufacturing processes, the disposal of unused or expired products, and excreta. The amount of pharmaceuticals and their bioactive metabolites being introduced into the environment is likely low. However, their continuous environmental input may lead to a high long-term concentration and promote continual, but unnoticed adverse effects on aquatic and terrestrial organisms. Numerous papers have reported levels of pharmaceuticals in wastewaters and aqueous and solid environmental matrices, the most ubiquitous ones in both surface and wastewaters being analgesics and antiinflammatories, antibiotics, lipid regulators, psychiatric drugs, β-blockers, and others such as ranitidine and omeprazole (ulcer healing), salbutamol (β-agonist), and iopromide (X-ray contrast media).

Different studies across Europe have reported medium concentrations for effluents of WWTPs from 0.3 to 2 $\mu g\ L^{-1}$, lipid regulators in the range from 0.1 to 2 $\mu g\ L^{-1}$, antipileptic drugs such as carbamazepine in medium ranges between 0.7 and 2 $\mu g\ L^{-1}$, and in lower concentrations antibiotics in the range of 0.05–0.3.

The human health and ecological significance of pharmaceutical residues in water resources is subject of on-going studies. To date, drug residues in drinking water have not been shown to adversely impact human health. There is concern, however, that antibiotics in surface water are playing a role in the ability of disease-causing pathogens to develop resistance to the drugs used to fight them. Additionally, numerous studies provide compelling evidence that hormonal drugs in surface water are causing adverse impacts to some fish populations.

Many believe that of all the emerging contaminants, antibiotics are the biggest concern because of the potential for antibiotic resistance [83–85]. The increasing use of these drugs in livestock, poultry production, and fish farming during the last five decades has caused a genetic selection of more harmful bacteria, which is a matter of great concern. However, other pharmaceutical compounds, especially polar ones, such as acidic antiinflammatory drugs and lipid regulators also deserve particular attention. Elimination of acidic pharmaceuticals in WWTPs has been found to be rather low and consequently sewage effluents are one of the main sources of these compounds and their recalcitrant metabolites. Because of their physico-chemical properties (high water solubility and often poor degradability) they are able to penetrate through all natural filtration steps and enter groundwater as well as drinking water.

4.7 Estrogens

Estrogens are mainly excreted as their less active sulfate, glucuronide and sulfo-glucuronide conjugates [164]. However, in WWTPs, as well as in the environment, these conjugates may suffer deconjugation and act as precursors of the corresponding free steroids. Thus, an appropriate evaluation of their

Table 11 Levels (ng L^{-1}) of free estrogens detected in wastewater

E2	E1	E3	EE	Reference
4.7–25	25–132	24–188	0.4–13	[106]
n.d.-21	10–57	27–220	n.d.	[113]
10–31	16–60	23–48	n.d.	[114]
3–22	8–52	n.a.	n.a.	[165]
n.d.-234	9.4–232	n.d.-108	2.4–138	[107]
17	<0.2–58	n.a.	<0.2–114	[166]

n.d. not detected; *n.a.* not analyzed

occurrence and impact requires the analysis of both free and conjugated estrogens.

Most of the studies dealing with the investigation of estrogens in wastewaters have been performed in WWTPs receiving urban/domestic discharges and concentrations reported have been most usually in the ng/L range. E2 and E1 have been the free estrogens most frequently found whereas E3 has been studied and detected only occasionally. However, E3 concentrations, when detected, have been usually higher than those of E2 and E1. In general, estrogen concentrations increase in the order E3 > E1 > E2 (see Table 11 for examples). Thorough revision of all data available situates mean and median concentrations in the range of 9–20 ng L^{-1} for E2, 20–55 ng L^{-1} for E1, and 45–75 ng L^{-1} for E3 [3].

The most studied synthetic estrogen, EE, has been either not detected [110, 113, 114] or detected at concentrations in general much lower than the other estrogens [100, 106, 108] (see Table 11). Levels higher than 100 ng L^{-1} have been only occasionally reported (e.g., 155 ng L^{-1} [167] and 138 ng L^{-1} [107]).

As regards conjugated estrogens, mostly sulfates and glucuronides of E1, E2, and E3 have been included as target analytes and detected at similar levels as the free estrogens [3].

5 Conclusions

LC–(tandem)MS is a very valuable tool for analysis of endocrine-disrupting compounds in wastewater since many classes of EDs are polar, nonvolatile compounds. Electrospray-tandem mass spectrometry with triple quadrupole analyzers operating in the SRM mode is by far the configuration most commonly used for quantitation of EDs in wastewater. The main advantages of this technique are sensitivity (achieving LOD in the ng L^{-1} or even pg L^{-1} range after sample preconcentration) and selectivity and confidence in compound identification (achieving a minimum of three identification points when acquiring two SRM transitions per compound). The main drawback, associated to the complexity of the samples, is the likely existence of matrix ionization effects; this makes recommendable the use of matrix-matched calibration, standard addition, or isotopically labeled surrogate standards for quantitation. The application of this methodology to elucidate the presence of EDs at the entrance and exit of WWTPs is primordial to know the efficiency of current wastewater treatments and check advanced, alternative ones for a better protection of the environment and human health from EDs effects.

Acknowledgments This work has been supported by the EU projects MODELKEY (GOCE 511237) and INNOVA MED (INCO-2006–517728) and by the Spanish Ministry of Science and Innovation (CTM2006–26227-E/TECNO, CTM2007–30524-E, and CGL2007–64551/HID). This article reflects the authors view alone. The EU is not liable for any use that may be made of the information contained in it. Cristina Postigo acknowledges the European Social Fund and AGAUR (Generalitat de Catalunya, Spain) for their economical support through the FI predoctoral grant.

References

1. IEH (2005) Web Report W20. Leicester, UK, MRC Institute for Environment and Health, available at http://www.le.ac.uk/ieh/
2. Directive 2008/105/EC. Official J Eur Union L 348:84–97
3. Petrovic M, Radjenovic J, Postigo C et al (2009) Emerging contaminants in wastewaters: sources and occurrence. In: Barcelo D, Petrovic M (eds) The handbook of environmental chemistry. Emerging contaminants from industrial and municipal wastewaters. Springer, Berlin
4. Borba da Cunha AC, Lopez de Alda M, Barceló D et al (2004) Multianalyte determination of different classes of pesticides (acidic, triazines, phenyl ureas, anilines, organophosphates, molinate and propanil) by liquid chromatography-electrospray-tandem mass spectrometry. Anal Bioanal Chem 378:940–954
5. Kuster M, Lopez de Alda M, Barceló D (2006) Analysis of pesticides in water by liquid chromatography-tandem mass spectrometric techniques. Mass Spectrom Rev 25:900–916
6. Kuster M, Lopez de Alda M, Barceló D (2008) Liquid chromatography–tandem mass spectrometric analysis and regulatory issues of polar pesticides in natural and treated waters. J Chromatogr A 1216:520–529
7. Fromme H, Kuchler T, Otto T et al (2002) Occurrence of phthalates and bisphenol A and F in the environment. Water Res 36:1429–1438
8. Staples CA, Peterson DR, Parkerton TF et al (1997) The environmental fate of phthalate esters: a literature review. Chemosphere 35:667–749
9. Tyler CR, Jobling SR, Sumpter JP (1998) Endocrine disruption in wildlife: a critical review of the evidence. Crit Rev Toxicol 28:319–361
10. Swan SH, Main KM, Liu F et al (2005) Decrease in anogenital distance among male infants with prenatal phthalate exposure. (Research: Children's Health). Environ Health Persp 113:1056–1061
11. Harris CA, Henttu P, Parker MG et al (1997) The estrogenic activity of phthalate esters in vitro. Environ Health Persp 105:802–811
12. van Wezel AP, van Vlaardingen P, Posthumus R et al (2000) Environmental risk limits for two phthalates, with special emphasis on endocrine disruptive properties. Ecotox Environ Safe 46:305–321
13. López-Jiménez FJ, Rubio S, Pérez-Bendito D (2005) Determination of phthalate esters in sewage by hemimicelles-based solid-phase extraction and liquid chromatography-mass spectrometry. Anal Chim Acta 551:142–149
14. Staples CA, Dorn PB, Klecka GM et al (1998) A review of the environmental fate, effects, and exposures of bisphenol A. Chemosphere 36:2149–2173

15. Brotons JA, Olea-Serrano MF, Villalobos M et al (1995) Xenoestrogens released from lacquer coatings in food cans. Environ Health Persp 103:608–612
16. Sonnenschein C, Soto AM (1998) An updated review of environmental estrogen and androgen mimics and antagonists. J Steroid Biochem 65:143–150
17. Rehmann K, Schramm K-W, Kettrup AA (1999) Applicability of a yeast oestrogen screen for the detection of oestrogen-like activities in environmental samples. Chemosphere 38:3303–3312
18. Christiansen LB, Pedersen KL, Pedersen SN et al (2000) In vivo comparison of xenoestrogens using rainbow trout vitellogenin induction as a screening system. Environ Toxicol Chem 19:1867–1874
19. Mehmood Z, Smith AG, Tucker MJ et al (2000) The development of methods for assessing the in vivo oestrogen-like effects of xenobiotics in CD-1 mice. Food Chem Toxicol 38:493–501
20. Van den Belt K, Verheyen R, Witters H (2003) Comparison of vitellogenin responses in zebrafish and rainbow trout following exposure to environmental estrogens. Ecotox Environ Safe 56:271–281
21. Suzuki N, Hattori A (2003) Bisphenol A suppresses osteoclastic and osteoblastic activities in the cultured scales of goldfish. Life Sci 73:2237–2247
22. Birkett JW and Lester JN (2003) Endocrine Disrupters in wastewater and sludge treatment processes. CRC, LLC, Florida
23. Gaido KW, Leonard LS, Lovell S et al (1997) Evaluation of chemicals with endocrine modulating activity in a yeast-based steroid hormone receptor gene transcription assay. Toxicol Appl Pharm 143:205–212
24. Rubin BS, Murray MK, Damassa DA et al (2001) Perinatal exposure to low doses of Bisphenol A affects body weight, patterns of estrous cyclicity, and plasma LH levels. Environ Health Persp 109:675–680
25. Elsby R, Maggs JL, Ashby J et al (2001) Comparison of the modulatory effects of human and rat liver microsomal metabolism on the estrogenicity of bisphenol A: implications for extrapolation to humans. J Pharmacol Exp Ther 297:103–113
26. Jonkers N, Knepper TP, De Voogt P (2001) Aerobic biodegradation studies of nonylphenol ethoxylates in river water using liquid chromatography – electrospray tandem mass spectrometry. Environ Sci Technol 35:335–340
27. Loyo-Rosales JE, Rice CP, Torrents A (2007) Fate of octyl- and nonylphenol ethoxylates and some carboxylated derivatives in three American wastewater treatment plants. Environ Sci Technol 41:6815–6821
28. Coldham NG, Dave M, Sivapathasundaram S et al (1997) Evaluation of a recombinant yeast cell estrogen screening assay. Environ Health Persp 105:734–742
29. Murk AJ, Legler J, van Lipzig MMH et al (2002) Detection of estrogenic potency in wastewater and surface water with three in vitro bioassays. Environ Toxicol Chem 21:16–23
30. Cespedes R, Petrovic M, Raldua D et al. (2004) Integrated procedure for determination of endocrine-disrupting activity in surface waters and sediments by use of the biological technique recombinant yeast assay and chemical analysis by LC-ESI-MS. Anal Bioanal Chem 378:697–708
31. Rutishauser BV, Pesonen M, Escher BI et al (2004) Comparative analysis of estrogenic activity in sewage treatment plant effluents involving three in vitro assays and chemical analysis of steroids. Environ Toxicol Chem 23:857–864
32. Cespedes R, Lacorte S, Raldua D et al. (2005) Distribution of endocrine disruptors in the Llobregat River basin (Catalonia, NE Spain). Chemosphere 61:1710–1719
33. Quiros L, Cespedes R, Lacorte S et al. (2005) Detection and evaluation of endocrine-disruption activity in water samples from Portuguese rivers. Environ Toxicol Chem 24:389–395
34. OECD (2002) Co-operation on existing chemicals-hazard assessment of perfluorooctane sulfonate and its salts. Environment Directorate Joint Meeting of the Chemicals Committee and the Working Party on Chemicals, Pesticides and Biotechnology, Organisation for Economic Co-operation and Development, Paris 21 November 2002

35. Zuccato E, Calamari D, Natangelo M et al (2000) Presence of therapeutic drugs in the environment. Lancet 355:1789–1790
36. Calamari D, Zuccato E, Castiglioni S et al (2003) Strategic survey of therapeutic drugs in the rivers Po and Lambro in Northern Italy. Environ Sci Technol 37:1241–1248
37. Heberer T (2002) Occurrence, fate, and removal of pharmaceutical residues in the aquatic environment: a review of recent research data. Toxicol Lett 131:5–17
38. Kuster M, Lopez de Alda MJ, Rodriguez-Mozaz S et al (2007) Comprehensive analytical chemistry. In: Petrovic M, D Barceló (eds) Analysis of steroid estrogens in the environment. Elsevier, Amsterdam
39. Viglino L, Aboulfadl K, Mahvelat AD et al (2008) On-line solid phase extraction and liquid chromatography/tandem mass spectrometry to quantify pharmaceuticals, pesticides and some metabolites in wastewaters, drinking, and surface waters. J Environ Monit 10:482–489
40. Baugros JB, Giroud B, Dessalces G et al (2008) Multiresidue analytical methods for the ultra-trace quantification of 33 priority substances present in the list of REACH in real water samples. Anal Chim Acta 607:191–203
41. Martinez Bueno MJ, Agüera A, Gomez MJ et al (2007) Application of liquid chromatography/quadrupole-linear ion trap mass spectrometry and time-of-flight mass spectrometry to the determination of pharmaceuticals and related contaminants in wastewater. Anal Chem 79:9372–9384
42. Freitas LG, Götz CW, Ruff M et al (2004) Quantification of the new triketone herbicides, sulcotrione and mesotrione, and other important herbicides and metabolites, at the ng/l level in surface waters using liquid chromatography-tandem mass spectrometry. J Chromatogr A 1028:277–286
43. Loos R, Hanke G, Eisenreich SJ (2003) Multi-component analysis of polar water pollutants using sequential solid-phase extraction followed by LC-ESI-MS. J Environ Monitor 5:384–394
44. Katsoyiannis A, Samara C (2004) Persistent organic pollutants (POPS) in the sewage treatment plant of Thessaloniki, northern Greece: occurrence and removal. Water Res 38:2685–2698
45. Fatta D, Canna-Michaelidou S, Michael C et al (2007) Organochlorine and organophosphoric insecticides, herbicides and heavy metals residue in industrial wastewaters in Cyprus. J Haz Mater 145:169–179
46. Pitarch E, Medina C, Portoles T et al (2007) Determination of priority organic micro-pollutants in water by gas chromatography coupled to triple quadrupole mass spectrometry. Anal Chim Acta 583:246–258
47. Öllers S, Singer HP, Fässler P et al (2001) Simultaneous quantification of neutral and acidic pharmaceuticals and pesticides at the low-ng/1 level in surface and waste water. J Chromatogr A 911:255–234
48. Li J, Cai Y, Shi Y et al (2008) Analysis of phthalates via HPLC-UV in environmental water samples after concentration by solid-phase extraction using ionic liquid mixed hemimicelles. Talanta 74:498–504
49. Castillo M, Alpendurada MF, Barceló D (1997) Characterization of organic pollutants in industrial effluents using liquid chromatography-atmospheric pressure chemical ionization-mass spectrometry. J Mass Spectrom 32:1100–1110
50. Cespedes R, Lacorte S, Ginebreda A et al (2006) Chemical monitoring and occurrence of alkylphenols, alkylphenol ethoxylates, alcohol ethoxylates, phthalates and benzothiazoles in sewage treatment plants and receiving waters along the Ter River basin (Catalonia, N. E. Spain). Anal Bioanal Chem 385:992–1000
51. Zafra-Gomez A, Ballesteros O, Navalon A et al (2008) Determination of some endocrine disrupter chemicals in urban wastewater samples using liquid chromatography-mass spectrometry. Microchem J 88:87–94
52. Petrovic M, Eljarrat E, Lopez de Alda MJ et al (2001a) Analysis and environmental levels of endocrine-disrupting compounds in freshwater sediments. Trends Anal Chem 20:637–648

53. Petrovic M, Eljarrat E, Lopez de Alda MJ et al (2002) Recent advances in the mass spectrometric analysis related to endocrine disrupting compounds in aquatic environmental samples. J Chromatogr A 974:23–51
54. Peñalver A, Pocurull E, Borrull F et al (2000) Determination of phthalate esters in water samples by solid-phase microextraction and gas chromatography with mass spectrometric detection. J Chromatogr A 872:191–201
55. Luks-Betlej K, Popp P, Janoszka B et al (2001) Solid-phase microextraction of phthalates from water. J Chromatogr A 938:93–101
56. Mauricio R, Diniz M, Petrovic M et al (2006) A characterization of selected endocrine disruptor compounds in a Portuguese wastewater treatment plant. Environ Monit Assess 118:75–87
57. Loos R, Hanke G, Umlauf G et al (2007) LC-MS-MS analysis and occurrence of octyl- and nonylphenol, their ethoxylates and their carboxylates in Belgian and Italian textile industry, waste water treatment plant effluents and surface waters. Chemosphere 66:690–699
58. Houde F, DeBlois C, Berryman D (2002) Liquid chromatographic-tandem mass spectrometric determination of nonylphenol polyethoxylates and nonylphenol carboxylic acids in surface water. J Chromatogr A 961:245–256
59. Petrovic M, Diaz A, Ventura F et al. (2001b) Simultaneous determination of halogenated derivatives of alkylphenol ethoxylates and their metabolites in sludges, river sediments, and surface, drinking, and wastewaters by liquid chromatography-mass spectrometry. Anal Chem 73:5886–5895
60. Careri M, Elviri L, Mangia A et al (2003) Liquid chromatography-electrospray ionization tandem mass spectrometry and liquid chromatography with fluorescence detection for determination of octylphenol and nonylphenol in municipal wastewater at trace levels. Chromatographia 57:321–327
61. Cohen A, Klint K, Bøwadt S et al (2001) Routine analysis of alcohol and nonylphenol polyethoxylates in wastewater and sludge using liquid chromatography-electrospray mass spectrometry. J Chromatogr A 927:103–110
62. Gonzalez S, Petrovic M, Barcelo D (2004) Simultaneous extraction and fate of linear alkylbenzene sulfonates, coconut diethanol amides, nonylphenol ethoxylates and their degradation products in wastewater treatment plants, receiving coastal waters and sediments in the Catalonian area (NE Spain). J Chromatogr A 1052:111–120
63. Di Corcia A, Costantino A, Crescenzi C et al (1998) Characterization of recalcitrant intermediates from biotransformation of the branched alkyl side chain of nonylphenol ethoxylate surfactants. Environ Sci Technol 32:2401–2409
64. Jeannot R, Sabik H, Sauvard E et al. (2002) Determination of endocrine-disrupting compounds in environmental samples using gas and liquid chromatography with mass spectrometry. J Chromatogr A 974:143–159
65. Clara M, Scharf S, Scheffknecht C et al (2007) Occurrence of selected surfactants in untreated and treated sewage. Water Res 41:4339–4348
66. Brossa L, Pocurull E, Borrull F et al (2004) Solid-phase extraction/high-performance liquid chromatography-electrospray mass spectrometry to determine endocrine disruptors in water samples. Chromatographia 59:419–423
67. Benijts T, Lambert W, De Leenheer, A (2004) Analysis of multiple endocrine disruptors in environmental waters via wide-spectrum solid-phase extraction and dual-polarity ionization LC-ion trap-MS/MS. Anal Chem 76:704–711
68. Cantero M, Rubio S, Perez-Bendito D (2006) Determination of alkylphenols and alkylphenol carboxylates in wastewater and river samples by hemimicelle-based extraction and liquid chromatography-ion trap mass spectrometry. J Chromatogr A 1120:260–267
69. Boulanger B, Vargo JD, Schnoor JL et al (2005) Evaluation of perfluorooctane surfactants in a wastewater treatment system and in a commercial surface protection product. Environ Sci Technol 39:5524–5530
70. Tseng C, Liu L, Chen C (2006) Analysis of perfluorooctanesulfonate and related fluorochemicals in water and biological tissue samples by liquid chromatography-ion trap mass spectrometry. J Chromatogr A 1105:119–126

71. Schultz MM, Barofsky DF, Field JA (2006) Quantitative determination of fluorinated alkyl substances by large-volume-injection liquid chromatography tandem mass spectrometry – Characterization of municipal wastewaters. Environ Sci Technol 40:289–295
72. Yamashita N, Kannan K, Taniyasu S et al (2005) A global survey of perfluorinated acids in oceans. Mar Pollut Bull 51:658–668
73. Gonzalez-Barreiro C, Martinez-Carballo E, Sitka A et al (2006) Method optimization for determination of selected perfluorinated alkylated substances in water samples. Anal Bioanal Chem 386:2123–2132
74. Richardson SD (2004) Environmental mass spectrometry: emerging contaminants and current issues. Anal Chem 76:3337–3364
75. Petrovic M, Hernando MD, Diaz-Cruz S et al (2005) Liquid chromatography-tandem mass spectrometry for the analysis of pharmaceutical residues in environmental samples: a review. J Chromatogr A 1067:1–14
76. Gros M, Petrovic M, Barcelo D (2006a) Multi-residue analytical methods using LC-tandem MS for the determination of pharmaceuticals in environmental and wastewater samples: a review. Anal Bioanal Chem 386:941–952
77. Heberer T, Dunnbier U, Reilich C et al (1997) Detection of drugs and drug metabolites in ground water samples of a drinking water treatment plant. Fresen Environ Bull 6:438–443
78. Ternes T, Bonerz M, Schmidt T (2001) Determination of neutral pharmaceuticals in wastewater and rivers by liquid chromatography-electrospray tandem mass spectrometry. J Chromatogr A 938:175–185
79. Miao XS, Koenig BG, Metcalfe CD (2002) Analysis of acidic drugs in the effluents of sewage treatment plants using liquid chromatography-electrospray ionization tandem mass spectrometry. J Chromatogr A 952:139–147
80. Marchese S, Gentili A, Perret D et al (2003) Quadrupole time-of-flight versus triple-quadrupole mass spectrometry for the determination of non-steroidal antiinflammatory drugs in surface water by liquid chromatography/tandem mass spectrometry. Rapid Commun Mass Spectrom 17:879–886
81. Hernando MD, Petrovic M, Fernández-Alba AR et al (2004) Analysis by liquid chromatography-electrospray ionization tandem mass spectrometry and acute toxicity evaluation for β-blockers and lipid-regulating agents in wastewater samples. J Chromatogr A 1046:133–140
82. Quintana JB, Reemtsma T (2004) Sensitive determination of acidic drugs and trichlosan in surface and wastewater by ion-pair reverse-phase liquid chromatography/tandem mass spectrometry. Rapid Commun Mass Spectrom 18:765–774
83. Hirsch R, Ternes TA, Haberer K et al (1998) Determination of antibiotics in different water compartments via liquid chromatography-electrospray tandem mass spectrometry. J Chromatogr A 815:213–223
84. Golet EM, Alder AC, Hartmann A et al (2001) Trace determination of fluoroquinolone antibacterial agents in urban wastewater by solid-phase extraction and liquid chromatography with fluorescence detection. Anal Chem 73:3632–3638
85. Reverte S, Borrull F, Pocurull E et al (2003) Determination of antibiotic compounds in water by solid-phase extraction-high-performance liquid chromatography-(electrospray) mass spectrometry. J Chromatogr A 1010:225–232
86. Castiglioni S, Bagnati R, Calamari D et al (2005) A multiresidue analytical method using solid-phase extraction and high-pressure liquid chromatography tandem mass spectrometry to measure pharmaceuticals of different therapeutic classes in urban wastewaters. J Chromatogr A 1092:206–215
87. Petrovic M, Gros M, Barcelo D (2006) Multi-residue analysis of pharmaceuticals in wastewater by ultra-performance liquid chromatography-quadrupole-time-of-flight mass spectrometry. J Chromatogr A 1124:68–81
88. Gros M, Petrovic M, Barcelo D (2006b) Development of a multi-residue analytical methodology based on liquid chromatography-tandem mass spectrometry (LC-MS/MS) for screening and trace level determination of pharmaceuticals in surface and wastewaters. Talanta 70:678–690

89. Schlüsener MP, Bester K (2005) Determination of steroid hormones, hormone conjugates and macrolide antibiotics in influents and effluents of sewage treatment plants utilising high-performance liquid chromatography/tandem mass spectrometry with electrospray and atmospheric pressure chemical ionisation. Rapid Commun Mass Spectrom 19:3269–3278
90. Yang S, Cha J, Carlson K (2005) Simultaneous extraction and analysis of 11 tetracycline and sulfonamide antibiotics in influent and effluent domestic wastewater by solid-phase extraction and liquid chromatography-electrospray ionization tandem mass spectrometry. J Chromatogr A 1097:40–53
91. Metcalfe CD, Koenig BG, Bennie DT et al (2003) Occurrence of neutral and acidic drugs in the effluents of Canadian sewage treatment plants. Environ Toxicol Chem 22:2872–2880
92. Vanderford BJ, Pearson RA, Rexing DJ et al (2003) Analysis of endocrine disruptors, pharmaceuticals, and personal care products in water using liquid chromatography/tandem mass spectrometry. Anal Chem 75:6265–6274
93. Vieno NM, Tuhkanen T, Kronberg L (2006) Analysis of neutral and basic pharmaceuticals in sewage treatment plants and in recipient rivers using solid phase extraction and liquid chromatography-tandem mass spectrometry detection. J Chromatogr A 1134:101–111
94. Van De Steene JC, Lambert WE (2008) Comparison of matrix effects in HPLC-MS/MS and UPLC-MS/MS analysis of nine basic pharmaceuticals in surface waters. J Am Soc Mass Spectrom 19:713–718
95. Schultz MM, Furlong ET (2008) Trace analysis of antidepressant pharmaceuticals and their select degradates in aquatic matrixes by LC/ESI/MS/MS. Anal Chem 80:1756–1762
96. Lacey C, McMahon G, Bones J et al (2008) An LC-MS method for the determination of pharmaceutical compounds in wastewater treatment plant influent and effluent samples. Talanta 75:1089–1097
97. Gros M, Pizzolato TM, Petrovic M et al (2008) Trace level determination of β-blockers in waste waters by highly selective molecularly imprinted polymers extraction followed by liquid chromatography-quadrupole-linear ion trap mass spectrometry. J Chromatogr A 1189:374–384
98. Quintana JB, Miro M, Estela JM et al (2006) Automated on-line renewable solid-phase extraction-liquid chromatography exploiting multisyringe flow injection-bead injection lab-on-valve analysis. Anal Chem 78:2832–2840
99. Salste L, Leskinen P, Virta M et al (2007) Determination of estrogens and estrogenic activity in wastewater effluent by chemical analysis and the bioluminescent yeast assay. Sci Total Environ 378:343–351
100. Koh YKK, Chiu TY, Boobis A et al (2007) Determination of steroid estrogens in wastewater by high performance liquid chromatography-tandem mass spectrometry. J Chromatogr A 1173:81–87
101. Grung M, Lichtenthaler R, Ahel M et al (2007) Effects-directed analysis of organic toxicants in wastewater effluent from Zagreb, Croatia. Chemosphere 67:108–120
102. Kim SD, Cho J, Kim IS et al (2007) Occurrence and removal of pharmaceuticals and endocrine disruptors in South Korean surface, drinking, and waste waters. Water Res 41:1013–1021
103. Farré M, Kuster M, Brix R et al (2007) Comparative study of an estradiol enzyme-linked immunosorbent assay kit, liquid chromatography-tandem mass spectrometry, and ultra performance liquid chromatography-quadrupole time of flight mass spectrometry for part-per-trillion analysis of estrogens in water samples. J Chromatogr A 1160:166–175
104. Schlüsener MP, Bester K (2008) Behavior of Steroid Hormones and Conjugates During Wastewater Treatment – A Comparison of Three Sewage Treatment Plants. CLEAN – Soil, Air, Water 36:25–33
105. Chen C-Y, Wen T-Y, Wang G-S et al (2007) Determining estrogenic steroids in Taipei waters and removal in drinking water treatment using high-flow solid-phase extraction and liquid chromatography/tandem mass spectrometry. Sci Total Environ 378:352–365
106. Baronti C, Curini R, D'Ascenzo G et al (2000) Monitoring natural and synthetic estrogens at activated sludge sewage treatment plants and in a receiving river water. Environ Sci Technol 34:5059–5066

107. Roda A, Mirasoli M, Michelini E et al (2006) Analytical approach for monitoring endocrine-disrupting compounds in urban waste water treatment plants. Anal Bioanal Chem 385:742–752
108. Zuehlke S, Duennbier U, Heberer T (2005) Determination of estrogenic steroids in surface water and wastewater by liquid chromatography-electrospray tandem mass spectrometry. J Sep Sci 28:52–58
109. Bacaloni A, Cavaliere C, Faberi A et al (2005) Determination of isoflavones and coumestrol in river water and domestic wastewater sewage treatment plants. Anal Chim Acta 531:229–237
110. Gomes RL, Birkett JW, Scrimshaw MD et al (2005) Simultaneous determination of natural and synthetic steroid estrogens and their conjugates in aqueous matrices by liquid chromatography/mass spectrometry. Int J Environ Anal Chem 85:1–14
111. Reddy S, Iden CR, Brownawell BJ (2005) Analysis of steroid conjugates in sewage influent and effluent by liquid chromatography-tandem mass spectrometry. Anal Chem 77:7032–7038
112. Ferguson PL, Iden CR, McElroy AE et al (2001) Determination of steroid estrogens in wastewater by immunoaffinity extraction coupled with HPLC-Electrospray-MS. Anal Chem 73:3890–3895
113. Komori K, Tanaka H, Okayasu Y et al (2004) Analysis and occurrence of estrogen in wastewater in Japan. Water Sci Technol 50:93–100
114. Laganà A, Bacaloni A, De Leva I et al (2004) Analytical methodologies for determining the occurrence of endocrine disrupting chemicals in sewage treatment plants and natural waters. Anal Chim Acta 501:79–88
115. Isobe T, Shiraishi H, Yasuda M et al (2003) Determination of estrogens and their conjugates in water using solid-phase extraction followed by liquid chromatography-tandem mass spectrometry. J Chromatogr A 984:195–202
116. Ingrand V, Herry G, Beausse J et al (2003) Analysis of steroid hormones in effluents of wastewater treatment plants by liquid chromatography-tandem mass spectrometry. J Chromatogr A 1020:99–104
117. D'Ascenzo G, Di Corcia A, Gentili A et al (2003) Fate of natural estrogen conjugates in municipal sewage transport and treatment facilities. Sci Total Environ 302:199–209
118. Gentili A, Perret D, Marchese S et al (2002) Analysis of free estrogens and their conjugates in sewage and river waters by solid-phase extraction then liquid chromatography-electrospray-tandem mass spectrometry. Chromatographia 56:25–32
119. Lopez de Alda M J, Barcelo D (2001) Determination of steroid sex hormones and related synthetic compounds considered as endocrine disrupters in water by fully automated on-line solid-phase extraction-liquid chromatography-diode array detection. J Chromatogr A 911:203–210
120. Lopez de Alda M J, Barcelo D (2000) Determination of steroid sex hormones and related synthetic compounds considered as endocrine disrupters in water by liquid chromatography-diode array detection-mass spectrometry. J Chromatogr A 892:391–406
121. Laganà A, Bacaloni A, Fago G et al (2000) Trace analysis of estrogenic chemicals in sewage effluent using liquid chromatography combined with tandem mass spectrometry. Rapid Commun Mass Spectrom 14:401–407
122. Mitani K, Fujioka A, Kataoka H (2005) Fully automated analysis of estrogens in environmental waters by in-tube solid-phase microextraction coupled with liquid chromatography-tandem mass spectrometry. J Chromatogr A 1081:218–224
123. Yu K, Krol J, Balogh M (2003) A fully automated LC/MS method development and quantification protocol targeting 52 carbamates, thiocarbamates, and phenylureas. Anal Chem 75:4103–4112
124. Peñalver A, Pocurull E, Borrull F et al (2001) Comparison of different fibers for the solid-phase microextraction of phthalate esters from water. J Chromatogr A 922:377–384
125. Cai Y-Q, Jiang G-B, Liu J-F et al (2003) Multi-walled carbon nanotubes packed cartridge for the solid-phase extraction of several phthalate esters from water samples and their determination by high performance liquid chromatography. Anal Chim Acta 494:149–156
126. Kim EJ, Lee JE, Gu MB (2002) Application of recombinant fluorescent mammalian cells as a toxicity biosensor. Water Sci Technol 46:51–56

127. Usami M, Mitsunaga K, Ohno Y (2002) Estrogen receptor binding assay of chemicals with a surface plasmon resonance biosensor. J Steroid Biochem 81:47–55
128. Zhao M-P, Li Y-Z, Guo Z-Q et al (2002) A new competitive enzyme-linked immunosorbent assay (ELISA) for determination of estrogenic bisphenols. Talanta 57:1205–1210
129. Matsunaga T, Ueki F, Obata K et al (2003) Fully automated immunoassay system of endocrine disrupting chemicals using monoclonal antibodies chemically conjugated to bacterial magnetic particles. Anal Chim Acta 475:75–83
130. Tschmelak J, Proll G, Gauglitz G (2004) Verification of performance with the automated direct optical TIRF immunosensor (River Analyser) in single and multi-analyte assays with real water samples. Biosens Bioelectron 20:743–752
131. Rodriguez-Mozaz S, Lopez de Alda M, Barceló D (2005) Analysis of bisphenol A in natural waters by means of an optical immunosensor. Water Res 39:5071–5079
132. Pedersen SN and Lindholst C (1999) Quantification of the xenoestrogens 4-tert.-octylphenol and bisphenol A in water and in fish tissue based on microwave assisted extraction, solid-phase extraction and liquid chromatography-mass spectrometry. J Chromatogr A 864:17–24
133. Sacher F, Lange FT, Brauch HJ et al (2001) Pharmaceuticals in groundwaters: analytical methods and results of a monitoring program in Baden-Wurttemberg, Germany. J Chromatogr A 938:199–210
134. Alder L, Deritz S, Lindtner K et al (2004) The ECHO technique – the more effective way of data evaluation in liquid chromatography-tandem mass spectrometry analysis. J Chromatogr A 1058:67–79
135. Farre M, Gros M, Hernandez B et al (2008) Analysis of biologically active compounds in water by ultra-performance liquid chromatography quadrupole time-of-flight mass spectrometry. Rapid Commun Mass Spectrom 22:41–51
136. Tan BLL, Hawker DW, Müller JF et al (2008) Stir bar sorptive extraction and trace analysis of selected endocrine disruptors in water, biosolids and sludge samples by thermal desorption with gas chromatography-mass spectrometry. Water Res 42:404–412
137. Fauser P, Vikelsøe J, Sørensen PB et al (2003) Phthalates, nonylphenols and LAS in an alternately operated wastewater treatment plant – fate modelling based on measured concentrations in wastewater and sludge. Water Res 37:1288–1295
138. Bolz U, Hagenmaier H, Korner W (2001) Phenolic xenoestrogens in surface water, sediments, and sewage sludge from Baden-Wurttemberg, south-west Germany. Environ Pollut 115:291–301
139. Aguayo S, Muñoz MJ, de la Torre A et al (2004) Identification of organic compounds and ecotoxicological assessment of sewage treatment plants (STP) effluents. Sci Total Environ 328:69–81
140. Lee H-B, Peart TE (2000) Bisphenol A contamination in Canadian municipal and industrial wastewater and sludge samples. Water Qual Res J Can 35:283–298
141. Körner W, Bolz U, Süßmuth W et al (2000) Input/output balance of estrogenic active compounds in a major municipal sewage plant in Germany. Chemosphere 40:1131–1142
142. Jiang JQ, Yin Q, Zhou JL et al (2005) Occurrence and treatment trials of endocrine disrupting chemicals (EDCs) in wastewaters. Chemosphere 61:544–550
143. Stasinakis AS, Gatidou G, Mamais D et al (2008) Occurrence and fate of endocrine disrupters in Greek sewage treatment plants. Water Res 42:1796–1804
144. Heemken O-P, Reincke H, Stachel B et al (2001) The occurrence of xenoestrogens in the Elbe river and the north sea. Chemosphere 45:245–259
145. Shin H-S, Park C-h, Park S-J et al (2001) Sensitive determination of bisphenol A in environmental water by gas chromatography with nitrogen-phosphorus detection after cyanomethylation. J Chromatogr A 912:119–125
146. Rodriguez-Mozaz S, Lopez de Alda M, Barcelo D (2004) Monitoring of estrogens, pesticides and bisphenol A in natural waters and drinking water treatment plants by solid-phase extraction–liquid chromatography–mass spectrometry. J Chromatogr A 1045:85–92
147. Focazio MJ, Kolpin DW, Barnes KK et al (2008) A national reconnaissance for pharmaceuticals and other organic wastewater contaminants in the United States – II Untreated drinking water sources. Sci Total Environ 402:201–216

148. Kuch HM, Ballschmiter K (2001) Determination of endocrine-disrupting phenolic compounds and estrogens in surface and drinking water by HRGC-(NCI)-MS in the picogram per liter range. Environ Sci Technol 35:3201–3206
149. Schonfelder G, Wittfoht W, Hopp H et al (2002) Parent Bisphenol A accumulation in the human maternal-fetal-placental unit. Environ Health Persp 110:703–707
150. Crescenzi C, Di Corcia A, Samperi R et al (1995) Determination of nonionic polyethoxylate surfactants in environmental waters by liquid chromatography/electrospray mass spectrometry. Anal Chem 67:1797–1804
151. Di Corcia A, Cavallo R, Crescenzi C et al (2000) Occurrence and abundance of dicarboxylated metabolites of nonylphenol polyethoxylate surfactants in treated sewages. Environ Sci Technol 34:3914–3919
152. Castillo M, Martinez E, Ginebreda A et al. (2000) Determination of non-ionic surfactants and polar degradation products in influent and effluent water samples and sludges of sewage treatment plants by a generic solid-phase extraction protocol. Analyst 125:1733–1739
153. Eichhorn P, Barceló D, Knepper TP (2000) Fate of surfactants and their metabolites in waste water treatment plants. Vom Wasser 95:245–268
154. Planas C, Guadayol JM, Droguet M et al (2002) Degradation of polyethoxylated nonylphenols in a sewage treatment plant. Quantitative analysis by isotopic dilution-HRGC/MS. Water Res 36:982–988
155. Ahel M, Molnar, E, Ibric, S et al. (2000) Estrogenic metabolites of alkylphenol polyethoxylates in secondary sewage effluents and rivers. Water Sci Technol 42:15–22
156. Terzic S, Senta I, Ahel M et al. (2008) Occurrence and fate of emerging wastewater contaminants in Western Balkan Region. Sci Total Environ 399:66–77
157. Jahnke A, Gandrass J, Ruck W (2004) Simultaneous determination of alkylphenol ethoxylates and their biotransformation products by liquid chromatography/electrospray ionisation tandem mass spectrometry. J Chromatogr A 1035:115–122
158. Tanghe T, Devriese G, Verstraete W (1999) Nonylphenol and estrogenic activity in aquatic environmental samples. J Environ Qual 28:702–709
159. Castillo M, Alonso MC, Riu J et al. (1999) Identification of polar, ionic, and highly water soluble organic pollutants in untreated industrial wastewaters. Environ Sci Technol 33:1300–1306
160. De Voogt WP, Kwast O, Hendriks R, Jonkers CCA (2000) Alkylphenol ethoxylates and their degradation products in abiotic and biological samples from the environment. Analysis 28:776–782
161. Vogelsang C, Grung M, Jantsch TG et al (2006) Occurrence and removal of selected organic micropollutants at mechanical, chemical and advanced wastewater treatment plants in Norway. Water Res 40:3559–3570
162. Key BD, Howell RD, Criddle CS (1998) Defluorination of organofluorine sulfur compounds by Pseudomonas sp. strain D2. Environ Sci Technol 32:2283–2287
163. Wang N, Szostek B, Folsom PW et al (2005) Aerobic biotransformation of C-14-labeled 8–2 telomer B alcohol by activated sludge from a domestic sewage treatment plant. Environ Sci Technol 39:531–538
164. Johnson AC, Williams RJ (2004) A model to estimate influent and effluent concentrations of estradiol, estrone, and ethinylestradiol at sewage treatment works. Environ Sci Technol 38:3649–3658
165. Lee H-B, Peart TE, Svoboda ML (2005) Determination of endocrine-disrupting phenols, acidic pharmaceuticals, and personal-care products in sewage by solid-phase extraction and gas chromatography-mass spectrometry. J Chromatogr A 1094:122–129
166. Pauwels B. Noppe H, De Brabander H et al (2008) Comparison of Steroid Hormone Concentrations in Domestic and Hospital Wastewater Treatment Plants. J Environ Eng-ASCE 134:933–936
167. Cui C, Ji S, Ren H (2006) Determination of steroid estrogens in wastewater treatment plant of a contraceptives producing factory. Environ Monit Assess 121:409–419

Index

MIX
Papier aus verantwortungsvollen Quellen
Paper from responsible sources
FSC® C105338

FSC
www.fsc.org

If you have any concerns about our products,
you can contact us on
ProductSafety@springernature.com

In case Publisher is established outside the EU,
the EU authorized representative is:
Springer Nature Customer Service Center GmbH
Europaplatz 3, 69115 Heidelberg, Germany

Printed by Libri Plureos GmbH
in Hamburg, Germany